Lecture Notes in Computer Science 891

Edited by G. Goos, J. Hartmanis and J. van Leeuwen

Advisory Board: W. Brauer   D. Gries   J. Stoer

T0205781

Claus Lewerentz   Thomas Lindner (Eds.)

Formal Development of Reactive Systems

Case Study Production Cell

Springer-Verlag

Berlin Heidelberg New York
London Paris Tokyo
Hong Kong Barcelona
Budapest

Series Editors

Gerhard Goos
Universität Karlsruhe
Vincenz-Priessnitz-Straße 3, D-76128 Karlsruhe, Germany

Juris Hartmanis
Department of Computer Science, Cornell University
4130 Upson Hall, Ithaka, NY 14853, USA

Jan van Leeuwen
Department of Computer Science, Utrecht University
Padualaan 14, 3584 CH Utrecht, The Netherlands

Volume Editors

Claus Lewerentz
Lehrstuhl für Software-Technik, Brandenburgische Technische Universität Cottbus
Karl-Marx-Str. 17, D-03044 Cottbus, Germany

Thomas Lindner
Forschungszentrum Informatik, Bereich Programmstrukturen
Haid-und-Neu-Str. 10-14, D-76131 Karlsruhe, Germany

CR Subject Classification (1991): D.2, C.3, D.1, D.3.1, J.6, J.7

ISBN 3-540-58867-1 Springer-Verlag Berlin Heidelberg New York

CIP data applied for

© Springer-Verlag Berlin Heidelberg 1995
Printed in Germany

Typesetting: Camera-ready by author
SPIN: 10479269   45/3140-543210 - Printed on acid-free paper

Preface

Mathematically precise methods play an increasingly important role in software development. To a growing extent, industry is utilizing this technology in areas where failure of software would result in injury to people or significant loss of money.

In one of the areas of computer science where a wide variety of methods confuse the practitioner, a comparative study can foster industrial acceptance by clarifying strengths and weaknesses, and elucidating the main ideas on a sample basis. This is the motivation and the goal of this book.

The book is based upon work done within the German KorSo project. We gratefully acknowledge the support given by the German Federal Ministry of Research and Technology (BMFT) with funding this project (grant number 01 IS 203). The editors would like to thank everyone who contributed by specifying, verifying, or implementing control programs. We thank the authors for their instructive contributions.

Special thanks are due to our colleague Emil Sekerinski, who brought this case study to our attention. We thank our colleagues from the Microcomputer Research Group at the Forschungszentrum Informatik, who provided us with a first version of a toy model of the production cell.

We hope that this work will stimulate many discussions and encourage everybody to try out their favourite method on the case study "production cell". We will keep on maintaining an ftp database for task description, simulation, and recording contributed work, be it published papers, reports, or specifications and implementations. Thus, all accumulated contributions may help in putting formal methods into industrial use by showing their benefits, explaining their use, and providing a roadmap for their selection and application.

Karlsruhe, September 1994

Claus Lewerentz
Thomas Lindner

Table of Contents

I. Introduction

Claus Lewerentz, Thomas Lindner

Forschungszentrum Informatik, Karlsruhe

1.1 Goal and Context of the Case Study

A recent report based on a questionnaire [1] authored by the British National Physical Laboratory (NPL) argued that one of the major impediments to formal methods gaining broader acceptance in industry is the lack of realistic, comparative surveys. With this case study we try to bridge this gap at least for the area of safety-critical reactive systems of moderate size.

The case study "Control Software for an Industrial Production Cell" was done as one of two major case studies of the KorSo project [2]. Both case studies have the common primary objective to show the usefulness of formal methods for critical software systems and to prove their applicability to real-world examples.

Whereas the HDMS-A [3] case study examines the application of an entire set of formal methods to a really large and complex information system, the "Production Cell" case study focuses on

- comparing different approaches of formal and semi-formal software construction with methods developed inside and outside the KorSo project, and

- checking their suitability for the class of problems represented by the production cell.

In the first phase of this case study, the results of which are presented in this book, 18 different approaches were used. In Chapters IV to X, the application of methods based on modelling with finite automata and using verification / validation techniques based on model checking is shown. In Chapter XI and XII modelling is shown based on the notion of streams and stream processing functions. The work reported in Chapters XII to XVI stresses verification by deductive methods and, in particular, the use of verification support tools. Approaches based on synthesis are presented in Chapters XVII and XVIII. The last section, i.e. Chapters XIX to XXI , is dedicated to object-oriented modelling methods and their integration with formal specification and verification techniques. Chapter III summarizes the final conclusions of all approaches and gives a comparative analysis.

1.2   Problem

The task for contributors to the case study is to develop the automation software to control a typical industrial production cell used in a metal processing factory. The example was taken from a real metal processing plant near Karlsruhe.

The production cell is composed of two conveyor belts, a positioning table, a two-armed robot, a press, and a travelling crane. Metal plates inserted in the cell via the feed belt are moved to the press. There, they are forged and then brought out of the cell via the other belt and the crane. Figure 1 shows a top view of the production cell.

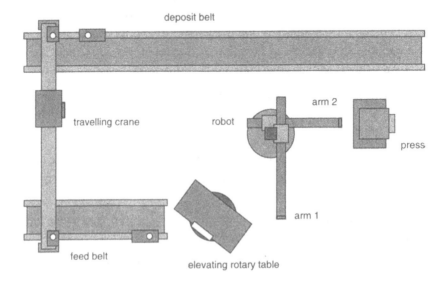

Figure 1   Top view of the production cell

On the bottom left the feed belt is shown which conveys the blanks to an elevating rotary table. This table has to be between the feed belt and the robot to bring the blanks into the right position for the robot to pick them up. To increase the utilization of the press, the robot is equipped with two arms – one only used for loading, the other one for unloading the press. Chapter II describes the physical system and the set of requirements for the control software in detail, and thus serves as the task description for contributors.

The problem addressed in the case study belongs to the category of *safety-critical systems*, as a number of properties must be enforced by the control software in order to avoid injury to people. It is a *reactive system*, as the control software has to react permanently to changes of the environment, i.e. the production cell. In principle a *real-time problem* has to be addressed, as the control software's reaction must be guaranteed within a certain interval of time[1]. The production cell can be modelled by a finite state automaton comprising in the order of at least 10^{12} states and, therefore, is a system of *moderate complexity*.

1.3   Required Properties

There are three kinds of properties required from the controlling software. The first and most important constraints are *safety requirements*. The controlling software

- must not allow a machine to collide with another one, in order to avoid the production cell damaging itself,

- must bar a machine from moving further than movability restrictions allow,

- must never allow blanks to be dropped outside "safe" areas (such as the press and the two belts) to avoid people being injured by falling metal plates,

- must guarantee that two consecutive blanks are always kept at a sufficient distance from one another to avoid having two blanks in the press simultaneously.

The controller should further guarantee the *liveness* of the system: each metal blank introduced into the system via the feed belt should eventually arrive at the end of the deposit belt and has been forged by the press.

---

1. All approaches reported in this book neglected this fact by simply assuming that the reaction of the control software is sufficiently fast to fulfil timing requirements. This simplification reduces the problem to be only based on causality.

In addition, some *other requirements* typical for industrial software development should be fulfilled: the design of the controller should be *flexible*, such that it would be easy to adapt it to similar production cells. The controller should finish putting the maximum number of blanks through the system, achieving the maximum flow of pieces. Finally, from a manager's point of view, it is important that the cost-benefit ratio is balanced.

1.4   Validation Environment: Toy Model and Graphic Simulation

FZI has a working toy model of the production cell (built with Fischer-Technik, cf. fig. 2 and 2) which can be controlled via a RS 232 serial line port. In order to perform demonstrations of the model, the production sequence should be able to run without an operator. The "forged" metal plates – which the press in the model does not actually modify – are therefore taken from the deposit belt back to the feed belt by a travelling crane, thus making the entire sequence cyclical.

The system is controlled by means of actuators using information received from sensors. The system has 13 actuators, for extending and retracting both robot arms independently, rotating the robot base, picking up metal plates with electro-

Figure 2   The toy model (top view)

Figure 2   The toy model (view on robot and press)

magnets installed at the end of each robot arm and other similar tasks. Commands given to the hardware are of the nature "switch on the motor for extending arm 1", and not "extend the robot arm". The controller has to take care that the motor is switched off at the right time.

There are 14 sensors providing the control program with information about the state of the system. Most of them return discrete values, like "press is open for loading" or "a blank arrived at the end of the feed belt". In addition, some potentiometers report on e.g. the rotation of the robot base. These sensors return real values, but as only a couple of these values are relevant, the production cell can be modelled as a finite automaton.

Experience shows that in addition to implementation errors, specification errors should be of major concern to the developer. Therefore, we had to provide all contributors with a validation environment. We built a simulation of the cell with a graphical visualization which runs under X Windows. Figure 3 shows this simulation together with a panel for manual control.

The simulation can be controlled with a program which reads sensor values from UNIX stdin and writes its control commands to stdout according to a standardized ASCII protocol. The same control program is used for operating the toy model as well.

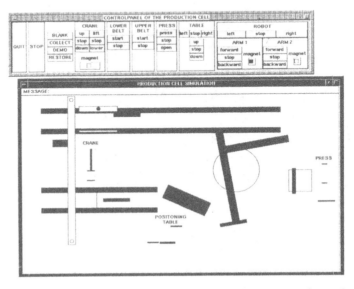

Figure 3   Snapshot of the simulation and the control panel

References

[1]   S. Austin and G. I. Parkin. Formal methods: A survey. Technical report, National Physical Laboratory, Great Britain, 1993.

[2]   M. Broy and S. Jähnichen, editors. *Korrekte Software durch formale Methoden.* Technische Universität Berlin, Franklinstraße 28–29, D-10587 Berlin, March 1993.

[3]   M. Löwe, F. Cornelius, J. Faulhaber, and R. Wessälly. Ein Fallbeispiel für KorSo — Das heterogene verteilte Managementsystem HDMS der Projektgruppe Medizin Informatik (PMI) am Deutschen Herzzentrum Berlin und an der TU Berlin — Ein Vorschlag. Technical Report 92–45, TU Berlin, 1992.

II. Task Description

Thomas Lindner

Forschungszentrum Informatik, Karlsruhe

Abstract

This chapter presents a case study in the field of control systems. The task consists of developing verified control software for a model representing a production cell installed in a metal-processing plant in Karlsruhe. The paper describes the functionality of the model, explains how the control program relies on the system's sensors, discusses the possibilities for driving the model with the help of various actuators, and finally defines the requirements that are to be fulfilled by the control software.

2.1 Description of the Production Cell

The Forschungszentrum Informatik has created a model of a production cell for mounting frames which was built as part of a study in microcomputer technology in 1989. This is not a model only in theory: it represents an actual industrial installation in a metal-processing plant in Karlsruhe.

The case study presents a realistic industry-oriented problem, where safety requirements play a significant role and can be met by the application of formal methods. The manageable size of the task allows for experimenting with several approaches.

The production cell processes metal blanks which are conveyed to a press by a feed belt. A robot takes each blank from the feed belt and places it into the press. The robot arm withdraws from the press, the press processes the metal blank and opens again. Finally, the robot takes the forged metal plate out of the press and puts it on a deposit belt (see figure 1).

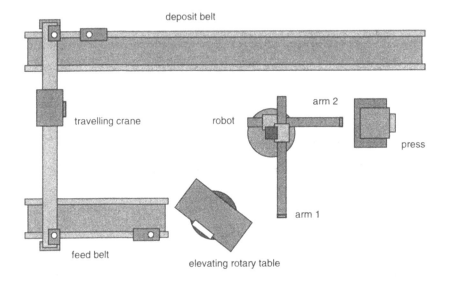

Figure 1   Top view of the model

This basic sequence is complicated by further details:

- To enhance the utilization of the press, the robot is fitted with two arms — thus making it possible for the first arm to pick up a blank while the press is forging another plate.

- The robot arms are placed on different horizontal planes, and they are not vertically mobile. This explains why an elevating rotary table has to be intercalated between the feed belt and the robot.

- Another consequence of the fact that the two robot arms are at different levels, is that the press has not only two, but three states: open for unloading by the lower arm, open for loading by the upper arm, and closed (pressing).

- In order to perform demonstrations with the model, the production sequence should be able to run without an operator. The "forged" metal plates — which the press in the model does not actually modify — are therefore taken from the deposit belt back to the feed belt by a travelling crane, thus making the entire sequence cyclical.

- A photoelectric cell at the end of the deposit belt informs the control program about the arrival of metal plates to be picked up by the travelling crane.

The general sequence (from the perspective of a metal plate) is the following:

1. The feed belt conveys the metal plate to the elevating rotary table.
2. The elevating rotary table is moved to a position adequate for unloading by the first robot arm.
3. The first robot arm picks up the metal plate.
4. The robot rotates counterclockwise so that arm 1 points to the open press, places the metal plate into it and then withdraws from the press.
5. The press forges the metal blank and opens again.
6. The robot retrieves the metal plate with its second arm, rotates further and unloads the plate on the deposit belt.
7. The deposit belt transports the plate to the travelling crane.
8. The travelling crane picks up the metal plate, moves to the feed belt, and unloads the metal plate on it.

This description of the system is of course rather simplified. First, the individual system components are not specified in detail. Secondly, the cell is configured so that several metal plates can be processed and transported simultaneously; this should allow an optimal utilization of the cell capacity.

2.1.1   Feed Belt

The task of the feed belt consists in transporting metal blanks to the elevating rotary table. The belt is powered by an electric motor, which can be started up or stopped by the control program. A photoelectric cell is installed at the end of the belt; it indicates whether a blank has entered or left the final part of the belt.

2.1.2   Elevating Rotary Table

The task of the elevating rotary table is to rotate the blanks by about 45 degrees and to lift them to a level where they can be picked up by the first robot arm. The vertical movement is necessary because the robot arm is located at a different level than the feed belt and because it cannot perform vertical translations. The rotation of the table is also required, because the arm's gripper is not rotary and is therefore unable to place the metal plates into the press in a straight position by itself.

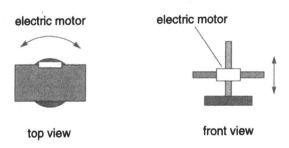

Figure 2   Elevating rotary table

2.1.3   Robot

The robot comprises two orthogonal arms. For technical reasons, the arms are set at two different levels. Each arm can retract or extend horizontally. Both arms rotate jointly. Mobility on the horizontal plane is necessary, since elevating rotary table, press, and deposit belt are all placed at different distances from the robot's turning center.

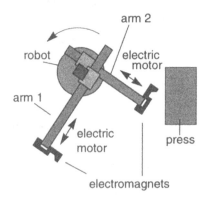

Figure 3   Robot and press (top view)

The end of each robot arm is fitted with an electromagnet that allows the arm to pick up metal plates. The robot's task consists in:

- taking metal blanks from the elevating rotary table to the press;

- transporting forged plates from the press to the deposit belt.

The robot is fitted with two arms so that the press can be used to maximum capacity. Below, we describe the order of the rotation operations the robot arm has to perform, supposed the feed belt to delivers blanks frequently enough. We presuppose that initially the robot is rotated such that arm 1 points towards the elevating rotary table, and assume that all arms are retracted to allow safe rotation.

1. Arm 1 extends and picks up a metal blank from the elevating rotary table.

2. The robot rotates counterclockwise until arm 2 points towards the press. Arm 2 is extended until it reaches the press. Arm 2 picks up a forged work piece and retracts.

3. The robot rotates counterclockwise until arm 2 points towards the deposit belt. Arm 2 extends and places the forged metal plate on the deposit belt.

4. The robot rotates counterclockwise until arm 1 can reach the press. Arm 1 extends, deposits the blank in the press, and retracts again.

Finally, the robot rotates clockwise towards its original position, and the cycle starts again with 1.

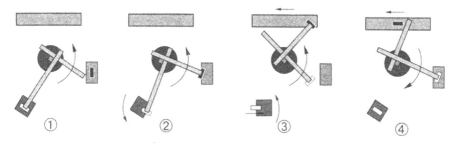

Figure 4   Order of the robot's actions

In order to meet the various safety requirements described in section 2.3, a robot arm must retract whenever a processing step where it is involved is completed.

2.1.4   Press

The task of the press is to forge metal blanks. The press consists of two horizontal plates, with the lower plate being movable along a vertical axis. The press operates by pressing the lower plate against the upper plate. Because the robot arms are placed on different horizontal planes, the press has three positions. In the lower position, the press is unloaded by arm 2, while in the middle position it is loaded by arm 1. The operation of the press is coordinated with the robot arms as follows:

1. Open the press in its lower position and wait until arm 2 has retrieved the metal plate and left the press.

2. Move the lower plate to the middle position and wait until arm 1 has loaded and left the press.

3. Close the press, i.e. forge the metal plate.

This processing sequence is carried out cyclically.

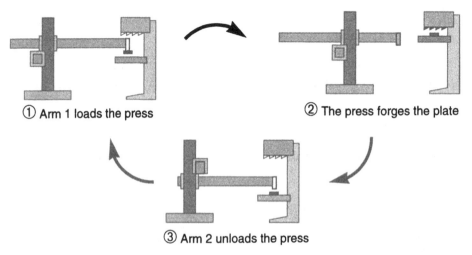

① Arm 1 loads the press   ② The press forges the plate

③ Arm 2 unloads the press

Figure 5   Robot and press (side view)

2.1.5   Deposit Belt

The task of the deposit belt is to transport the work pieces unloaded by the second robot arm to the travelling crane. A photoelectric cell is installed at the end of the belt; it reports when a work piece reaches the end section of the belt. The control program then has to stop the belt. The belt can restart as soon as the travelling crane has picked up the work piece.

The system designer is free to decide if the belts are to run continuously and should be stopped only when necessary, or if they should stand still and move only when necessary.

2.1.6   Travelling Crane

The task of the travelling crane consists in picking up metal plates from the deposit belt, moving them to the feed belt and unloading them there. It acts as a link be-

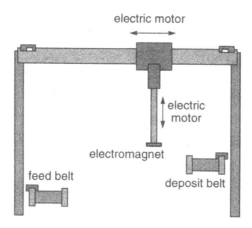

Figure 6   Travelling crane

tween the two belts that makes it possible to let the model function continuously, without the need for an external operator. In a more realistic setting, the travelling crane could unload the metal plates into a container, or link the production cell to a further manufacturing unit.

The crane has an electromagnet as gripper which can perform horizontal and vertical translations. Horizontal mobility serves to cover the horizontal distance between the belts, while vertical mobility is necessary because the belts are placed at different levels. The typical operation of the crane is as follows:

1. After the signal from the photoelectric cell indicates that a work-piece has moved into the unloading area on the deposit belt, the gripper positions itself through horizontal and vertical translations over the deposit belt and picks up the metal plate.

2. The gripper transports the metal plate to the feed belt and unloads it there.

Efficiency considerations may lead a system designer to move the travelling crane back to the deposit belt at the end of this sequence so that incoming plates can be transported immediately.

2.2   Actuators and Sensors

In the previous section, the system and its operation have been described from an "object-oriented" perspective — in the broadest possible sense of the term. In this section, additional information is given from a different perspective, that of the control program responsible for driving the model.

2.2.1 Actuators

The system can be controlled using the following actions:

1. move the lower part of the press (electric motor);
2. extend and retract 1st robot arm (electric motor);
3. extend and retract 2nd robot arm (electric motor);
4. pick up and drop a metal plate with 1st arm (electromagnet);
5. pick up and drop a metal plate with 2nd arm (electromagnet);
6. rotate robot (electric motor);
7. rotate elevating rotary table (electric motor);
8. move elevating rotary table vertically (electric motor);
9. move gripper of travelling crane horizontally (electric motor);
10. move gripper of travelling crane vertically (electric motor);
11. pick up and drop a metal plate with gripper of travelling crane (electro-magnet);
12. activate and deactivate feed belt (electric motor);
13. activate and deactivate deposit belt (electric motor).

2.2.2 Sensors

The control program receives information from the sensors as follows:

1. Is the press in its lower position? (switch)
2. Is the press in its middle position? (switch)
3. Is the press in its upper position? (switch)
4. How far has 1st arm been extended? (potentiometer)
5. How far has 2nd arm been extended? (potentiometer)
6. How far has the robot rotated? (potentiometer)
7. Is the elevating rotary table in its lower position? (switch)
8. Is the elevating rotary table in its upper position? (switch)
9. How far has the table rotated? (potentiometer)
10. Is the travelling crane positioned over the deposit belt? (switch)

11. Is the travelling crane positioned over the feed belt? (switch)

12. What is the current vertical position of the gripper? (potentiometer)

13. Is there a metal plate at the extreme end of the deposit belt? (photoelectric cell)

14. Is there a metal plate at the extreme end of the feed belt? (photoelectric cell)

Both photoelectric cells switch on when a plate intercepts the light ray. Just after the plate has completely passed through it, the light barrier switches off. At this precise moment, the plate is in the correct position to be picked up by the travelling crane (sensor 13 of the deposit belt), respectively it has just left the belt to land on the elevating rotary table — provided of course that the latter machine is correctly positioned — (sensor 14 of the feed belt).

While light barriers and switches provide a go/no-go kind of information, the potentiometer returns a value — which, in the case of rotation for instance, is proportional to the angle.

2.3 Requirements

In a reactive system, one typically distinguishes between safety and liveness requirements. Obviously, the safety requirements are most important in this setting: if a safety requirement is violated, this might result in damage of machines, or, even worse, injury of people. The safety requirements are described in section 2.3.1, liveness properties are discussed in section 2.3.2, and the last section discusses other properties interesting in this context.

The requirements listed below should be viewed as a pool of ideas. This case study allows for evaluating methods and approaches according to a wide spectrum of requirements, but not all properties can be formally proved to hold. We encourage contributors to prove representants of the single classes of properties, or discuss whether or how certain kinds of properties can be expressed or verified using the method under consideration.

2.3.1 Safety requirements

The control program must make sure that various safety requirements are met. Each safety requirement is a consequence of one of the following principles:

- the limitations of machine mobility: the robot, for instance, would destroy itself if rotated too far; the press would damage itself if opened too far;

- the avoidance of machine collisions: the robot, for instance, would collide with the press arm 1 would extend too far while pointing towards the press;

- the demand to keep metal blanks from being dropped outside safe regions: the robot, for instance, may deposit blanks only at some, few places, the feed belt has to make sure that the table is in the right position before transporting the blank too far;

- the necessity to keep the metal blanks sufficiently seperate: light barriers, for instance, can distinguish two consecutive blanks only, if they have a sufficient distance.

Restrict machine mobility!

The electric motors associated with the actuators 1-3 and 6-10 (cf. section 2.2.1) may not be used to move the corresponding devices further than necessary. In detail:

- the robot must not be rotated clockwise, if arm 1 points towards the elevating rotary table, and it must not be rotated counterclockwise, if arm 1 points towards the press,

- both arms of the robot must not be retracted less than necessary for passing the press, and they must not be extended more than necessary for picking up blanks from the press,

- the press must not be moved downward, if sensor 1 is true, and it must not be moved upward, if sensor 3 is true,

- the elevating rotary table must not be moved downward, if sensor 7 is true, and it must not be moved upward, if sensor 8 is true,

- the elevating rotary table must not be rotated clockwise, if it is in the position required for transfering blanks to the robot, and it must not be rotated counterclockwise, if it is in the position to receive blanks from the feed belt,

- if the crane is positioned above the feed belt, it may only move towards the deposit belt, and if it is positioned above the deposit belt, it may only move towards the feed belt,

- the gripper of the crane must not be moved downward, if it is in the position required for picking up a work piece from the deposit belt, and it must not be moved upward beyond a certain limit.

To fulfil these restrictions, the certain constants must be known. We refer to Appendix A.

Avoid machine collisions!

A couple of possible collisions are already avoided by simply obeying the above-mentioned restrictions on machine mobility. We do not mention these collisions in the following. Additionally, collision is possible and has to be avoided between the press and the robot, and between the crane and the feed belt:

- the press may only close when no robot arm is positioned inside it,
- a robot arm may only rotate in the proximity of the press if the arm is retracted or if the press is in its upper or lower position,
- the travelling crane is not allowed to knock against a belt laterally (this would happen if the travelling crane moved from the deposit belt to the feed belt without a simultaneous vertical translation),
- the travelling crane must not knock against a belt from above.

Again, we refer to Appendix A for the corresponding constants.

Do not drop metal blanks outside safe areas!

Metal blanks can be dropped for two reasons:

- the electromagnets of the robot arms or the crane are deactivated,
- a belt transports work pieces too far.

To avoid this, it suffices to obey the following rules:

- the magnet of arm 1 may only be deactivated, if the arm points towards the press and the arm is extended such that it reaches the press,
- the magnet of arm 2 may only be deactivated, if its magnet is above the deposit belt,
- the magnet of the crane may only be dactivated, if its magnet is above the feed belt and sufficiently close to it,
- the feed belt may only convey a blank through its light barrier, if the table is in loading position,
- the deposit belt must be stopped after a blank has passed the light barrier at its end and may only be started after the crane has picked up the blank.

Keep blanks sufficiently distant!

Errors occur if blanks are piled on each other, overlap, or even if they are too close for being distinguished by the light barriers. To avoid these errors, it suffices to obey the following rules:

- a new blank may only be put on the feed belt, if sensor 14 confirms that the last one has arrived at the end of the feed belt,

- a new blank may only be put on the deposit belt, if sensor 13 confirms that the last one has arrived at the end of the deposit belt,

- do not put blanks on the table, if it is already loaded,

- do not put blanks into the press, if it is already loaded

- do not move the loaded robot arm 1 above the loaded table, if the latter is in unloading position (otherwise the two blanks collide).

2.3.2   Liveness properties

A very strong liveness property for this system is satisfied, if the following requirement is fulfilled:

Every blank introduced into the system via the feed belt will eventually be dropped by the crane on the feed belt again and will have been forged.

There are many weaker forms of this liveness requirement.

2.3.3   Other requirements

Efficiency. It might be required that no blank is longer than a certain amount of time in the production cell. The best result would be to prove that the implemented controller achieves minimum possible time. To prove these properties it is necessary to remove the crane from the part of the system where time is measured.

Additionally, it can be required that the controller takes care that there are never less then a certain number of work pieces in the system, provided that there are enough blanks available.

Flexibility. The control software has to be as flexible as possible. The effort for changing the control software and proving its correctness must be as small as possible, when the requirements or the configuration of the cell change.

Questions

Several contributors found it helpful to consider the following question during their work or while writing the documentation:

- Which properties have been proved?
- Have assumptions about the architecture or the behavior of the production cell been made explicit and are they documented?
- How long, how complicated is the description? Is it understandable without deep knowledge of the method? Can it be discussed with a potential customer?
- How much effort was spent? Is the cost-benefit ratio balanced?
- Is it easy to change the controller? Can proofs be reused, or does a change in one part of the cell invalidate all proofs?
- How efficient is the controller? Does it achieve maximum possible throughput?
- Is it possible to draw conclusions on how the hardware design of the production cell could be improved? Would it be easier to prove certain properties, if additional sensors would be added? Would it be easier to control the cell, if any other additional hardware would be provided?

Acknowledgements

The author thanks Eduardo Casais for various suggestions and for providing the production cell clip-art from which all figures are adapted. Several contributors detected various errors and inconsistencies in earlier versions of this case study. Finally, thanks are due to Jochen Burghardt, who found the classification of safety requirements presented in section 2.3.1.

III. Comparative Survey

Summary and Evaluation of the Case Study
"Production Cell"

Claus Lewerentz, Thomas Lindner

Forschungszentrum Informatik, Karlsruhe

Abstract

This chapter summarizes the knowledge gained through work with the different modelling, specification, verification, and validation approaches to the Production Cell problem. Each of the 18 contributions is briefly presented and discussed according to a set of evaluation criteria. It turns out, that it is not easy to directly compare the different contributions, because different aspects of the same problem have been modelled, formally specified, or verified. The section on evaluation summarizes the most important conclusions concerning the suitability of the different approaches to tasks of which the Production Cell is representative.

It is a very challenging task to compare different approaches to the construction of reliable control software. One difficulty is that the different methods focus on particular aspects of the same problem and not all contributions effectively address the same problem. Some contributions model the behavior of the Production Cell as a whole, others give specifications of the control program, and a third group offers specifications for both the environment and the control program.

The criteria used as a basis to evaluate and to compare the different approaches are presented in section 3.1. For each contribution the control program properties addressed in the solutions and the assumptions about the system that were necessary to guarantee these properties are summarized. Furthermore measurements are made, like the lenghts of the specifications, the number of proof steps, and the time needed for the development effort. Finally, re-usability and extendability of the presented solutions are questioned.

3.1 Criteria

The main emphasis of our study is on the evaluation of applied *methods*, not on the skill the developer shows in applying the method. Our main criterion is the quality of the resulting software. Therefore, we formulate criteria which try to make our notion of quality measurable.

3.1.1 Properties

The first and traditionally most important quality criterion of a software product is *correctness*. In this case study, correctness primarily refers to *safety*. In order to avoid injury to people, the program must include a number of safety requirements. These are listed in Part II, Section 2.3.1.

Liveness is another prerequisite for correctness. A control program which would not control the Production Cell so that every metal blank is processed, would not be satisfactory. Therefore, we ask whether liveness can be proved or not.

There are other properties especially important from a manager's point of view: Does the control program achieve the maximum possible throughput of blanks through the system? This property is very difficult to prove. We are convinced that it is important to show that the application of formal methods does not result in a less efficient code, but also that it is not important to prove that a program is the most efficient possible.

In our opinion, a case study like this can only prove that certain kinds of properties can be guaranteed with a certain methodology. The opposite is beyond the scope of our evaluation: if a treatment fails to show or express one property, it is possible that the persons doing the modelling failed to do it the right way.

3.1.2 Assumptions

A property of the controller can only be proved by relying on certain assumptions. There are two classes of assumptions which influence the validity of a proof:

- As proofs are only possible in a formal system, reality has to be translated into a model and all properties can only be proved for this model. However, one is interested in statements about the behaviour of the real world.

One has to assume that the model covers all "relevant" aspects of the part of the real world under consideration. As it is never possible to model every detail of the real world, one has to assume that the model is sufficient. We call the amount of detail, which the model has with respect to reality, the *granularity* of the model.

- Proofs about the control program can hardly be carried out without assuming certain *environmental behaviour*. In our example the environment of the Production Cell consists of the processor which executes the control program, human interaction with the cell, and so on.

The granularity of the model is a key point in modelling: if it is too fine, the model is very complex and it may be very hard to prove anything.[1] If the model has a very coarse granularity, stronger assumptions are made: The mapping of different states or behaviours onto one single "grain" in the formal model assumes that these states or behaviours are not important to distinguish between the formal treatment. This assumption has to be made explicit and be justified.

As the choice of the right level of granularity is mainly a question of the skill of the modeller, we are interested here only whether the applied method gives the modeller a reasonable choice.

Our main focus is on the explicitness of the assumptions on the environment. It is not really important whether one proof makes more assumptions than another one, important from our point of view is that these assumptions would be explicit. Applied to our case study the questions are: Can one detect in the formal specification the assumptions which are presupposed, or were these assumptions mainly made in the developer's mind? Does the method encourage the developer to make these assumptions explicit?

3.1.3 Measurements

Even if a formal method provides one with the best possible software, it is unemployable for practical use if the costs in time and money for achieving this quality are too high: formal methods must be cost effective.

It was very difficult to analyse the question of cost-benefit-ratios. Benefits depend heavily on the market situation and on whether there are national or international standards which require the use of formal methods. This makes the

---

1. Please note that a finer granularity does not *necessarily* increase the set of properties which are provable. It is of course possible to construct models which are both compact and allow for proving many properties.

calculation of benefits difficult if not impossible within the scope of our analysis. The actual costs are also difficult to calculate, as measurements of the effort involved, the lines of specification or program written, or the number of proof steps carried out provide only poor points of reference: Many contributions to this survey were carried out by experts, many by novices. All of the contributions differed according to the number and kind of properties which were proven. Many of the researchers spent time on tool enhancement while working on the case study. Thus the amount of time spent working on the case study has to be seen in relation to the many factors which are hard to quantify. The lines of code written provide an equally rough estimate.

Nevertheless, we present these figures, as far as they are known to us. We stress that these figures must be seen in relation to the context they were taken from.

The question whether a property can be proven has to be seen into relation to the measurements for the following reason: In most languages there is a way to express almost every imaginable property. Though a "provable" property in our understanding, is a property which can be expressed and proven with a "reasonable" amount of complexity and effort. The best available measurements (besides inspecting specifications and proofs themselves) for both complexity and effort are measurements of lines of code and time spent.

3.1.4　Flexibility

Under realistic circumstances a project is not likely to be finished after the (verified) program has been delivered to the customer. Usually the detection of errors (in the specification, if not in the implementation), the change of requirements, and other factors force one to change the software to further meet customers demands.

What then counts is the ease of extending or correcting the software. Ideally, small changes in the software should only require little work on proving that the corrected program still fulfils the same properties as before.

3.2　Contributions

Most participants in the case study did not produce a formal treatment which was complete in every respect. This was not necessary for finding out which kind of properties can be guaranteed under which assumptions, nor for pointing out the main advantages and disadvantages of the respective method. Therefore the scope of each contribution is slightly different.

Accordingly, we start our summary of each contribution by naming its *goal*. We then characterize the approach chosen by both a short *description of the method* applied and a summary of *how the method was used* for this particular problem. The *outcomes* are named (like specifications, implementations, proofs). Finally, we *evaluate* the contribution according to the criteria given in Chapter 3.

Table 1 gives a short characterization of each of the applied methods.[1]

Method	Description
CSL	controller: declarative synchronous specification language; properties: temporal logic (PTL,CTL)
Esterel	imperative synchronous programming language
Lustre	controller: declarative synchronous programming language; properties: simple temporal logic
Signal	declarative synchronous programming language
StateCharts/STD	controller/environment: visual formalism describing synchronous state machines; properties: graphical variant of temporal logic
TLT	controller: unity-like specification language; properties: temporal logic
SDL	visual formalism describing processes communicating asynchronously via buffered channels
Focus	system described as set of asynchronously communicating agents
Spectrum	embedding CSP-like programming constructs in an algebraic specification language using streams of values; and (on a more abstract level) observing sequences of events via stream-processing functions
KIV	first order predicate logic (for specification) and dynamic logic (for verification)
Tatzelwurm	controller: first order predicate logic with equality;properties: Hoare-like assertions

Table 1   Short characterization of the specification mdethos as applied in this case study

Method	Description
HTTDs and HOL	controller: hierarchical timed transition diagrams, properties: real time temporal logic
RAISE	communicating sequential processes (CSP)
Deductive Synthesis	first order predicate logic
Symbolic Timing Diagrams	graphical variant of temporal logic
LCM/MCM	declarative, object-based specification language
Modula-3	some properties expressed as class invariants
TROLL *light*	interacting parallel objects constrained by event sequences

Table 1   Short characterization of the specification mdethos as applied in this case study

3.2.1   CSL

The goal of this contribution is to specify and automatically synthesize a correct control program for the Production Cell, to verify it and connect it to the graphic simulation. The specification should make use of the modular structure of the cell to enable re-use of components in similar cells.

CSL is a declarative specification language which adopts the synchronous paradigm. A CSL specification consists of a component library and a structure description. To write the specification, the relevant plant component types are first identified. Secondly, the constraints placed on the controller by an instance of a type are defined for each type. The specification is completed by describing the structure of a particular plant.

For CSL, a compiler is available which transforms a specification into a Finite State Automaton (FSA). These FSAs can then be used as input for both a code generator and the System Verification Environment (SVE) [11], a symbolic model checker developed by Siemens.

When applied to the Production Cell example, a CSL specification was written. It was decided to model arms and conveyer belts as special component types and instantiating both types twice. The other devices are transformed into component types straightforward.

1. Please note that this table is not supposed to be understandable by the beginner. It rather tries to give a quick overview for the more experienced reader.

The synthesized controller was then verified using SVE. In order to do this, the environment was modelled by formalizing the assumptions of physical plausibility ("a sensor signal will only arise when the corresponding device is actually moving towards the position indicated by the sensor") and fairness ("when a device keeps moving towards some position, it will eventually arrive there and be detected by the appropriate sensor"). Fairness assumptions were only used for liveness proofs.

Liveness was proven component for component, meaning that for each component "when a part is in an initial position, it will eventually arrive in a final position", and "the component will always return to its initial position".

Evaluation

All essential safety and liveness properties were specified and verified. The assumptions about the environment are explicit. The specification is nine pages long. It took three days to design, code, and debug it. Within one week it was connected to the simulator. The safety properties were specified and verified in less than one week. Liveness proofs took about two weeks.

The model checker tool had no problem verifying the safety properties. For the verification of the liveness requirements, the SVE developers assisted and introduced a few optimizations (precomputation of the set of reachable states, dynamic reordering of the BDD variables). The model featured about 50 million reachable states.

3.2.2 Esterel

This contribution was aimed at designing a working controller for the Production Cell.

Esterel [16] is an imperative programming language for reactive systems developed by Gérard Berry at Sophia-Antipolis. Like Lustre (cf. section 3.2.3), it is based on the hypothesis of perfect synchronization: it is assumed that input signals are synchronous with output signals, and that the program is executed on an infinitely fast machine. Other members of the family of synchronous languages are Signal and StateCharts (cf. section 3.2.4).

Communication in Esterel is based on signals. They can be emitted and tested for presence or absence. Internal communication with local signals looks and behaves like communication with the environment. Signals are broadcasted: if a signal is emitted, it is visible at that instant of time and cannot be removed. In the next instant it is forgotten.

Esterel provides the programmer with constructs for

- guarded statements triggered by a certain signal
- statements checking the absence or presence of signals
- parallel execution of statements
- modularization on the level of macro-expansion

An Esterel program describes a finite state machine. It is possible to generate implementations in C and an automaton description which can be graphically visualized or used as input for an automatic verification tool, called AUTO. The C programming language is used as a host language, dealing with all the matters which are outside the scope of Esterel.

The Production Cell was modelled in Esterel by decomposing it into five modules, mostly resembling the physical structure of the cell. It turned out that a single automaton describing the entire system would have too many states to be useful for code generation and automized proof. Therefore, each module was taken as a separate development unit and compiled onto an automaton of its own. Thus, it is easy to distribute the control program onto several hardware modules. On the other hand, the signals used for cooperation between the modules had to be propagated between the automata.

The outcome from this contribution was an Esterel program and a generated C program which controlled the simulation. No proofs were carried out.

Evaluation

Since no proofs were carried out, we could not evaluate the suitability of Esterel for ensuring properties. As a consequence, no assumptions were made. It took about three days to write the Esterel program, which is about 400 loc long, and another day to implement the C interface. Due to the usage of object-oriented design pattern, and due to the decision to model the devices as autonomous automata, the program is reusable and flexible. Reuse of proofs is imaginable on the basis of the automata; this can be done by using them via the particular protocol defined by their signals to the environment. A disadvantage in reusing Esterel programs is the unpredictable size of the generated automaton: small changes in an Esterel program can result in significant changes of the size to the automaton.

3.2.3 Lustre

The goal of the Lustre contribution was to develop a controller for the simulation and to formally verify the safety requirements of the task description.

Lustre [16] is a declarative synchronous programming language for reactive systems developed by N. Halbwachs. Pure Lustre, that is Lustre without data types, allows for writing control programs which react to boolean inputs by generating boolean outputs.

Programs are structured in *nodes*. Each node has a set of input parameters and may have local variables. For each output parameter there has to be exactly one assignment with an expression built from input and local variables. One may use some built-in simple temporal logic operators and user-defined functions to construct such expressions.

Lustre has a discrete time model. A Lustre program is supposed to run forever, receiving input at each instance of time and calculating simultaneously the corresponding outputs. The *synchronization hypothesis*, on which all verification is based, assumes that the reaction of the control program is "quick enough", or formally, that the program's execution does not consume any time.

Lustre programs can the be translated into both hardware and software code. The Lustre compiler constructs a finite automaton out of each node which can be translated both to C code and to hardware descriptions (VHDL). The user has to provide the interface of the so-called *main node* with the environment by writing some additional functions.

The verification of pure Lustre programs is supported by *Lesar*, a symbolic model checker. Lesar analyses a Lustre node with a single output variable and checks automatically whether this output variable is constantly true. Assertions may be included in the Lustre node to provide Lesar with necessary information on the environment.

The Production Cell was programmed in Lustre as a set of nodes most resembling the physical structure of the cell. For instance, there is one node for controlling the press, but three nodes for controlling the robot. The two feed belts do not appear as autonomous nodes, they are rather integrated in the nodes controlling their neighbour machines. Similar behaviour of several machines was extracted and put into a generic node called "moving item".

All safety requirements have been verified. The resulting controller is able to control the simulation by handling five blanks simultaneously.

Evaluation

Proving safety properties was very convenient with the verification tool Lesar. It could be done fully automatically and with short running time (< 1 minute). Liveness was outside the scope of this contribution. All assumptions which had to be made for the proofs are explicitly named in the assumptions clauses inside the Lustre programs. A closer look at the proof goals shows that some implicit knowledge about the Production Cell was used in formulating the safety requirements.

It took about three days to implement the first version of the controller, and two weeks to correct and verify the program.

Reuse of Lustre programs and proofs is only possible on the level of nodes, the only structuring mechanism provided by the language.

3.2.4 Signal

The aim of the Signal [16] contribution was to formally specify, verify and implement a controller for the Production Cell.

Signal was developed at IRISA/INRIA (Rennes, France) and belongs to the family of synchronous programming languages. Like Lustre, Signal is data flow oriented and has a discrete time model.

Signal manipulates unbounded series of typed values (called *signals*). With each signal, a *clock* is associated denoting the set of instants when values are present. Special signals are *events*, which can only be present or absent. Systems can be specified by relating signals, i.e. their values and their clocks, to to each other. These relations, denoted in *equations*, can be collected, and together they constrain the reactive program.

The Signal programming environment features a diagram representation of the language; a compiler, which performs the analysis of the consistency of the system of equations and checks for the validity of synchronization constraints; and a proof system, called Sigali [10], which can be used to verify dynamic properties of programs. The compiler is also able to generate executable code in Fortran or C, if the specification is deterministic.

Applied to the Production Cell example, Signal was used to describe both safe and erroneous states of the environment. This specification was composed with a specification of safety rules, which say how to avoid dangerous positions by sending commands to the environment. The resulting specification was finally enriched with a scheduler, linking the machines and guaranteeing the flow of metal plates.

Evaluation

A number of safety properties were proved using Sigali. Liveness was not addressed. Verification in Sigali requires one to specify the assumptions about the environment by describing them as a process which is put parallel to the control program. This guarantees that assumptions about the environment are explicit.

In total, four weeks were spent for the specification (2), implementation (1), and verification (1). The Signal code, automatically generated from the graphical representation, is about 1700 lines long. Compilation of the program takes some seconds. The executable file is about 90 K.

During specification and verification, care was taken to make the control program and the proofs as reusable as possible. In the specification this was achieved by a loose coupling between the transformational and the transport part of the production cycle, and by the use of generic procedures at several places. The required properties were formalized using generic functions as well, making them fairly reusable for similar applications.

3.2.5   Statecharts with Timing Diagrams

This contribution aimed at the formal specification of both the control program and the Production Cell, and the verification of safety and liveness properties of the composed system.

The control program and the behaviour of the Production Cell were specified using a sub-language of StateCharts, introduced by D. Harel [18]. This is a visual formalism capable of describing state transition systems. In addition, one may describe states hierarchically (a state being allowed to have sub-states) and express parallelism by allowing a system to be in two or more states at the same time. State transitions are enabled by events, which may additionally depend on conditions (boolean expressions over states) and which are allowed to initiate other events.

Safety and liveness properties are also expressed in a graphical language, called Symbolic Timing Diagrams (SDTs). These have been introduced by Schlör and Damm [31] and allow for expressing temporal logic formulas in a graphical way, similar to the timing diagrams known through telecommunication or signalling systems. These diagrams can, in principle, be checked automatically against the StateCharts model by translating the latter into a so-called Binary Decision Diagram (BDD), transforming the former in temporal logic, and finally checking the validity of the logical formulas by symbolic model checking [6].

StateCharts and SDTs were applied to the case study, a description of the application follows. First, the interface between Production Cell and control program was defined. It was decided to model both actuators and sensors as states. Then the StateCharts model of both the controller and the cell were developed independent-

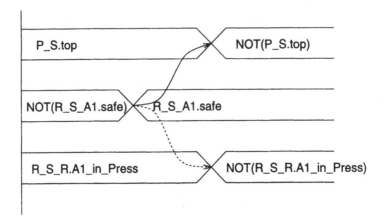

Figure 1   A simple STD expressing a (conditional) safety requirement

ly from the formalization of the safety and liveness requirements expressed in temporal logic using STDs. It turned out that the generated model had about 10^{19} states, and that the smallest representation found was a BDD comprising about 72000 nodes. Thus, due to this size, only some properties could be checked on the full model, an ad-hoc approach to compositional model checking was chosen which allowed the verification of the other requirements.

Evaluation

All safety and liveness properties could be conveniently expressed using STD. To verify them the model checking had to be restricted to specific parts of the model. All assumptions were explicitly described in a StateChart model of the Production Cell. The visual languages for both model and properties contributed to the comprehensibility of the specification.

It took about 10 days to specify controller, cell, and safety and liveness properties, resulting in 760 loc in the textual representation of StateCharts and 18 timing diagrams. The effort for the complete verification is estimated to be in the order of 20 days, supposed that more efficient versions of the tools (currently in prototype state) would be available. The model checker ran between one and five minutes for each property proved (Sparc 10 with 95 MB core memory).

3.2.6   TLT

In this contribution a distributed controller for the production cell was specified and the correctness of the controller was verified with respect to safety and liveness properties. The controller was tested in a simulation environment connected to the graphic simulation of the cell.

The Temporal Language of Transitions (TLT) is a framework for the design and verification of distributed systems. It was developed at Siemens Corporate Research in Munich [3]. TLT is based upon the ideas of Chandy and Misra's language UNITY [8] and Lamport's Temporal Logic of Actions [25].

It contains a first order logic, state-transition based language together with a temporal logic for expressing properties, and an operational semantics for TLT programs. Decomposition and composition mechanisms like modules, interface constraints, and explicit modelling of component environments are provided. System components are cooperating by both, explicit synchronization and asynchronous message passing between such synchronization points. TLT specifications consist of instructions written using state predicates as guards and transition predicates. Parallel composition of instructions expresses simultaneous execution of actions.

The application of TLT to the production cell consists of the following steps:

1. the behavioural properties of the entire system, i.e. the controller with its environment, are specified on a rather abstract level;

2. safety and liveness properties were specified and proved to hold for this system specification;

3. the system was decomposed and refined into a centralized controller and an environment part. Explicit synchronization between these system parts is introduced;

4. the controller part is further refined to get a distributed solution, i.e. small local controllers for groups of cooperating physical machines in the system. The properties proven in the previous step are preserved in the distributed solution by the refinement and composition process. Additionally, the properties were proved independently through model checking;

5. the distributed controllers are further refined to optimize the system behaviour (e.g. to increase the number of plates that can be handled safely).

Evaluation

TLT was used to write a complete specification of a controller. All relevant safety and liveness properties were proved using the model checking environment. The specification and the verification was done in about 30 hours. Specification steps, particularly refinement, and verification steps were interleaved. Each design step was immediately checked using the SVE tool for symbolic model checking. The times for automatically proving particular properties with the model checker took only few seconds to some minutes of machine calculation. It was proved that the controller is able to handle up to eight plates, whereas with nine plates it would deadlock.

As there does not exist a compiler for TLT yet, no code was generated from the specifications. Instead, some simulation runs of the specification were done in connection with the graphic simulation environment.

The structure of the abstract specification is quite independent from the details of the actual production cell. Parts of it can be reused for a variety of similar cells as long as the link constraints are preserved.

3.2.7   SDL

The goal of the SDL contribution was to specify both the control program and the Production Cell in order to prove properties of the system.

SDL, the Specification and Description Language standardized by CCITT, is an asynchronous programming language for distributed communicating systems. On the process level systems are specified by describing their behaviour and their communication via graphical notation resembling flowcharts. On the system level one can structure the system with blocks, which can either contain new blocks or processes. As SDL is, in its latest standardized version, a fully object-oriented language, inheritance, instantiation, and encapsulation techniques are usable on some levels like blocks, processes, and procedures.

Processes run parallel to each other and communicate with signals sent via asynchronous, buffered channels. Processes are extended finite state machines; they always have a current state and may only change their state when receiving a signal. While changing a state the process may update local variables, call procedures, and send signals to other processes, and the like.

There are several commercial tools on the market to support this language which is very popular for telecommunications applications. For our purposes, the SDT tool from TeleLOGIC was used. It supports the verification by allowing the performance of some reachability analyses. One can specify message sequence charts (MSCs), a very simple form of timing diagrams, and verify a SDL specification against them. SDT proved to be useful for debugging and systematic testing, but is not really well suited for formal verification.

Therefore, translating SDL specifications into transition systems and doing symbolic model checking of temporal logic formulas [8] is being considered. This work is still being carried out, and due to the lack of tools, it is mainly of interest from a research point of view.

Currently, the main outcome of the work is a controller generated from SDT[1]. A variant of a liveness property was derived by verifying an MSC against the specification. Simulations showed that the safety properties hold as well, but there is no formal proof for this so far.

Evaluation

SDL, when supported by currently available proof support, is only suitable for semi-formal simulations of the system under construction. No mathematically precise proofs are possible. The assumptions can be made explicit by modelling the environment in SDL as well, but the user is not really encouraged to do so. Reuse of program code is supported by the availability of object-oriented constructs, reuse of proofs has not been addressed.

It took about one month to write and validate the specification, which is about 2000 lines in length (SDL textual notation generated from diagrams).

3.2.8   Focus

The object of this contribution is to specify the Production Cell system in a formal framework called Focus and implement the specification in Concurrent ML. In addition, some liveness and safety properties were proven.

Focus [5] is a design methodology for distributed systems developed by M. Broy at the Technische Universität München. A system is designed in three phases:

In the requirement specification phase one describes the behaviour of the system in a trace logic as a set of action traces.

In the design specification, the second phase, the system is modelled by agents communicating asynchronously via channels. The communication history of a channel is viewed as a finite or infinite sequence of actions. The agents are modelled as a set of functions over these sequences. Some refinement techniques allow for translating descriptive specifications into constructive ones. Another set of re-

---

1. This controller could not be connected with the simulation due to the lack of appropriate I/O libraries, which were too expensive to buy just for this case study.

finement techniques allow the specification to be modularized without changing the semantics of the specification.

In the third phase the specification is translated into either a functional or an imperative programming language.

For the Production Cell example the emphasis is put on the design level of the specification. An abstract specification is split into specifications of the single machines. The specifications of the control procedures are refined into constructive specifications, which are finally implemented in Concurrent ML. All refinement steps were proven on a sample basis. It was also proven on the design level that the system is deadlock-free.

Evaluation

No safety properties were proven. The system specified is deadlock-free; this is guaranteed by the contruction method, if the lower-level components acknowledge each control signal in finite time. As all proofs carried out only address properties of the specification which do not depend on properties of the system environment, no assumptions about the latter are necessary.

All of the work was done by a student within half a year. The high level specification was made as compact as possible, after some iterations resulting in a specification of 80 lines.

Focus allows for modular proofs, therefore, if just the architecture of the components need to be rebuilt, the implementation of the components would be reusable including the proofs.

3.2.9   Spectrum

This contribution describes the Production Cell on two levels of abstraction:

CSP Specification

The aim of this specification is to formally describe the behaviour of the Production Cell using a sub-language of CSP. CSP (Communicating Sequential Processes) [23] is a language used for the description of systems operating concurrently, developed by C. A. R. Hoare at Oxford. Communicating processes are used to describe the behaviour of systems. Processes can be specified in an imperative style (using communication primitives, parallel and sequential composition, e.g.) and in a descriptive style (using predicates over event traces), as well. For this case study, a sub-language of the more imperative style of CSP was used for specifying the possible behaviour of the Production Cell. For the construction of the specification so-called *event diagrams* (cf. Fig. 7) were used, illustrating the event flow in the system and the synchronization between the terms. In event diagrams, nodes represent events and the connecting edges represent sequentially occurring events or events which two processes have in common, respectively. A pseudo-interpreter for the chosen sub-language of CSP was implemented in SML. The program can work in interactive and automatic mode and helps in ensuring that the specification models reality correctly. It is also used to search for deadlocks or livelocks of the system.

Stream-based Specification

The aim of this work was to establish a formal specification of the Production Cell, that allows its behaviour to be described by axiomating its possible runs (infinite) as sequences of observable actions. This specification was used to formalize the safety properties given in the informal task description. Streams, as found in Spectrum [5] or Focus [3], both developed at the Technische Universität München by M. Broy, are finite or infinite sequences of observable actions. They can be used to describe runs of a distributed system. Predicates over streams describe valid or invalid runs of the system. Streams were used to construct different views of the Production Cell: a data flow, a buffer, a protocol, and an action oriented view were given. A future goal is the formalization of the semantic relationship between both specifications using Spectrum as a common framework. Towards this end, both the CSP and the stream-based specification were embedded in Spectrum.

Evaluation

The stream logical specification provides an abstract specification of the Production Cell. It formalizes the safety requirements of the informal task description. There are no liveness requirements described. The stream logical specification is structured horizontally as well as vertically. Horizontal structure: for each component of the Production Cell there exists a separate specification module. Synchronization between interacting components is described in separate specification modules as well. Vertical structure: several views to the system describe it on different levels of abstraction. The CSP description is a model of the stream logical specification (i.e. each of the traces in its semantics fulfils the stream logical specification). So far, no proofs have been carried out.

The CSP description comprises about 100 actions and is about 200 lines in length. It took about two weeks to write it, three days were spent implementing the pseudo-interpreter. The stream-based specification describes about 130 actions and is composed out of 50 modules. It is about 10 pages in length, developed in about four weeks.

3.2.10 KIV

This contribution focuses on studying how distributed systems can be modelled within first-order logic, and which requirements for the correctness of a central control program can be expressed and verified using the KIV system.

The KIV (Karlsruhe Interactive Verifier) [19][20] system was designed for the development of correct software systems. The system supports structured first order specifications and a tactical theorem proving approach for program verification. Specifications are written in first order predicate logic and refined towards an implementation in a PASCAL-like programming language. Verification is carried out using dynamic logic, an extension of first order logic.

For the Production Cell case study the specification consists of three parts. On the bottom level there are two specifications for every machine, describing the events and states of the individual devices. At a second level there is a specification for every device that describes the restrictions imposed by the use of the device in the context of the Production Cell. On the top level the system specification is defined by combining the device specifications and the interaction between them.

Starting from this specification, a complete formal development was carried out including implementation and verification. Safety properties were shown by proving that an initially valid assertion holds after each control command. Liveness was shown by proving that in each state reached by the system there is still an expected event for which not all devices will be stopped.

Both liveness (in the understanding of the notion used here) and the safety properties could be expressed in pure propositional logic, which in principle, is decidable. Due to the extremely large state space the theorem prover failed to do these proofs automatically. To tackle this problem, the necessary case distinctions in the proofs often had to be done interactively.

Evaluation

All safety properties were expressed in the specification, but not completely proven. A weak form of liveness, basically saying that at every instance a device is moving, has been proven. The more complete liveness property, saying that each blank will be processed and will appear at the end of the deposit belt, seems to be inexpressable at the language level chosen. The specified safety properties make use of implicit assumptions about the behaviour of the Production Cell, which is not explicitly specified.

It took two weeks to come up with a first version of specification, implementation, and some lemmata. During the proof of the liveness property, which lasted about two more weeks, and during which about 30 errors were detected, 1000 lemmata were formulated. The proof of the safety properties resulted in about 50 cases, of which eleven were carried out, taking about eleven days. The specification contains about 600 axioms and is about 2000 loc long. The program has a length of 160 lines.

One of the three specification levels is fairly reusable. Liveness proofs seem not to be reusable at all, but the excellent proof replay mechanism of KIV makes the correction of specification and implementation easier.

3.2.11 Tatzelwurm

The goal of this contribution was to come up with a complete formal development for the control software of the Production Cell, enforcing a number of safety and liveness conditions.

The verification system *Tatzelwurm* was designed at the Universität Karlsruhe [24] to develop and verify sequential programs written in an imperative language. A Hoare-like calculus is used to generate proof obligations sufficient for the correctness of the program. For the proof of the validity of the verification conditions a theorem prover specializing in program verification is available.

The system accepts a subset of Pascal as a programming language. An order-sorted first order logic with equality is used as the specification language. The input for the verification system consists of a program containing annotations on the specification. The prover uses the analytic tableaux developed by Smullyan [32]. The set of rules is enlarged by rules for equivalence and for a generalization of the modus ponens. The prover can be used automatically without user interaction.

For this case study a program controlling the Production Cell was implemented. To prove the validity of safety and liveness requirements both the behaviour of the Production Cell, as well as the required properties have been specified. A discrete, explicit model of time was used. The assumption that the execution speed of the program and the reaction of the devices was quick enough to stop the machines was made. The state of the Production Cell was described by a set of functions, predicates, and state variables corresponding to the position and motion of each device.

The final version of the implemented control program is able to handle several blanks synchronously. It works by repeatedly (1) reading the sensor values, (2) computing the corresponding actuator events using a decision table, and (3) sending the computed output to the cell. Some safety requirements were shown almost automatically after having stated some lemmata valid for the decision table mechanism. Although in principle regarded to be easy, liveness was not formally proved.

Evaluation

All safety properties were specified. For a simpler version of the control program (CP1, capable to handle only one blank in the system) all properties were proven. For the complete version of the control program (CP2) only some proofs were carried out. Liveness was not addressed. The assumptions about the environment used for the proofs are explicit and separated from the specification of the control program. As the model does not specify the whole automaton, but only the relevant part, proving properties was not so hard.

After getting familiar with the problem, CP1 was implemented in three days and verified within three weeks. About 14 days were spent to implement CP2.

3.2.12 Timed Transition Diagrams and HOL

The object of this contribution was to illustrate the use of a graphical specification language and an interactive theorem prover for the formal development of the Production Cell controller. The contribution is distinctive in three ways: *real-time behaviour* of controller and environment are modelled, the specification *generalizes* the task as given in Chapter II, allowing many different geometries and component speeds, and verification is done using *partially automated deductive proofs*.

The control program was specified using Timed Transition Diagrams (TTD)[21][22]. This is a graphical notation for state transition systems where the state transitions depend on enabling conditions and timing constraints. The enabling condition states a necessary condition for the transition. The timing constraint gives a lower limit for how long the enabling condition must hold before the transition may be taken, and an upper limit for how long at most the enabling condition may be true without taking the transition. Hierarchical Timed Transition Diagrams (HTTD) also allow the use of Hierarchical Transitions (HT), which are here sequences of timed transitions.

All requirements were specified in Real Time Temporal Logic (RTTL). In addition to the standard safety and liveness requirements as stated in the task description, this contribution also comprises some successfully proven real-time liveness properties. An example for a real-time liveness property is "if arm1 is about to unload the table, then eventually withing a given time the press will be loaded."

Proofs were carried out using the HOL theorem prover [15]. HOL is a generic higher order logic theorem prover. A special instantiation of HOL for HTTD was used. The specification was written as a HOL-HTTD theory, some samples of proofs have been successfully carried out on this theory.

Evaluation

All requirements within the task description were specified. All properties were proven, samples of these properties were proof-checked using the HOL system. Additional real-time properties were specified and proven. Necessary assumptions on the environment were made explicit by listing them in a distinct process.

Approximately two person months was spent on this contribution. Most of this time was spent on developing proof support in HOL and identifying a specification style which led to faithful implementations. With hindsight, and the current state of the tools, the time necessary for a similar problem is estimated to be about four weeks, including complete verification with HOL-HTTD.

During specification and verification, errors or ambiguities were uncovered. From the graphic simulation, a number of incorrect assumptions made in the specification were discovered.

3.2.13 RAISE

The objective of the RAISE contribution was to specify the production cell system, to prove properties of the system at a high level of abstraction, to generate an implementation, and prove that the implementation fulfils the specification.

RAISE stands for "a Rigorous Approach to Industrial Software Engineering". RAISE is a software development method along with a specification language (RSL) and tools supporting the development method. The development method is characterized by stepwise refinement of the specification (until an implementation is obtained) and by *rigorous* proofs (for example the proof of an implementation relation). Rigorous means: by default formal, but not necessarily formal. RAISE provides a tool to support rigorous proving: the justification editor. RSL is a VDM-like specification language with many enhancements, the main addition being CSP-like constructs allowing for process-oriented specification.

In this project the stepwise refinement process consisted of the following steps:

- description *ProductionCell_1* of the machine's states,
- specification *ProductionCell_2* of the safety conditions with the CSP-like RSL-constructs by refinement (extension) of *ProductionCell_1*,
- axiomatic specification *ProductionCell_3* of the control processes by extension of *ProductionCell_1*,
- proof that *ProductionCell_3* fulfils the safety conditions of *ProductionCell_2*,
- explicit recursive definition *ProductionCell_4* of the control processes (refinement of *ProductionCell_3*),
- formal proof that *ProductionCell_4* is an implementation of *ProductionCell_3* (proof of the *implementation relation*),
- iterative implementation *ProductionCell_5* of the control processes,
- rigorous proof that *ProductionCell_5* is an implementation of *ProductionCell_4*,
- generation of executable code by the RAISE Tools (as soon as the RAISE Tools allow for).

The main outcomes of the project at the current time are (1) specifications at different levels of abstractions of the press, the robot and the production cell hardware interface processes, (2) an implementation of the press, the robot and the production cell interface processes, and (3) a rigorous proof that the implementation satisfies the safety conditions.

An significantly improved version of this contribution was published in [11].

Evaluation

The set of expressible properties was restricted by the language. One had to put in a lot of implicit knowledge about the production cell in order to ensure safety properties, which are only properties of the control program. These are – as the designer implicitly knows – sufficient to guarantee the intended properties of the system. Liveness properties are – on the language level chosen – not expressible.

RSL, being a wide-spectrum language, provides the user with a vast amount of possibilities to model the system. The choice among them was really difficult. The size of the language has another consequence: automation of proofs is nearly impossible. The support tool for proofs is a justification editor which still imposes the problem to the user to find the right theory to apply. Both reusability of specification and proof seems to be quite restricted.

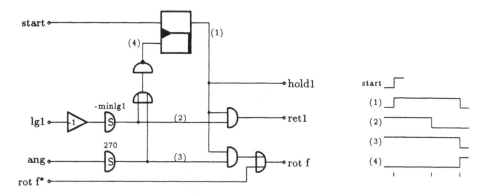

Figure 2 Circuitry for transporting a blank from the elevating rotary table into the press (without safety conditions)

The single specifications described above are all between 100 and 200 lines in length. A couple of weeks was spent for writing these specifications. It took about one day to carry out the proofs.

3.2.14 Deductive Synthesis

The aim of this contribution was to write a formal specification of the Production Cell that was as close as possible to the informal requirements description and showed how a verified TTL-like circuitry could be constructed from this description, using Deductive Program Synthesis.

The deductive program synthesis approach developed by Manna and Waldinger [26] is a method for program development in the small. It concentrates on deriving one algorithm from a given specification and some given axioms of background knowledge. Axioms and specifications are given as first order predicate logic formulas. One tries to prove the specification formula, thereby simultaneously constructing a correct functional program from the answer substitutions arising from unification.

A suitable terminology was defined for the case study. The creation of a formal language level was attempted, in which the informal description given could be expressed almost "1:1" – in order to defuse the problem of introducing errors while formalizing the requirements. Time was modelled by explicit parameters, space described by three-dimensional Cartesian coordinate vectors. After this, a collection of obvious facts about the behaviour of the machines was formalised.

The formal specification consists of four parts: the description of the behaviour required from each machine, the description of the behaviour required from each control circuit, background facts from geometry, arithmetics, and physics, and the actual specification of the goal of the Production Cell.

The control circuitry was not really "synthesized" in the sense that an actual intermediate proof goal would provide many hints about which program or circuitry constructs to insert. Instead, a previously constructed circuitry was in fact verified. During the proof some higher level constructs were introduced which turned out to be valuable in lifting specification and proofs to a higher level of expressivity. Two logical operators, *unless* and *leads to*, borrowed from Misra/Chandy's language *Unity* were defined. Not all proofs were carried out completely.

Evaluation

Both safety and liveness properties were expressible at the language level chosen. Liveness was even the main proof goal and thus addressed immediately. All assumptions necessary for the proofs are explicit to a high degree, both geometry and time are modelled with extraordinarily fine granularity. This makes it possible to implement a very efficient control program, which – due to its good knowledge about the geometry – is actually able to ensure the safety properties paying the minimum possible price. For instance, the robot's arms need not to be retracted completely if one has the necessary data.

The specification comprises 80 axioms and is about 600 lines in length. The circuitry synthesized (verified) so far treats only a part of the production cycle: it is able to move a blank from the beginning of the feed belt into the press. In the first attempt, it took 1–2 months to verify this circuitry, after having gained more experience and using the *unless* and *leads to* operators, this time was shortened to 1–2 days.

Flexibility is not the strength of this approach, as deductive program synthesis is aimed at programming in the small. We think this technique is an interesting approach to HW/SW-Codesign, if one suitably extends the set of available constructs for synthesizing the system according to a given goal.

3.2.15 Symbolic Timing Diagrams

The goal of this contribution is the synthesis of a control program directly from requirement specifications. Unlike the current approaches for hardware synthesis, this work starts from a non-operational description. This specification is given with so-called Symbolic Timing Diagrams (STD) [31], a graphical notation, used to write down temporal logic formulae (cf. Section 3.2.5).

An STD specification is a set of symbolic timing diagrams associated with an interface description which can be compared to an entity declaration in VHDL. Each STD represents a particular safety or liveness requirement. These diagrams are used to synthesize a controller using the tool and associated method of the ICOS^2 tool. This is done in two steps, the decomposition phase and the synthesis phase.

In the decomposition phase global system requirements are separated into local ones which are guaranteed by the system components. The correctness of this decomposition is supported in the ICOS^2 system by testing as well as automatic formal verification.

In the synthesis phase the decomposed specification is first checked for consistency (i.e. satisfiability and compositionality). The specification is then checked for input determinism (i.e. the whole specification determines a unique output for a given input). Finally, ICOS^2 synthesizes a controller by generating a set of finite state machines. These are then translated to C-code.

Evaluation

It is the very nature of this approach, that system requirements are the starting point. Therefore, these properties are guaranteed by the contruction approach taken. As all requirements have been used for the synthesis, all of them are guaranteed to hold for the controller.

The approach also implies that assumptions about the environment are made explicit. Implicit knowledge of the designer is only introduced if additional symbolic timing diagrams are added due to the lack of input determinism. All assumptions are explicitly represented by STDs.

This contribution re-used work completed in the StateCharts/STD solution (cf. 3.2.5). Using ICOS^2, a student designed the Production Cell controller within 30 hours. The size of single automata varies between 66 nodes and 347 edges (for the automaton controlling the feed belt, for instance) and 151 nodes and 716 edges (robot). Both the translation from STDs to automata and from automata to C-code was automatically done in only a few seconds.

3.2.16 LCM/MCM

The goal of the MCM/LCM contribution was to model and to formally specify a production cell controller. Verification aspects were discussed and it was explored how proofs of safety properties would be performed. No implementation of a control program was derived from the specification.

LCM (Language for Conceptual Modelling) is a formal modelling language based on dynamic logic and process algebra. It allows to specify applications as systems of synchronously communicating, dynamic objects. MCM is a design method for LCM that gives guidance in finding informal and formal system models and in validating the latter against informal system descriptions. LCM and MCM were developed by Feenstra and Wieringa at the Vrije Universiteit Amsterdam in the context of object oriented information systems and databases [12]. This background and the general view on structural and behavioural system modelling taken in LCM is similar to the TROLL *light* approach (cf. section XXI).

The application in this case study shows that the language concepts of LCM are also suitable for modelling control applications, whereas the design strategy of MCM had to be adapted with respect to so-called control objects. The design process started with informally defining objects of the system and a description of local events within objects as well as communication events between these objects. Parts of the physical environment like sensors and actuators were modelled as virtual device objects. Two device objects communicate by participating in a synchronous events. The control function of the system, i.e. the control program, is modelled by control objects. They encapsulate device communications and enforce the required behaviour of the devices. The structure of objects is given in an algebraic style like abstract data types. The behaviour of objects is described by a recursive process specification (cf. [2]).

Safety constraints are written as static integrity constraints in the form of logic predicates. Such constraints serve as an invariants which have to be true in all observable states of a model. Verification of particular constraints is done by explicitly proving this invariant for all relevant state transitions.

Evaluation

In this contribution the main focus was on modelling and specification and on the question how safety and liveness properties could be expressed in LCM and be verified using theorem proving techniques. Due to the lack of an appropriate theorem prover no detailed proof of safety and liveness constraints was done.

The LCM specification makes a number of assumptions about the behaviour of the production cell, which simplifies the control task considerably. It is assumed that the conveyor belts are moving continuously and that the feed belt delivers blanks to the table at the right moments, i.e. there is always left enough time for the table and the robot to return to the required positions. These assumptions are explicitly expressed in the specification.

Through three iterations an informal system model was built within about 12 hours. This model was validated by informal examination of use-cases. It took one more day to write a formal specification of about 400 lines.

3.2.17 Modula-3

The object of this contribution was to examine how the benefits of object-oriented design and parallel programming can be combined by modelling and implementing a controller of the Production Cell in the programming language Modula-3 [17]. Also, safety requirements were verified for the model obtained.

Modula-3 adopts the techniques of structured and modular programming and adds constructs for object-orientation and parallel programming. Concurrent programming is supported in Modula-3 by so-called *threads* which allow multiple points of execution. Concurrency is *lightweight*, since all threads in a program live in the same address space. To guarantee mutual exclusion when entering critical sections, Modula-3 provides mechanisms for the synchronization of threads.

All devices of the Production Cell are modelled as individual objects. They have to fulfil their tasks autonomously, and communicate with each other to ensure the correct processing. Therefore, the control of each device is organized in parallel threads that communicate via signals.

To avoid active waiting, which would result in the purely sequential controlling of the entire cell, a particular sensor control object dealing with all sensor values was introduced. The process associated with this object cyclically reads all sensors and, if a sensor returns a significant value, sends a signal to the device that needs to know this sensor value.

The resulting control program showed that object-oriented modelling profits from the parallel constructs allowing for more intuitive modelling. The concurrency model implemented in Modula-3 (threads as lightweight processes) was easily applicable in this case study.

Safety requirements of the Production Cell were expressed as class invariants. Parallel constructs were incorporated into the verification by additionally assigning a signal invariant to each communication signal. Such an invariant describes a condition which holds if a process sends a signal, and which another process can assume to hold when receiving this signal. Using the weakest precondition calculus, several safety requirements were verified with comparatively little effort.

Evaluation

Some safety requirements were expressed as class invariants and successfully proven. The modelling allows for the reuse of single components as well as larger aggregates. The Modula-3 program comprises about 24 classes, having about 1000 lines of code. Modelling and implementation were done by a student in about 3 months.

3.2.18 TROLL *light*

As the focus of TROLL *light* is the construction of correct information systems, the objective of this contribution was not the proof of the suitability of the method for constructing a correct implementation. Instead, the Production Cell was specified as a community of concurrently existing and interacting objects.

TROLL *light* is a language for the conceptual modelling of information systems and has been developed by the Technische Universität Braunschweig [14]. Systems are specified as a community of interacting objects, the main specification concept being object descriptions, called *templates*. Templates describe static and dynamic properties (i.e. valid states and valid state evolutions of a prototypical object). An object description may depend upon predefined data types, other objects, and may make use of other objects as sub-objects. Templates have visible attribute states and can constrain the possible event sequences occurring to an object. They describe the effect of events on attributes as well as the synchronization of events in different objects.

TROLL *light* was used to describe the possible states and state evolutions of the Production Cell. In the first step the objects have been identified and placed into an object hierarchy, the template *ProductionCell* being the highest one. Secondly, the attributes and events as well as the effects of the events on attributes were described. By restricting possible attribute states using constraints and possible event sequences by behaviour patterns in the third step, it was possible to express some safety properties. Finally, the synchronization of events in different objects was specified using interaction rules.

Evaluation

It was possible to restrict object behaviour in order to ensure some of the safety properties, but TROLL *light* has no means to describe stipulation of particular object behaviour. Therefore, the specification does not describe the Production Cell completely. Especially, liveness properties are not expressible within the language. Properties have not been proved, therefore, no assumptions about the cell's behaviour were used. The specification consists of 400 loc, and it took less then five days to complete it. We assume that parts of the specification code could be reused for an extended or modified version of the Production Cell; the contribution allows no conclusion about the reuse of proofs.

3.2.19 Other contributions

Other formal developments have been finished using

* the telecommunication specification language LOTOS,

* the Duration Calculus [29], a real-time specification language developed at the DTH Lyngby,

* and the object-oriented extension to the specification language Z, Object-Z [1].

Additionally, an object-oriented framework for the rapid construction of control programs for a wide range of different production cells has been developed and successfully instantiated for this production cell [7]. The framework was implemented in an concurrent variant of the language Eiffel [27].

3.3   Conclusions

3.3.1   Guaranteed properties

This case study shows how formal methods allow for guaranteeing properties of systems consisting of software, hardware, and mechanics. The resulting contributions can be used as a roadmap both for selecting and applying a method to a similar problem. As the role of software in life critical applications is becoming more and more important, we stress that in this case study, formal methods proved to be the only viable way to achieve a high level of safety.

3.3.2 Errors in requirements and specifications

The case study was carried out by several partners all over Europe. We tried to come up with an unambiguous, complete requirements description. Even though the Production Cell is a system which is relatively easy to understand, our first task description was both incomplete and erroneous. During the specification process errors were detected and corrections of the task description were submitted by many contributers. The same occured during the construction of the simulation with the graphical visualization. The Esterel contribution, for instance, revealed that the feed belt must have a certain minimum length, if the controller is supposed to put two blanks on it. After being stopped with one blank at the one end, there must be space enough at the other end to put another blank there.

It is well known that the transition from the informal to the formal description does not only detect errors in the informal requirements, but also is likely to introduce errors. The probability of these errors grows with the difficulty of the formal notation chosen.

Generally, a formal specification can be inadequate for several reasons:

- it can repeat an error already contained in the informal task description,
- it can be the consequence of a wrong assumption which closes a gap which remained open in the informal description, or
- it can be a misinterpretation of the informal description, which, due to the impreciseness of informal text can never be avoided.

This implies that one has to address the problem of formalization with the same rigor as implementation. We see a need for intuitive and adequate formalisms that allow for building models which closely simulate reality and support discussions with the customer. Additionally, formal specification must be combined with traditional techniques like tests, reviews, walk-throughs, and rapid prototyping.

Formal methods can help in detecting errors in the informal description, but it is unlikely that the formal description achieved at first is perfect.

3.3.3 Reuse of proofs

As described above, some control programs which were specified and verified formally proved to be incorrect upon connection with the simulation. This was because of errors typical to the software development process, not due to bad management or insufficient diligence. We regard these mistakes as unavoidable: formal methods maybe able to eliminate errors in the final product, but they cannot deter errors from occurring during the development. Thus, specifications and implementations should be flexible so that invested proof work, typically the most expensive in the whole software life cycle, can easily be reused.

We do not believe that proof automation can replace suitable proof re-use mechanisms. This case study shows that even problems of modest real world complexity are just within the current limits of available automatic proof tools and model-checkers. Reuse of proofs is likely to be achieved by suitable modularization techniques, the most promising currently being those of object-oriented approaches.

3.3.4   Scalability

The complexity of the case study was underestimated in the beginning by almost all contributors. Both current formal verification techniques, namely (at least partially) automated logical deduction and symbolic model-checking cannot be applied to problems much larger than this case study without further additional efforts. In our opinion it is necessary to combine both techniques, as done in the StateChart (section 3.2.5) contribution, and planned in the SDL work (section 3.2.7).

3.3.5   Assumptions about the environment

Assumptions about the environment of the control program can either be put into the formal model or can remain implicit in the developer's mind. The granularity of the models chosen by the contributions differs very much (in the contribution described in Section 3.2.14 it is extraordinarily fine), and its impacts on the ease of proofs is obvious: a very simple model allows for easy, automatic proofs, but their results depend on stronger assumptions, often not explicitly justified.

Having fixed and justified the granularity of the model, there are still some alternatives about how explicit the assumptions about the environment must be. Using interactive simulation techniques (possible e.g. for SDL or StateCharts) bears the danger that assumptions play their main role in the mind of the person interacting with the simulator. Among others, sections 3.2.5 and 3.2.14 show very nice examples of how an explicit model of the environment was used for verification.

3.3.6   Hardware-Software Co-design

At the time the case study was launched there was no sensor installed at the end of the feed belt. During formalization we recognized that one would be able to prove much stronger properties using much weaker assumptions *with* such a sensor installed. The task of writing a control program would become much simpler. Thus we decided to install the sensor.

An abstraction of this particular problem would immediately lead into the field of hardware-software co-design. It turns out that formal methods can give useful hints for the hardware design of a machine. Of course, there is a mutual dependency between the system and the formal method applied: methods capable of dealing with real-time constraints would rather use information about the time it takes a blank to get from the begin of the feed belt to the elevating rotary table.

It would be quite handy to have local microprocessors for each actuator which could help in issuing commands to the machines at a higher level. For instance, instead of switching the motor of the press on and switching it off after the press reached the upper position, then one could simply say "move the press to upper position". This would make the task of writing a control program much easier, and the programs running on the local microprocessors could easier be verified.

3.3.7   Outlook

We are aware that the present "comparative survey" is rather a survey than a comparison. The case study provides us with a lot of material we can learn from. The authors will carry on trying to evaluate this example.

Acknowledgements

The authors want to thank all people who contributed to this case study. These are too numerous to name them all. We thank Eduardo Casais and Hardi Hungar for their valuable comments. We thank Bradley Schmerl for proof-reading an earlier version of this article, and Laura Olson for assistance on language and presentation of the current version.

References

[1]   C. Albers. Spezifikation und Verifikation einer industriellen Fertigungszelle mit Object-Z. Diploma thesis, 1994, Forschungszentrum Informatik, Haid-und-Neu-Straße 10-14, 76131 Karlsruhe, Germany. In German language.

[2]   J. C. M. Baeten, W. P. Weijland. Process Algebra. Cambridge Tracts in Theoretical Computer Science 18, Cambridge University Press, 1990

[3]   D. Barnard, J. Cuellar, M. Huber. A Tutorial Introduction to TLT –Part I: The Design of Distributed Systems; – Part II: The Verification of Distributed Systems, Technical report Siemens ZFE BT SE 11, 1994

[4]   M. Broy, F. Dederichs, C. Dendorfer, M. Fuchs, T. F. Gritzner, and R. Weber. The design of distributed systems – an introduction to Focus. Technical Report SFB 342/2/92, Technische Universität München, 1992.

[5]   M. Broy, C. Facchi, R. Grosu, R. Hettler, H. Hußmann, D. Nazareth, F. Regensburger, and K. Stølen. The requirement and design specification language Spectrum, an informal introduction. Technical Report TUM-I9140, Technische Universität München, 1992.

[6]   J. R. Burch, E. M. Clare, K. L. McMillan, D. L. Dill, and J. Hwang. Symbolic model checking: 10^{20} states and beyond. In *Proceedings of the Fifth Annual Conference on Logic in Computer Science*, pages 428–439, 1990.

[7]   E. Casais. An Experiment in Framework Development. Technical Report, 29 pp., Forschungszentrum Informatik, Haid-und-Neu-Straße 10-14, 76131 Karlsruhe, Germany. Submitted for publication.

[8]   K. M. Chandy, J. Misra. Parallel Program Design - A Foundation. Addison-Wesley, 1988

[9]   E. M. Clarke, E. A. Emerson, and A. P. Sistla. Automatic verification of finite state concurrent systems using temporal logic specifications. In *Proceedings of the 10th ACM Symposium on Principles of Programming Languages*, pages 117–126, 1983.

[10]   B. Dutertre. *Spécification et preuve de systèmes dynamiques*. Ph.D. thesis, University of Rennes 1, France, December 1992. (In French)

[11]   F. Erasmy, E. Sekerinski. Stepwise Refinement of Control Software — A Case Study using RAISE. Proceedings of the FME 94, Barcelona, Spain, LNCS, Springer Verlag, to appear.

[12]   R. B. Feenstra, R. J. Wieringa. LCM 3.0: a language for describing conceptual models. Technical Report IR-344, Faculty of Mathematics and Computer Science, Vrije Universiteit, Amsterdam, December 1993

[13]   Th. Filkorn, H.-A. Schneider, A. Scholz, A. Strasser, P. Warkentin, *SVE System Verification Environment,* to appear

[14]   S. Conrad, M. Gogolla, and R. Herzig. TROLL *light*: A core language for specifying objects. Informatik-Bericht 92-02, Technische Universität Braunschweig, 1992.

[15] M. J. C. Gordon, T. F. Melham. *Introduction to the HOL System*, Cambridge University Press, March 1994

[16] N. Halbwachs. *Synchronous Programming of Reactive Systems*. Kluwer Academic Publishers, 1993.

[17] S. P. Harbison. *Modula-3*. Prentice Hall, 1992.

[18] D. Harel. A visual formalism for complex systems. *Science of Computer Programming*, 8:231–274, 1987.

[19] M. Heisel, W. Reif, W. Stephan: *A Dynamic Logic for Program Verification*. "Logic at Botik" 89, Meyer, Taitslin (eds.), Springer LNCS 1989.

[20] M. Heisel, W. Reif, W. Stephan: *Tactical Theorem Proving in Program Verification*. 10th International Conference on Automated Deduction, Kaiserslautern, FRG, Springer LNCS 1990.

[21] R. W. S. Hale, R. M. Cardell-Oliver, J. M. J. Herbert, An Embedding of Timed Transition Systems in HOL. *Formal Methods in System Design*, 3(1&2),pages 151-174, Kluwer, September 1993

[22] T. A. Henzinger, Z. Manna, A. Pnueli, Temporal proof methodologies for real-time systems. In *Proceedings of the 18th Symposium on Principles of Programming Languages*. ACM Press, 1991.

[23] C. A. R. Hoare. *Communicating Sequential Processes*. Prentice Hall, 1985.

[24] T. Käufl. The program verifier *Tatzelwurm*. In H. Kersten, editor, *Sichere Software: Formale Spezifikation und Verifikation vertrauenswürdiger Systeme*. 1990.

[25] L. Lamport. The Temporal Logic of Actions. Technical Report, Digital Systems Research Center, 1991

[26] Z. Manna and R. Waldinger. A deductive approach to program synthesis. *ACM Transactions on Programming Languages and Systems*, 2(1):90–121, Jan. 1980.

[27] B. Meyer. *Object-oriented Software Construction*. Prentice Hall, 1988.

[28] O. Nierstrasz, S. Gibbs, and D. Tsichritzis. Component-oriented software development. *Communications of the ACM*, 35(9):160–165, Sept. 1992.

[29] J. L. Petersen. Specifying a computer controlled forging machine. Diploma Thesis, Technical University of Denmark, Lyngby, 1994.

[30] J. L. Peterson. *Petri Net Theory and the Modelling of Systems*. Prentice Hall, 1981.

[31] R. Schlör and W. Damm. Specification and verification of system-level hardware designs using timing diagrams. In *The European Conference on Design Automation with the European Event in ASIC Design*, pages 518–524, 1993.

[32] R. M. Smullyan. *First Order Logic*. Berlin Heidelberg New York, 1968.

IV. CSL

Controller Synthesis and Verification: A Case Study

Klaus Nökel, Klaus Winkelmann,

Siemens AG, Corporate Research and Development

Abstract

CSL (Control Specification Language) is a declarative language for specifying structure and behavior of finite-state systems. We generated a controller for the production cell which was verified using SVE (System Verification Environment), a symbolic model checker. We were able to verify all demanded safety properties, and in addition a substantial set of liveness properties. Model checking also revealed a subtle error in the original design of the controller.
From the good performance results of our experiments we conclude that current techniques can handle realistic problems of complexity well beyond the one represented by the production cell.

4.1   Goals of the CSL Contribution

Our goal was to specify and automatically synthesize a correct controller for the production cell, verify it and connect it to the simulation.

The specification should make use of the modular structure of the cell so that re-use of components in a similar cell would be possible with reasonable effort. The suitability of CSL for the specification of production cell controllers was to be assessed, in particular the following questions had to be addressed:

- Can the specification be written in an application-oriented style?

- How easy is it to express interesting technological requirements?

- Which aspects of a specification can be expressed naturally, and which less naturally in CSL?

4.2 A First Look at CSL

4.2.1 Design goals

CSL was conceived as a declarative specification language that bridges the gap between two conflicting goals in the design of large discrete controllers. On the one hand control tasks often form families in the following sense: the controlled plant or process is composed of distinct interacting physical objects and exists in many different configurations of these objects. Each plant configuration is associated with the problem of designing a controller that is adapted to the particular configuration. Examples occur in building automation, traffic light controllers, railway interlocking systems, and industry automation. In fact, the production cell can be viewed as just one instance of the more general task of designing controllers for any arrangement of belts, presses, robots, etc. The problem calls for a compositional specification method that separates the description of the plant component types from the description of an individual plant structure.

On the other hand the synchronous approach to reactive system design possesses a number of important advantages [1]. From the implementation perspective, synchronicity is most easily guaranteed in a centralized controller, for example in a finite state automaton (FSA) that receives plant events and computes the control actions simultaneously for all plant components given complete information about the plant status without further communication. However, for control tasks of industrial size FSAs tend to grow extremely large (the *state explosion problem*), so that manual construction quickly becomes impossible. Therefore FSA synthesis must be automated.

With CSL we attempt to reconcile a compositional specification style with the automated synthesis of a centralized controller.

4.2.2 Specifications in CSL

Our approach to compositional specification has been inspired by similar work in model-based reasoning (MBR) about physical systems [4] [5]. In MBR the physical behavior of aggregates is inferred from the given physical behaviors of components that are considered primitive and from structural information about the aggregate (interactions between the primitive components).

Similarly, a CSL specification consists of a component library and a structure description. An intuitive representation of the plant structure for MBR purposes enumerates the component instances and defines how they are interconnected. Since the controller depends on the particular plant configuration, we use the same structure description (i.e. in terms of *plant* components) as a part of the controller specification. The structural description may be hierarchically organized.

The key idea in CSL is that the presence of a physical component in the plant not only influences the aggregate plant behavior, but also constrains the functionality of the controller. Stated in FSA terms, each physical plant component contributes its share to the controller state space and partially defines the transition and output functions of the controller. Hence, a CSL component library is organized by physical plant components, but contains type-specific constraints on the controller behavior[1]. These constraints are instantiated according to the structure description and from these the state transition and output functions of the finite automaton are computed.

Writing a CSL specification consists of three steps:

- identify the relevant plant component types,
- for each type, define the constraints placed on the controller by an instance of the type,
- describe the structure of individual plants.

The component library is simply a collection of component type descriptions. Each type is given as a 6-tuple (N, P, S, I, O, T) where

- N is the name of the type;
- P is a finite set of engineering parameters which differ from instance to instance but which are constant over time;
- $S = \{ s_1, ..., s_n \}$ is a finite set of controller state variables, each with a finite domain $dom(s_i)$;
- I is a finite set of input symbols;
- O is a finite set of output symbols;
- T is a finite set of so-called transitions (see below).

---

1. This is in contrast to MBR where the component library contains a model of the physical behavior of the primitive component types themselves.

A structural description of a plant consists of

- a finite set of typed instances;

- a value assignment to the type-specific engineering parameters for each instance;

- optionally, a neighborhood relation on the instances given as a labelled graph[1].

The state space of the controller is defined as the cartesian product of the state spaces contributed by the plant component instances which in turn are the cartesian products of the type-specific dom(s_i). Likewise, the set of controller input (output) symbols is the union over all instances of their type-specific input (output) symbols.

The transition and output functions of the controller FSA are determined by the transitions. Each transition is a quadruple (*name, input, condition, output*) The language for the conditions is propositional and contains as literals all formulas of the form "$s_i(x) = a_{ij}$" , for $a_{ij} \in$ dom(s_i) and some component instance x. In addition to the boolean operators the language contains a (one-step) next operator, denoted by "**". Informally, transition conditions denote relations on two successive control states.[2] The semantics of a transition (*name, input, condition, output*) is as follows: if the controller receives *input* in state *s* and there is a future control state *s'* such that *condition(s,s')* is satisfied, then one such successor state is selected and *output* is emitted. Transitions should not be confused with rules; first, multiple transitions for the same input are applied simultaneously and secondly, transitions may be nondeterministic.

The CSL compiler takes as input a component library and a structural description and synthesizes the appropriate FSA. Whenever the instantiated constraint system has no solution, the compiler pinpoints the transitions involved in the conflict. If the system is under-constrained, the compiler arbitrarily chooses one deterministic FSA that satisfies all constraints. The resulting FSA is represented as a data structure that is efficiently interpreted by a universal FSA interpreter. The interface between plant and controller consists of a set of C routines corresponding to the input and output symbols which are linked to the FSA interpreter.

---

1. not needed in our specification of the production cell.
2. Furthermore the language contains syntactic sugar to "quantify" over sets of instances. These "quantifiers" allow for concise formulations, but since all domains are finite, they are merely abbreviations for certain conjunctions / disjunctions. A macro facility supports naming and reuse of complex formulas.

4.3 Modelling the Production Cell

4.3.1 Cell layout

The structure of the production cell can in a natural way be mapped to a collection of CSL components. Obviously, each component like the press or the crane becomes one type in CSL. There is yet some freedom left in this mapping; to make maximal re-use possible, we chose to define a robot arm as a separate type, and a belt as a type. Then we get two instances of the arm type, and two of the belt type. Of the other types, which are robot (i.e. robot base), press and crane, we have one instance each.

The CSL engineering parameters ("parameters") are used to specialize the generic types, by defining which other components the instance "knows". The table, for instance, knows the feed belt, because there is a dependency between those; namely, the controller has to assure that no part is transported beyond the end of the feed belt if the table is not ready for it. Likewise, the press knows the robot arms and so on. Loosely speaking, every component gets as parameters its neighbors in a logical sense, which often coincide with physical neighbors.

In the general specification of one type, we need not state what type its neighbor is, but can for example just require that it has particular attributes or macros, like "part_present" or "ready_to_feed".

4.3.2 Component specification

For each component type, we have to describe (as explained in section 4.2) the name, engineering parameters, inputs, outputs, internal states and transitions.

We present the complete CSL code for one type, the elevating rotary table, interleaved with some commentary text. We start with the name and engineering parameters:

```
declTyp(rot_table, Self,
     [parameters: [feeder: belt],
```

The *internal states* represent the physical attributes of a component. Each state variable ("zustand") has finitely many values, based on a discretization of the state space:

```
states:[
       vertical_position: [low, moving_up, high, moving_down] default low,
       rotary_position: [0, moving_plus, 45, moving_minus] default 0,
       part_on_table: bool default no],
```

Here, the "default" construct specifies the initial values.

Inputs correspond to sensors. Any event that is relevant for the control is assumed to be signalled by an input symbol. The event set needed for this application is also finite; although CSL allows for real parameters to be part of an input, we did not make use of this feature, but have used a discretized model, assuming that for the analogous sensors, only the relevant milestones are signalled to the controller.

```
inputs: [low(rot_table), not_low(rot_table),    /* sensor 7: */
        high(rot_table),                        /*sensor 8 */
        rot_0(rot_table), rot_45(rot_table)],   /*sensor 9 */
```

The *outputs* directly correspond to the control commands:

```
outputs:
        [up(rot_table), down(rot_table), stop(rot_table),
        rot_plus(rot_table), rot_minus(rot_table),
        stop_rot(rot_table)],
```

A very useful facility to write a legible and well structured specification are the *macros* of CSL. We use them to express higher-level properties of the state of a component, defined by logical expressions over the internal states, for example the condition for the robot for picking up a part from the table, and the condition for the feed belt for feeding a new part:

```
macros: [ready_for_pickup(Self) :=
        part_on_table(Self) = yes
        ∧ vertical_position(Self) = high
        ∧ rotary_position(Self) = 45,
    ready_for_new_part(Self):=
        part_on_table(Self) = no
        ∧ vertical_position(Self) = low
        ∧ rotary_position(Self) = 0],
```

After defining the static part of a component (inputs, outputs, states and macros) we come to the dynamic part, i.e. the *transitions*. Each transition is a quadruple of

```
[transition name, input symbol, condition, output]
```

The following transitions are normally activated cyclically in the given order, apart from the pair "start feeder/stop feeder", which can occur in any order.

```
transitions:
```

When a part disappears from the feeding belt, it is assumed to be on the table. The table is lifted, and the belt keeps moving until stopped by the transition stop_feeder.

```
[start_up, part_disappears(feeder(Self)),
vertical_position(Self) = low
    ∧ ** part_on_table(Self) = yes
    ∧ ** vertical_position(Self) = moving_up,
[up(Self)]],
```

Rotation of the table starts only when a safe level is reached[1].

```
[rotate, not_low(Self),
vertical_position(Self) = moving_up
    ∧ ** rotary_position(Self) = moving_plus,
[rot_plus(Self)]],

[ stophigh, high(Self),
** vertical_position(Self) = high, [stop(Self)] ],

[ stop45, rot_45(Self), ** rotary_position(Self) = 45,
[stop_rot(Self)] ],
```

Here, the robot is assumed to take away the part and set the variable "part_on_table to "no".

```
[rotate_back, empty,
vertical_position(Self) = high ∧ part_on_table(Self) = no
    ∧ ** rotary_position(Self) = moving_minus,
[rot_minus(Self)]],
```

To guarantee that the table does not rotate in the low position, we must first rotate back before we start moving down.

```
[start_down, rot_0(Self),
rotary_position(Self) = moving_minus
    ∧ ** rotary_position(Self) = 0
    ∧ ** vertical_position(Self) = moving_down,
[down(Self), stop_rot(Self)]],
```

Arriving at the bottom concludes the cycle:

```
[stoplow, low(Self), ** vertical_position(Self) = low,
    [stop(Self)] ],
```

---

1. The task description that we used stated that "the table must not rotate in the low position", a requirement which was weakened later, but is met by our solution. To achieve it, we had to introduce the input "not_low(table)" indicating when a sufficiently high position is reached.

The feeder is restarted when it is not moving and the table is ready for a part, and stopped when a part is about to arrive and the table is not ready. Note that in the following, "feeder(Self)" refers to the device serving as the feeder of "this" table, which in the current layout happens to be the feed belt:

```
[start_feeder, empty,
ready_for_new_part(Self)
        ∧ moving(feeder(Self)) = no ∧ ** moving(feeder(Self)) = yes,
[forward(feeder(Self))]],
[stop_feeder, empty,
part_at_end(feeder(Self)) = yes
        ∧ not(ready_for_new_part(Self))
        ∧ moving(feeder(Self)) = yes ∧ ** moving(feeder(Self)) = no,
[stop(feeder(Self))]]]
```
]].

This concludes the specification of the table. In this style, each component is specified. The total specification of the six types is nine pages of CSL code, and took two days to write and debug. For debugging, apart from the CSL compiler we used a simulation tool which is part of the CSL environment. We started to formally verify the specification by model checking only when simple simulations indicated that a plausible solution was arrived at.

4.3.3   Implicit assumptions in the modelling

In every formal model, some implicit assumptions are necessary. They influence the choice of the parameters and variables of the model itself, and cannot be expressed within the model. An example of an implicit assumption which hardly anyone would question is

"at a given time, the press is in at most one position".

Another one which we made and which turned out to be less certain, refers to the

Capacity of the devices:

We assume: "Each of the magnets, as well as the table and the press can hold at most one part at a time".

This resulted in a binary variable for each of these places indicating the presence of a part, thus we cannot directly express the requirement that there is only one part for example on the table. Yet, as will be demonstrated, it is no problem to prove that no part will be moved to the table if there is already one. A similar problem occurs with the

Capacity of the belts:

When analyzing the original task description and the simulation we noted the sensors did not provide sufficient information for deciding when it is safe to put a part on a belt (both the feed and deposit belt). In general, there are several options at this point:

- One might demand additional sensors and modify the cell layout. This seemed not within the rules of the case study.

- One can take the most cautious approach and (make the controller) deposit a part only when the previous part has arrived at the end of the belt. This seemed unsatisfactory for the following reasons:

  — Even this solutions has to presuppose some minimal requirements of the relative lengths of the parts and the belts, which are strictly speaking not given.

  — When the parts are small, compared to the length of the belts, this solution would heavily under-utilize the belts, which in practice may well serve as buffer capacities.

  — In a realistic setting, i.e. without the crane to close the loop, some external agent would have to put the parts on the feed belt anyway, without control by the cell controller.

- Therefore we chose a third, pragmatic option: we assume that the belts move fast enough that two consecutive parts are sufficiently separated, and that new parts are put on the feed belt by the external agent only at appropriate times.

With the current simulation, the controller resulting from this assumption can handle only four parts, because, as the relative device speeds happen to be, the fifth part would still wait on the initial position when the crane deposits the first part at that same position. Making the crane wait in that situation would indeed allow us to handle a total of five parts, but, taking into account that a typical objective in industrial automation is to *minimize*, rather than maximize "work in progress", this did not seem worthwhile to us.

The deposit belt turns out not to be a problem because it is significantly faster to move a part away than it takes the robot and press to provide the next part.

4.3.4   Connecting the simulation

When connecting the CSL controller to the simulation or more generally, embedding it in an environment by means of the inputs and outputs, one has to translate between the different ways to handle events; on the input side, CSL expects an input symbol (event) whenever something relevant has changed (e.g. the table has arrived at its upper position). The simulation, on the other hand, provides the whole state as a status vector.

We wrote a small C program to convert the state changes to CSL input events. This is obvious for the binary sensors. For the real-valued sensors this interface program generates events whenever the value is crossing certain threshold ("milestone") values. For the output, as the simulation does expect the commands in the style provided by CSL(i.e. only changes are note) the interface does nothing but translate the names from CSL style to the format understood by the simulation, for example:

forward(feed_belt) is translated to belt1_start

rotate_minus(robot) is translated to robot_right, etc.

Defining such an interface for CSL is done by just customizing some well-defined routines which are called by the automaton interpreter. In that way the controller can be connected to various environments easily.

4.4   Verification Using Model Checking

The finite automaton generated by the CSL compiler was examined by the System Verification Environment (SVE), a very efficient symbolic model checker developed by Siemens ([2], [3]). Converting the compiler output to the SVE format is a standard feature of CSL.

To express the properties to be checked SVE uses a language (model checker language or "MCL") with the following essential elements:

- CSL expressions are used as an embedded language to refer to static program properties (any CSL expression without the next operator "**", including macros),

- temporal aspects can be expressed either in propositional linear-time temporal logic (PTL) or in computation tree logic (CTL),

- inputs and outputs can be referred to within MCL, which contains a predicate logic for expressing properties about states.
- boolean operators are available in MCL as well as in embedded CSL.

Each call to the model checker identifies

1. The automaton to check (in our case the production cell controller),
2. The commitment (i.e. the property to prove), and
3. Optionally, an assumption to use in the model checking. The assumption is expressed in PTL and defines which executions of the automaton are considered in the verification - i.e. the commitment is checked for the automaton restricted to the execution sequences defined by the assumption.

As the formulae needed for (2) and (3) are often rather long, a convenient macro facility allows to pre-define abbreviations in a file.

For the production cell we had SVE examine the complete controller. It was not necessary to split it in its components for efficiency reasons, although SVE offers a feature for such splitting (projections).

Environment model

To prove any non-trivial property of the production cell controller, we must make explicit the assumptions about the behavior of the environment, in other words, the process model. These assumptions are of two types:

1. physical plausibility: "a sensor signal will only arise when the corresponding device is actually moving towards the position indicated by the sensor", and
2. fairness: "when a device keeps moving towards some position, it will eventually arrive there and be detected by the appropriate sensor".

For proving safety properties, we used only physical plausibility assumptions, some of which are listed below. Such assumptions were defined for each device and their conjunction was used for all safety proofs.

```
a_press :=
    (input = lower(press) ==> (position(press) = moving_ul))
    ∧ (input = middle(press) ==> (position(press) = moving_lm
                    ∨ position(press) = moving_ul))
    ∧ (input = upper(press) ==> (position(press) = moving_mu)).

a_robot := (input = pos1(robot) ==> (position(robot) = moving31))
    ∧ (input = pos2(robot) ==> (position(robot) = moving12
                    ∨ position(robot) = moving31))
```

∧ (input = pos3(robot) ==> position(robot) = moving23).

It is interesting to note that these assumptions make only very sparse use of temporal operators, which makes them easy to write and read. Essentially only one "globally"-operator ranging over the whole conjunction is needed. The only exception is caused by our special treatment of the sensor "not_low(table)"

The overhead for handling the process model assumption is needed only once, as SVE is able to save for future computations the finite state machine described by the model plus an assumption.

Safety properties

In formalizing the safety properties, it was possible to restrict ourselves to properties of the simple form "globally (i.e. in every reachable state), some static property holds". In other words our safety properties describe subsets of the state space; they make only trivial use of temporal logic and are therefore easy to understand.

This is so because all relevant aspects of the cell are directly modelled as CSL variables, and also because of our above-mentioned simplifying assumption on the belt capacity.

Outputs are in this case considered to be part of the state in which they occur.

Formalizing the safety requirements is for most cases straightforward as is illustrated by the following examples.

```
/*------ SAFETY REQUIREMENT 1 ----------------------------------*/
/* the press closes only when no arm is inside*/
safe_prop1 := ag(
    press_closes ==> (not arm1_inpress ∧ not arm2_inpress)
    ).
press_closes :=
    position(press) = moving_lm ∨ position(press) = moving_mu.
arm1_inpress :=
    (position(robot) = pos3) ∧ not(extension(arm1) = 0).
arm2_inpress :=
    (position(robot) = pos2) ∧ not(extension(arm2) = 0).


/*------ SAFETY REQUIREMENT 9 ----------------------------------*/
/* feed belt only moves a blank to table when table is ready for it */
safe_prop9 := ag(ready_belt ==> ready_table).
ready_belt := (moving(feed_belt) = yes ∧ part_at_end(feed_belt) = yes).
ready_table := (vertical_position(table) = low ∧
        rotary_position(table) = 0∧ part_on_table(table) = no).
```

Less elegant is the formalization of a more dynamic property such as "all motors are stopped at the boundaries", for it must be translated as "the motor must stop when the device *arrives* at the boundary" rather than "... when it *is* at a boundary". This is not a problem in MCL, as shown below (mobility restriction for the crane), but a more natural treatment would be possible if the original model had been more explicit in this aspect by introducing additional values "out_of_boundary" for each position attribute.

```
/*------ SAFETY REQUIREMENT 4 ------------------------------------*/
/* all motors are stopped at the boundaries */
/* (crane:) */
safe_prop4_crane_h := ag(
    (((input = over_dep_belt(crane)) V (input = over_feed_belt(crane)))
    ==> 'stop_horizontal(crane)')).

safe_prop4_crane_v := ag(
    (((input = height1(crane)) V (input = height2(crane)))
    ==> 'stop_vertical(crane)' )).
safe_prop4_crane := safe_prop4_crane_h ∧ safe_prop4_crane_v.
```

Liveness properties

We aimed at a compositional treatment of the liveness of the cell. For each component we formalized and checked properties of the type:

"when a part is in an initial position A, it will eventually arrive in a final position B", and

"the component will always return to its initial position".

As parts are not identified (e.g. by part numbers) in our model, it is not strictly possible to express the first property above, but only something like "... some part will eventually arrive in position B". Given the additional knowledge that the components (table, each robot arm, press) can hold at most one part, the desired property follows.

Obviously, to prove liveness properties, fairness assumptions have to be made. For each desired liveness property we used just the necessary fairness assumptions, because of performance reasons; model checking becomes the more complex the more fairness assumptions are taken into account.

For the table, the necessary fairness assumption looks like this:

```
fair_table :=
    g(f((vertical_position(table) = moving_down) ==> (input = low(table))))
    ∧
```

g(f((vertical_position(table) = moving_up) ==> (input = high(table))))
∧
g(f((rotary_position(table) = moving_plus) ==> (input = rot_45(table))))
∧
g(f((rotary_position(table) = moving_minus) ==> (input = rot_0(table)))).

With this, we can prove a simple progress property, namely "the table will, once a part has arrived to it, move to the upper position":

life_table_12 :=
    ag((part_on_table(table) = no) ==>
        ax((part_on_table(table) = yes)

            ==>

            af((vertical_position(table) = high) ∧ (rotary_position(table) = 45)))).

For proving the converse (i.e. the table will move down again), we need the additional assumption that the robot will eventually remove the part:

robot_takes := g(
    ready_for_pickup(table) ==> f(part_on_table(table) = no)).

With these assumptions (a_proc_OK, fair_table and robot_takes), we can now also prove that the table will always return to its initial position:

life_table :=
    ag(af(vertical_position(table) = low)).

The assumption about the robot that was used above can in turn be proved separately. We call it therefore a lemma. To prove it we need in addition a fairness assumption for robot arm 1 as well as another lemma stating that the robot will always return to its initial position:

life_robot :=
    ag(af(position(robot) = pos1)).

To prove life_robot the fairness assumptions about both arms as well as the press have to be included. In this way we have grounded all necessary lemmas on plausible fairness assumptions. We could instead of using the lemmas directly put all fairness assumptions into the proof of the top level goal (in this example life_-table_21), but the compositional approach used has two advantages:

- it gives better insight into the dependencies of the properties, and

- it reduces run-times.

Almost all the expected progress properties could be proved in this way. Probably the most interesting experience of our verification attempts occurred when we repeatedly failed to prove the following claim: "whenever the press holds a part eventually the robot will go to the position for unloading the press",

life_robot2 :=

   g((part_present(press) = yes) ==> f(position(robot) = pos2)).

Even with adding more and more fairness assumptions about robot, arms, press and even table, this did not become true. Finally we had to accept that it is, in fact *not true* for our controller:

Following the task description we had built a controller which makes the robot pick up a part from the table before it unloads the press. This is indeed the most efficient way to do the task, but what if there is no further part on the table?!

The model checking has uncovered this problem with our controller, which can be stated as: the last part (of a lot or a day production) will not be unloaded! Obviously, testing the controller with the closed loop setting would never uncover this problem.

4.4.1   Performance of model checking

When we started using SVE the safety properties did not pose any performance problem; they were handled in well below one minute each. Introducing fairness assumptions initially exploded the computation times to the 30 minute range or more, making efficient testing with several iterations practically impossible. Yet when the SVE developers looked at the problem they were able to introduce a few optimizations which resulted in drastic improvements:

- The assumptions are only computed once (as explained above).

- The set of reachable states is precomputed, which takes a few minutes. As an aside, the number of reachable states is about 50 million for our model. The set of reachable states is then used as a don't-care set in the model checking algorithms.

- The most efficient improvement resulted from dynamic reordering of the BDD variables. It is well known that the BDD sizes are very sensitive to the variable order. Variables are reordered during the model checking process to keep the intermediate BDDs small.

With these improvements all considered liveness properties were handled in less than 300 seconds each on a Sparc 2 work-station, most of them in less than 30 seconds.

4.4.2   Experience

Although here only the final positive results have been given, the way to these results was of course not quite so straight. At many points an initial proof attempt
failed and we could make excellent use of the debugging feature of SVE. When a
property cannot be proven, a counter-example is provided; that is a sequence of
states which leads to violation of the property; in case of a liveness property it is a
cyclic sequence.

The counter-examples often led to additional assumptions that were necessary
in the proof.

But we also encountered more dangerous bugs in our own formulae. At some
points a property was proven even without assumptions, or in an extremely brief
time. This often was a symptom of a faulty formalization. In case of a commitment
of the form g(A ==> B) we performed in such cases a simple plausibility check,
for example by trying g(not A); if that is true, g(A ==> B) is very likely not to be
what one wanted to express.

Extreme care has to be taken when formalizing properties. One example illustrates how intuitive names can easily suggest meaning that must not be relied on;
in one version, we proved that "moving(table)" is false in a certain situation. Yet
as that variable was neither used nor ever set by (this particular early version of)
the specification, this has no connection whatsoever to the real world. Double-
checks like the one suggested above can help to avoid such fallacies.

Writing the necessary assumptions was straightforward in most cases, as with
the examples given above. Only when the state space of a component was modelled in an ad hoc non-standard fashion this led to barely comprehensible assumptions, as in the following example: For the robot arms, the state space was defined
as

states: [extension: [0, pos_put, pos_get, moving01, moving10] default 0, <etc.>

So when the arm is "moving" in or out the variable carries only the direction
but not the complete information about the interval (is it between 0 and the put-
position or between the put- and the get-position?). This turns out to be sufficient
for the controller. But for model checking this information has to be re-constructed
using history, which makes the formulae unnecessarily complicated:

```
fair_arm1 :=
    g(f((extension(arm1) = moving10) ==> (input = ext0(arm1))))
    ∧ g((extension(arm1) = 0)
        ==> x(
```

f((extension(arm1) = moving01) ==> (input = ext_put(arm1)))))
∧ g((input = ext_put(arm1))
    ==> x(
    (f((extension(arm1) = moving01) ==> (input = ext_get(arm1))))
    ∨ (extension(arm1) = pos_put))).

The verification of the safety requirements, including the formalization of the physical plausibility assumptions, took less than one week. The liveness properties were handled by a student who had never before heard of model checking but had a good background in mathematics. It took her about one week to learn using the system, and two weeks to verify all the properties reported here (and some more) including thinking of and formalizing the necessary assumptions.

4.5   Evaluation

Here we address the questions posed by Lindner/Lewerentz in the task description for the case study.

We have proved all safety requirements demanded in the original version of the task description. For the general mobility restriction (any device must be stopped at the boundaries of its operating region) we have proved only some instances because not only the duplication seemed uninteresting but these properties can be easily seen from the specification.

We have proved an essential set of liveness properties whose conjunction implies the correct operation of the cell. We have in these proof attempts uncovered a situation which would lead to an undesired blocking in a real production cell (without the artificial closing of the loop using the crane), viz. in the case of no more blanks arriving from the feed belt.

The assumptions about the correctness of each component are divided in plausibility and fairness assumptions. All of them were coded in temporal logic. They are all of similar form and most of them can be well understood. It turned out yet, in the case of the arms' extensions, that the assumptions are tricky to write and understand if the modelling of the state space in the specification itself is not well structured.

The specification takes 9 pages of code including comments. The static part is easy to understand, while reading the transitions requires a well documented specification as well as some background in the CSL semantics. This is mostly due to the fact that the inherent sequential structure of the problem is not directly mirrored in a transition system. We have meanwhile developed polished-up versions of the specification which make this structure more visible by introducing constructs to specify action sequences directly. CSL's macro facility allows such extensions with very little effort.

The efforts spent are summarized in table I (person days and person weeks):

activity	effort	code size
designing, coding and debugging CSL specification	3 days	9 pages (CSL)
connecting simulation	1 week	7 pages (C)
verifying safety	< 1 week	5 pages (MCL)
verifying liveness	2 weeks	5 pages (MCL)

Table 1　Total Efforts for CSL Solution

When taking into account that it was the first task of its type that we addressed and that the liveness verification was done by an inexperienced student, we consider these figures very satisfactory. With some experience and the specification even better adapted to the verification needs, we expect the verification effort to come down considerably for future applications.

These figures should be compared with the typical effort in programming a conventional industrial controller for a similar task. According to experience at Siemens this would take the order of two to six weeks including test and debugging if a language like Step 5/ Graph 5 is used. Thus we can safely conclude that, even with today's state of the art, CSL competes favourably with a conventional approach - keep in mind that if we verify only safety and not liveness, we have already a more reliable system than with a conventional approach.

We have not actually experimented with changing the controller but the CSL separation of cell structure and component library provides a sound basis for adaptations. As long as the state variables stay the same (e.g. in an optimized version of the controller) the formulae for model checking could be re-used. The question of re-using proofs itself is not applicable to a model checking approach as this runs fully automatic.

As far as we can see, our controller achieves maximum throughput; the press, which turns out to be the bottle-neck, is always busy. Yet, as explained above, not more than four parts can be handled simultaneously, and we have not fixed the above-mentioned "last part" problem.

4.6   Conclusion

It has been shown that CSL and model checking are powerful tools for synthesizing a provably correct controller for a small production cell.

Further, from the moderate resources consumed by our solution, we conclude that the method can handle a complexity well beyond that of the considered production cell. In other words, realistic problems are within the scope of the method and we have indeed started using CSL on several industrial control tasks.

Experience has shown that formal verification is indeed of practical use in developing a controller. On the other hand, we have also learned that care is necessary in writing the properties to be proved and in interpreting the results. We believe that more research should be invested in ways to make the mapping between the program variables and the reality less prone to the kind of errors outlined above.

We will end with indicating a direction of future improvement which applies not just to CSL but to most specification languages. Although we call the code in these languages "specifications", it is semantically very close to an actual program, in the sense that we specify not only what the controller should do but also how it is to be done. Obviously there is plenty of room for more abstract specifications as indicated by the following example. Instead of prescribing which output signals to activate in what situation, a specification might just say

- what position a device should be moved to, and

- what the safety constraints (legal state combinations) are.

Lifting the specification to this abstract level would result in an even better reuse as well as easier adaptation, for example to modified safety requirements.

We are working on enhancing CSL with that kind of declarative mechanism.

Acknowledgements

We are grateful to Thomas Lindner for actively supporting our participation in the case study. Thomas Filkorn and Peter Warkentin provided valuable support and powerful improvements in SVE. Ina Kühnel has been of great help in proving a substantial number of liveness properties.

References

[1]   A. Benveniste, G. Berry: *Real Time Systems Design and Programming*, in Proc. of the IEEE, vol. 9, no. 9, September 1991

[2]   J.R. Burch, E.M. Clarke, K. L. McMillan, D. L. Dill, L.J. Hwang, *Symbolic Model Checking: 10^{20} States and Beyond*, Information and Computation June 1992, vol. 98, pp 142-170.

[3]   Th. Filkorn, H.-A. Schneider, A. Scholz, A. Strasser, P. Warkentin, *SVE System Verification Environment*, to appear

[4]   W. Hamscher, L. Console, J. de Kleer (eds.): *Readings in Model-Based Reasoning*, Morgan Kaufmann, 1992

[5]   D.S. Weld, J. de Kleer (eds.): *Readings in Qualitative Reasoning about Physical Systems*, Morgan Kaufmann, 1990

V. ESTEREL

Applied to the Case Study Production Cell

Reinhard Budde

GMD Birlinghoven

Abstract

The aim of this contribution was to design a control program for the production cell using Esterel, an imperative language for reactive programming. Esterel is based on the perfect synchronization hypothesis and relies on signal broadcasting. The Esterel program is executable. It may either be simulated, or, interfaced with simple C-functions, run as a UNIX-process, or cross-compiled to be executed by a micro-controller. Some properties of the program are proven by an Esterel-compatible model-checker.

The contribution explains the language and the synchronous paradigm, describes the integration of Esterel into a complete embedded system, discusses design style and use of compiler, theorem prover and graphic tool.

The first section introduces to the language, the second to the design of the production cell. The third section discusses in more detail the ROBOT_PRESS module and how properties are proven, the fourth the interfacing of Esterel with the "real" environment. In the last section experiences are reported.

5.1   A Quick Introduction to Esterel

Esterel is a member of the family of synchronous programming languages, which also comprises Lustre, Signal and StateCharts. This family spans a wide range of modelling styles: StateCharts are well-suited as means for specification, Lustre and Signal are declarative languages based on a dataflow model, Esterel is an imperative language. Synchronous programming languages are well-suited for reactive programming [1].

5.1.1   Basic concepts

Reactive programming

A reactive program has to respond to stimuli from the environment. To achieve this, the program is thought to be connected with input and output lines to it. Signals can flow in and out using these lines. Signals have to be declared:

```
input     FEED_BELT_OCCUPIED, TABLE_CAN_TAKE_PLATE;
output    FEED_BELT_MOTOR_ON, FEED_BELT_MOTOR_OFF;
% Two input and two output signals are declared
```

The program does not react continuously to the environment. The following operational model explains this. Two different phases are distinguished: the program is reacting or it is idle.

When the program is idle, arriving signals on input lines (e.g. FEED_BELT_OCCUPIED) are collected, but not yet propagated to the program. The change from the idle phase to the reacting phase is effected by an activation event (in hardware design one would call this a clock). Whether this is done periodically, using a quartz or a timer subsystem of the underlying hardware, or depending on conditions, is not modelled in the Esterel program. Whenever the activation event happens, the gathered signals together with a special signal tick make up the input event and are provided to the program.

The program may test presence or absence of signals, emit signals, start parallel branches etc. All output signals emitted during the reaction are part of the output event. The reaction is complete if the program halts (see the halt-statement, later). Then all signals of the output event are made available to the environment and the program is idle again.

Usually the lines of the output signals are connected to actuators or display units and effect changes in the environment. These changes may produce input signals, which are collected, again. Then, if the clock ticks, the next reaction step is initiated.

Each reaction step is called an instant. A program reacting is logically disconnected from the environment, i.e. it is impossible to add input signals to the input event during a reaction.

The following example of a simple reactive program is explained below.

```
loop                                                                   %(1)
    await   [FEED_BELT_OCCUPIED and not TABLE_CAN_TAKE_PLATE];          %(2)
    emit    FEED_BELT_MOTOR_OFF                                         %(3)
end loop
```

- Comments start with a %

- The loop (1) never terminates.

- If the loop is activated in an instant, the await (2) commits the loop to halt. Nothing else is done in that instant.

- In each future instant it is checked whether the signal FEED_BELT_OCCUPIED is present and the signal TABLE_CAN_TAKE_PLATE is absent. If this is the case, in the same instant, i.e. instantly, the await (2) terminates and statement (3) emits the motor-off-signal. The emit statement terminates instantly and the loop body is executed again. As described above, the await (2) commits the loop to halt.

There is no rendezvous-like concept in Esterel. Sender, which emit signals, will never be blocked. This would be the case in asynchronous languages, where sender are blocked if no corresponding receiver awaits the signal. Further, Esterel has no concept of a channel. Signals are instantaneously broadcasted to the program, and they are *not* consumed in an await statement. This facilitates simultaneous reactions to the same signal, as the example shows:

```
output DISPLAY(string);                                                      %(1)
signal PLATE_IN_DANGER in                                                     %(2)
    [                                                                        %(3)
        loop
            wait [FEED_BELT_OCCUPIED and not TABLE_CAN_TAKE_PLATE];           %(4)
            emit PLATE_IN_DANGER
        end loop
    ||  every PLATE_IN_DANGER do                                             %(5)
            emit FEED_BELT_MOTOR_OFF
        end every
    ||  every PLATE_IN_DANGER do                                             %(6)
            emit DISPLAY("kb-28-10-89: jam on feed belt")
        end every
    ]
end signal
```

- In (1) a valued output signal is declared. It can emit a string to the environment. The string is displayed on an output device.

- A local signal PLATE_IN_DANGER is declared in (2) and may be used in the scope of the signal block.

- The statement [... || ...] (3) executes its constituents in parallel.

- The local signal PLATE_IN_DANGER is emitted if the more complex await condition of (4) becomes true.

- The signal PLATE_IN_DANGER is recognized at the *same time* in (5) and (6) and effects two different (sub-)reactions.

The perfect synchronization hypothesis

The paradigm underlying synchronous languages is the perfect synchronization hypothesis. Reactions are instantaneous, i.e. the output event is computed from the input event without any delay. This is equivalent to:

- input signals are synchronous with output signals.

- the program is considered to be executed on an infinitely fast machine.

- reactions are atomic. Different reactions cannot interfere. During a reaction input lines cannot influence the program.

This seems to be an unnatural and unreasonable paradigm. One could argue, that real computations use real time (>0), and that no engineer can build an infinitely fast machine. But let us take another position. *Infinitely fast* is defined as faster than any given limit. From a constructive point of view, one has to select an arbitrary test value for the limit and then to conduct an experiment to check whether the program reacts faster. However, for any system there exists an upper bound for these test values. Consider, for example, a system that uses only one A/D-converter as a sensor to receive input signals and only one motor as actuator controlled by an output signal. Typically the A/D-converter delivers input with at most 50 Hz and the motor can handle at most one command in 0.1 sec. In this configuration a test value for infinity higher than 100 Hz makes no sense. In a nutshell: the granularity of time in the environment is coarser than 100 Hz.If the system reacts always in less than 1/200 sec, one cannot conduct an experiment in *this* environment to detect the system's finite execution time.

The consequence of this hypothesis is an obligation for the controller programmer. He or she has to prove, that the time between two instants (this time depends on the clock event, which is originated in the environment) is large enough to compute any of the possible reactions. If this is guaranteed, then — relative to the environment — the program, which of course needs time to compute a reaction, can be considered either to be infinitely fast, or atomic, or able to compute output instantaneous when the input arrives.

The hypothesis simplifies the design. The specification of real time behaviour (based on perfect synchronisation) and implementation (has to guarantee perfect synchronisation) are clearly separated.

5.1.2   Notions and programming style

To understand the Esterel design of the production cell, some basic constructs are explained. Readers interested in a more complete overview should read [1], those interested in the semantics of the language should read [2]. All constructs are introduced using examples from the production cell. Simple statements of Esterel are:

- emit for emitting signals.
- halt for freezing further computation of a reaction in this instant.

Compound statements of Esterel:

- present to test for presence or absence of signals.
- [.. || .. || ..] for parallel execution.
- loop for executing a loop-body forever.
- watching for hard preemption of a guarded statement.
- trap for weak preemption.

Further statements of Esterel:

- signal to introduce local signals.
- module for structuring in the "large".

Watching

The basic statement for programming reactions is watching. It executes a statement guarded by a watchdog:

```
do
    loop                                        %(1)
        emit ROBOT_CAN_TAKE;
        await tick
    end loop
watching ROBOT_HAS_TAKEN
```

The guarded statement is the loop (1). Without the watching it would never terminate. It emits the signal ROBOT_CAN_TAKE and then awaits the next instant (remember that the signal tick is part of every input event).[1] The watching statement can be understood in different ways:

---

1. For the whole loop the abbreviation sustain ROBOT_CAN_TAKE can be written. In Esterel many of such abbreviations are defined, based on a small number of primitive statements. This small kernel makes the semantics easy to comprehend.

- a watchdog is attached to a body. If the watchdog signal is emitted else-where, the body is preempted immediately, i.e. has no chance to react fur-ther (hard preemption).
- a priority between signals is defined. ROBOT_HAS_TAKEN has higher priority than ROBOT_CAN_TAKE. Watching statements can be nested to build more complex priority structures.

If a reaction is needed when the watchdog terminates a watching, a timeout-clause is used. The full syntax is: do Stmt1 watching SignalExpr timeout Stmt2 end

Present

The present statement tests for the absence or presence of signals. Consider the feed-belt of the production cell. If a plate is detected by the sensor at the end of the belt, the motor has to be switched off, except for the case that the table signals the plate can be moved onto the table. The present statement is used to avoid a motor-off immediately followed by motor-on.

```
await S_FEED_BELT_OCCUPIED;
present FEED_BELT_CAN_DELIVER then
    nothing
else
    emit A_FEED_BELT_OFF;
    await FEED_BELT_CAN_DELIVER;
    emit A_FEED_BELT_ON
end present;
```

Halt

The halt statement never terminates. It freezes the program component at this very point. Without a preemption statement like watching or trap a halt blocks perma-nently. The statement await X is an abbreviation of do halt watching X.

Parallel execution

The parallel operator ‖ makes statements execute in parallel. The compound state-ment [A ‖ B ‖ C] terminates after all branches A, B, C have terminated. Communi-cation between branches is only possible by emitting of and watching for signals.

Trap

The watching statement is a hard preemption of a statement from outside. Weak preemption from inside uses the trap construct. It defines a context, which is left instantaneously if a corresponding exit statement is executed. Parallel branches have a chance to continue until they halt (a "last will").

```
loop                                                                     % (1)
    %   press is free, table has to deliver
    await ROBOT_can_TAKE_from_TABLE;
    %   press is loaded by the first robot arm
    trap PRESS_IN_USE in                                                 % (2)
        loop                                                             % (3)
            await PRESS_CAN_BE_UNLOADED;
            % press is unloaded by the second robot arm.
            % the first robot arm may or may not take from the table (depends on the table)
            present ARM_1_HAS_A_PLATE then                               % (4)
                % press is loaded again, PRESS_IN_USE remains valid
            else
                exit PRESS_IN_USE                                        % (5)
                % preemption from inside: press is idle, wait for the next plate from the table
        end loop
    end trap
end loop
```

In this example the trap (2) denotes a state, in which the press is operating continuously controlled by the loop (3), i.e. is loaded immediately by the first arm after
the second arm has unloaded it. Whether this state remains valid or not, is checked
inside the load-unload-loop in (4) and thus needs a weak preemption in (5). This
terminates the trap and restarts the loop (1).

Signal

Details of communication between program components can be hidden from the
outside by introducing local signals. These signals are like input and output signals, they belong to an instant, cannot be consumed etc. They are broadcasted like
all other signals, only the scope in which they are visible is restricted.

```
signal   PRESS_CAN_BE_LOADED, PRESS_CAN_BE_UNLOADED,
         PRESS_LOADED, PRESS_UNLOADED in
             [ % implementation of the robot
             II % implementation of the press
             ]
end signal
```

The robot and the press act independent from each other, expressed by parallel
execution. Their synchronisation needs four signals. These signals are made local
to exclude undesired interference from other parts. If no other components uses
these synchronisation signals, one could move them to global scope without
changing the semantics of this example (but for reasons of information hiding, one
should desist from this, of course).

Modules

Modules can be defined for grouping a program, but they do not define name spaces, i.e. scopes for signals etc. Modules are used in run statements like macros as shown below:

```
module FEED_BELT:                              module PRODUCTION_CELL:
    output  FEED_BELT_HAS_DELIVERED,               % ...
            CRANE_CAN_DROP;                        [ run FEED_BELT
    input   FEED_BELT_CAN_DELIVER,                 || run TABLE
            CRANE_HAS_DROPPED;                      || run ROBOT_PRESS
    loop                                           || run DEPOSIT_BELT
        do  sustain CRANE_CAN_DROP                 || run CRANE
        watching CRANE_HAS_DROPPED;                ]
        % ...                                  end module
    end loop
end module
```

The compiler replaces a run statement by the list of statements contained in the module definition. Like with macro expansion, renaming of signals and constants is allowed. This module concept bears all problems known from macro techniques: re-use of module is hard and readability is bad; information hiding is achieved by conventions only.

Nondeterminism and Causality errors

Many program designer accept the following dependencies as inevitable: If they want deterministic behaviour of a program, they have to confine themselves to sequential programming languages, but, if they require nonsequential features, they have to master nondeterminism. With respect to this, Esterel is quite uncommon: the language combines deterministic behaviour and nonsequential features. Both is desirable for embedded systems: Active components and their coordination are described realistically in a nonsequential model. To master the technical complexity of these systems, a reproducible deterministic behaviour is required, however.

To achieve this in Esterel causal dependencies between signals and to a weaker degree between data are checked. Only if there exists a unique "reason" for each reaction, the program is considered legal.

Consider the signals first. A reaction is the set of emitted signal. If in a fixed instant a signal is added to this set by emit, it cannot be removed within this instant (the reactive effect of a program can be understood as a fix-point of this set). For any set of input signals and any given state of the program, this set has to exist and to be unique. The compiler checks this.

Error (a)   Error (b)

```
signal X in                          signal X in
  present X then                       present X then
    emit X                               nothing
  else                                 else
    nothing                              emit X
  end present                          end present
end signal                           end signal
```

The present statement tests whether the local signal X is emitted in the current instant. In (a) the set of output signals is not unique. Both, {} and {X} are valid outputs. In (b) the set of output signals does not exist. If X is a member, it is not emitted, if it is not a member, it is emitted.

In both cases the compiler detects so-called causality errors. Sometimes it is hard to fix such a problem, because complex dependencies between signals may produce it (A more complex example is given in the ROBOT-PRESS object).

In Esterel some basic datatypes are built-in. Their dataflow is analysed by the compiler to exclude race conditions.

Error (c)   Example (d)

```
var a : integer in                   do
  [ await X;                            await X;
      a := a+1                            a := a+1
  || await Y;                          watching Y
      b := a                          timeout
  ]                                     b := a
end var                              end
```

Program (c) is rejected: the compiler checks for hidden communication, which could produce a nondeterministic behaviour. By introducing priorities (d) between signals this conflict can be resolved.

5.2 The Production Cell in Esterel

5.2.1 Object-based partitioning

An object-based partitioning technique was used to design the program. This can be done in many different ways. One possibility is to concentrate on the moving plate and describe its state changes. But as the design should facilitate re-use, it makes sense to use the manipulation devices for partitioning the program, as they

are the physical units reused in similar production tasks. Thus, the production cell
was divided into five objects:

- the feed-belt,
- the table,
- the robot-press (which contains two objects as its parts, the robot and the
  press),
- the deposit-belt and
- the crane.

In Esterel each object is designed as a module.

5.2.2   Signals used to communicate with the environment

All modules are engaged in communication with the environment. Data is sent to
actuators and received from sensors. Signals relevant for the actuators in the environment are pure, i.e. they don't carry values. For instance, if the signal
A_FEED_BELT_ON is present, it has to trigger a hardware-component to switch
the motor of the belt on, if the signal is absent, nothing happens. A second signal
A_FEED_BELT_OFF is used to stop the belt. The reduce the number of signals,
one could map the two signal to one, that takes a value from the enumeration
{on,off} as parameter. This is not used, because the theorem prover can only deal
with pure signals. By convention actuator signals are prefixed with A_.

To get feedback from the environment, most actuator signal are supplemented
by a sensor signal. These input signals deliver boolean information about the state
of the environment. This is realized as pure signals, signalling true by their presence and false by their absence. This is a feasible solution for the photo-electric
cells, e.g. For the robot, the crane and the table, however, the interpretation is as
follows: if the sensor signal is part of the input event, then the actuator command
issued in a previous instant terminated successfully. By convention sensor signals
are prefixed with S_

A typical communication sequence with the environment reads like

```
[   emit A_ARM_1_TO_TABLE;
    await S_ARM_1
||  emit A_ROBOT_TURN_LEFT;
    await S_ROBOT
]
```

The parallel statement terminates, if both actuator commands have been issued
and terminated successfully.

5.2.3 Cooperation characteristics of modules

In the production each object acts as consumer to one and as producer to the other neighbour. The crane consumes plates from the deposit belt and produces plates for the feed-belt.[1]

In each producer-consumer cooperation one device is engaged in an active, the other-one in a passive role:

- the crane is active in both cooperations (deposit belt, feed belt).
- the deposit belt is passive in both cooperations (robot, crane).
- the feed belt is active in one (table) and passive in the second cooperation (crane)

A concept that supports re-use of components, is based on four abstractions:

- the active producer (the crane related to the feed belt)
- the passive producer (the table related to the robot)
- the active consumer (the crane related to the deposit belt)
- the passive consumer (the table related to the feed belt)

Now a naming scheme could be developed, behaviour could be cast into modules and these modules could be instantiated in the objects for the production cell. The design is based on such a model, but the abstract naming and structuring of the modules hinders the readability of the program. Thus easy-to-read names like CRANE_can_TAKE_from_DEPOSIT_BELT are used for cooperation signals. This should not be taken as an argument against the re-usability of this design (saying for instance: "the design is not re-usable, because the crane knows that a deposit belt is its neighbour"). The names are used for didactic reasons only.

To reduce the strength of the coupling between the modules, no assumptions are made about the relative speed of the devices controlled by the modules. If for example all devices are fast and the feed-belt is very slow, one could assume, that always if the feed-belt can deliver a plate the table is ready to accept it. A bad design would mingle this knowledge into the program's structure, making the program inflexible and not re-usable.

To decouple the modules, hand-shaking is needed. This is easily realized with rendez-vous-like constructs, which are excluded from Esterel. Two signals are needed to realize the blocking of a producer until a consumer is willing to communicate. Take the following example: If the crane and the deposit belt are interfaced,

---

1. Producing and consuming are used in an abstract sense.

the crane is an active consumer, while the deposit belt is a passive producer. The crane is only allowed to take a plate, when it is available. The availability of a plate is a state the deposit belt is in. This state is terminated, after the crane has taken the plate. Thus, the cooperation structure can be expressed in the following program:

```
module CRANE
    input CRANE_can_TAKE_from_DEPOSIT_BELT;
    output CRANE_has_TAKEN_from_DEPOSIT_BELT;
    loop
        await CRANE_can_TAKE_from_DEPOSIT_BELT;                      %(1)
        % take it                                                   %(2)
        emit CRANE_has_TAKEN_from_DEPOSIT_BELT;                     %(3)
        % consume                                                   % (4)
    end loop
end module

module DEPOSIT_BELT
    output CRANE_can_TAKE_from_DEPOSIT_BELT;
    input CRANE_has_TAKEN_from_DEPOSIT_BELT;
    loop
        % produce it                                                % (5)
        do
            sustain CRANE_can_TAKE_from_DEPOSIT_BELT 1              %(6)
            watching CRANE_has_TAKEN_from_DEPOSIT_BELT;            %(7)
        % is taken
    end loop
end module
```

- the state, that a plate is available, is modelled by the sustained signal CRANE_can_TAKE_from_DEPOSIT_BELT. The state is left, after the watchdog CRANE_has_TAKEN_from_DEPOSIT_BELT preempts the sustain.

- The cooperation is insensible to the number of instants, which are used to consume (4) and produce (5) the plate. This is due to the watchdogs (1) and (7), which are triggered by the cooperating partner in (6) and (3). The cooperation is also insensible to the number of instants, that are needed to take the plate, with the following exception:

---

1. sustain S emits a signal S repeatedly in the current and all further instant. It is defined as: loop emit S; await tick end

- the taking-action (2) must not terminate instantly, i.e. it *must* contain an halt. Otherwise the program is not well-formed: the emitted signal (6) may terminate the await (1). If now (2) would terminate instantly, the signal in (3) triggers the watchdog (7), which in turn kills the sustain (6). The result is, that the presence of a signal in (6) effects its absence in (6). This causality error is detected by the compiler, fortunately.

5.3 Structure and Properties of the Production Cell

In this section we outline how the production cell is modelled by Esterel modules. Using the modules ROBOT and PRESS as examples we discuss the mechanisms for communication and coordination between objects. As Table 1 shows, this

Module	States	Actions	Calls
Feed belt	6	11	26
Table	10	17	76
Deposit Belt	7	11	41
Crane	10	20	76
Robot/Press	53	24	376

Table 1   Complexity of the modules

module, combining the robot and the press, is the most complex one. The number of states is relatively small compared to asynchronous approaches. This is an impressive consequence of the synchronous paradigm due to the fact that signals are totally ordered in time.

The Esterel code for the module ROBOT-PRESS is shown in an appendix, for the full description of the design and a discussion of the complete source code the reader is referred to [4].

In Figure 1 the overall structure of the communication and the cooperation between the modules is shown. For each pair of modules the communication signals they exchange are depicted, and the style of cooperation (i.e. passive/active) is indicated. The Esterel modelling of the modules and their communication according to these patterns as discussed in section 5.2.2 was done systematically. An important step for each module is to define the proper initial conditions that have to hold,

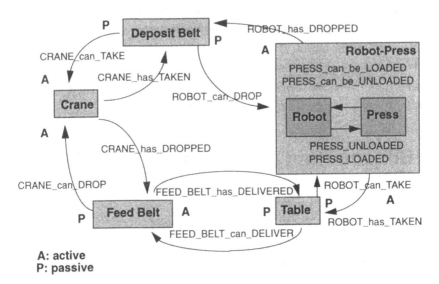

Figure 1   Overall communication structure

i.e. to find the right initialization of the program. In the sequel these issues are discussed for the modules ROBOT and PRESS.

The modules ROBOT, PRESS and ROBOT-PRESS

To achieve a safe initialization, the robot first retracts both arms, then turns to the table. Then it extends its first arm to the table. It is assumed that this never effects a collision with a plate on the table.

Note, that retracting the arms is done concurrently. This is safe, but what about turning and extending an arm? A first version was:

```
[   emit A_ROBOT_TURN_LEFT; await S_ROBOT
||  emit A_ARM_1_TO_TABLE; await S_ARM_1
];
```

Now the program reads:

```
emit A_ROBOT_TURN_LEFT; await S_ROBOT;
emit A_ARM_1_TO_TABLE; await S_ARM_1
```

Only the latter solution avoids a possible collision with the press. Note, how unpleasant the geometry of the physical parts dictates the *structure* of the program. Such dependencies are easily overlooked in code-reviews.

The main structure of the robot controller reflects two cases:

1. The press is idle. The robot waits for a plate or loads the press. This is programmed as the first part of the loop outside the trap construct.

2. The press is productive. The robot unloads the press and puts the plate onto the deposit belt. This is a loop inside a trap PRESS_IN_USE. If at unload time the table is empty, the robot commits to pure unloading and the first arm moves forth and back without a plate.[1] The press is not loaded in this case. Later on this will effect the preemption of the trap by executing an exit statement and resumption of the outer loop. If the table was not empty, the first arm takes the plate, and the inner loop surrounded by the PRESS_IN_USE-trap continues.

The press is trivial. The combination of both modules to describe the robot-press aggregate is trivial, too. The press is encapsulated by the robot and not vice versa. This encapsulation is done by declaring all cooperation signals between press and robot local signals, and defining the remaining cooperation signals of the robot as cooperation signals of the robot-press.

Thus the cooperation between press and robot is hidden inside the object ROBOT_PRESS. This is done to localize safety requirements. Otherwise a requirement like "the robot arms should not be destroyed by the press" has to be proven as a property of the whole program. This would be a not compositional, not re-usable design. Esterel is well suited to apply information hiding between objects: Actuators, sensors as well as communication signals are hidden in the ROBOT_PRESS. Perfect synchronization guarantees, that no reactions from other objects can disturb (dead-lock, decrease speed etc.) the ROBOT_PRESS.

Informal techniques like code inspection can be used to validate the safety requirement, that the arms are never destroyed by the press:

Viewed from the press, before pressing the signal PRESS_LOADED is awaited, after pressing the (sustained) signal PRESS_CAN_BE_UNLOADED is emitted. Only between these signals pressing is done.

The robot emits PRESS_LOADED twice. In each case a short inspection shows, that the arm positioned in front of the press is retracted before. It remains to be checked, that the arms remain retracted until the press emits PRESS_CAN_BE_UNLOADED. Both emissions of PRESS_LOADED are immediately followed by awaiting the PRESS_CAN_BE_UNLOADED signal, thus are safe.

---

1. A local signal is used to remember whether the loading arm (ARM_1) is carrying a plate. The signal is sustained in that case in a loop at the end of the ROBOT-module.

The safety requirement can also be formally proven by an automaton-based verification tool. Remember, that an Esterel-program reacts in an instant to a input event by generating a unique output event and becoming then idle (again). The idle state can be characterized by all those halts (for instance effected by await-statements), which are effective in this situation. Thus an Esterel-program is equivalent to a finite state machine, whose states are subsets of all halts of the program and the transitions are labelled by signal-sets (precondition for this transition to occur) and an action-table (signals to emit, e.g.). Such FSM can be analysed and manipulated by the verification tool AUTO [3]. On demand the Esterel compiler produces an automaton in AUTO-format. The main support of this tool is a semantically grounded abstraction facility to reduce the finite state machine and checking equivalence of two FSM's by bi-simulation.

In our example the FSM has to be reduced to a FSM exhibiting only the signals PRESS_LOADED PRESS_CAN_BE_UNLOADED and all the signals related to extending and retracting the both arms. The reduced automaton can be checked easily by exhaustive testing. It is also possible, to describe a FSM in terms of the process algebra Meije, that behaves safe, i.e. cyclically retracts the arms, then presses the plate, then extends the arms. This FSM can be checked for equivalent behaviour w.r.t. the FSM derived from the Esterel program. Often drawing a reduced automaton assures the programmer that a desired predicate holds. The AUTOGRAPH tool allows to visualize a FSM, produced by the AUTO-tool.

Unfortunately only pure control (i.e. no data) can be considered. For control-dominated, small examples like the production cell this is not a problem, but larger applications are beyond the scope of pure control.

5.4   Remark About Collisions and Systems Engineering

The production cell needs two views: the continuous view of the moving physical units and the discrete view of its controller. Problems to mediate between these views have gained much interest in the last years. They extend the focus from program design to systems design. They state a lot of very interesting problems, but are outside the scope of this presentation of Esterel. Thus the term *system design* was avoided and the term *program design* was used. Nevertheless some remarks are necessary.

It is hardly possible to avoid implicit assumptions during programming. But it is a bad design style to scatter implicit assumptions over a whole program, when they concern well-known problems of an engineering discipline and are likely to be invalidated in variants or later versions. A typical problem in the production cell are assumptions about geometry and relative speeds. Consider two typical problems:

Geometry

The feed belt is equipped with one sensor, answering to the question "Is there a plate at the end?", realized as a photo-electric cell. We consider this as bad *system* design. The belt should be equipped with a second photo-electric cell "Is there space for a plate at the begin?". Otherwise an inexperienced programmer could make the implicit assumption that if only one plate is on the belt and the plate is at the end, then another-one may be put at the other end. This assumption is based on the assumption, that the belt is much longer than the plates.

A programming design that avoids implicit assumptions of this kind has to be extremely defensive. This reduces concurrency and the throughput, and may create bottlenecks. It seems very hard to avoid these implicit assumptions. It is not sure, whether all or most of these dependencies are removed from the Esterel program.

A similar problem is related to the crane. When may the deposit belt move a plate to its end? This depends on the presence of the crane signal CRANE_has_TAKEN_from_DEPOSIT_BELT. But what is the precondition for the crane to emit this signal? The environment of the controller doesn't provide a sensor "Is the gripper out of the collision area w.r.t. the deposit belt?". Thus the safest solution is to emit the signal when the crane has approached the feed belt. This is the most distant position and can be sensed by a photo-electric cell. But even this contains the implicit assumption, that the plates are "enough" smaller than the reach of the crane.

Relative speed

When the robot was programmed, a decision about relative speed of the components had to be taken. That decision will influence the programs structure and, if the programmer is not aware of this problem, will remain implicit and hinder re-use.

If a plate is pressed and no unpressed plate is available at the table, the programmer has to decide,

- whether the robot unloads the press immediately
- or waits for some time to give the table a chance to deliver a plate.

If the robot's speed is the system's bottleneck, the first alternative is inferior to the second. Safety problems won't occur, but efficiency will be influenced. If the objects are re-used in different configuration, one has to design objects, which can be parametrized with speed assumptions and behave safe for all assumptions.

By the way, the analysis of relative speeds revealed a further problem of the systems design. No device had been provided to introduce new plates into the production cell's loop. At least a button for the operator is needed to announce, that he or she will put a new plate onto the feed belt. Otherwise consider the situation, that all components operate very rapidly: the operator could be injured by a plate. An implicit assumption underlying the case-study is, that either the loop is closed or the devices are operating at a speed not too fast for human beings.

5.5   From Esterel Programs to Executable Code

We assumed that all parts of the controller for the production cell have to be realized in software, if that is possible at all. An important question is how to interface an Esterel program to the "real world" environment.

For this case study it is assumed that the production cell should be controlled by one or more micro-controllers (i.e. micro-processors combined with some peripheral devices like A/D converters, timers, etc. on one chip) executing the Esterel-programs. Each micro-controller is connected to the sensors and actuators.

Take for example the module FEED_BELT. It is translated by the Esterel compiler to a C program representing a finite state automaton that implements the behaviour of the Esterel module. The compiler generates a collection of C-functions:

- For each input signal S a function FEED_BELT_I_S() is generated. If the function is called, S is added to the input event for the next reaction step.
- FEED_BELT(). If this function is called, one reaction step is carried out.

The technique of interfacing Esterel with C and hardware is the following: Sensors are connected to input pins of the micro-controller and may trigger an interrupt. To each sensor a interrupt routine is connected. Assume the program is idle. If a sensor S triggers an interrupt, the interrupt routine stores the signal by calling FEED_BELT_I_S() and then returns immediately from the interrupt. A timer of the micro-controller is used to produce periodically the reaction. If the timer elapses, the micro-controller is made non-interruptible. Then the reaction is computed by calling the automaton FEED_BELT(). If during a reaction an output signal A_MOTOR_OFF is emitted e.g., the Esterel generated function FEED_BELT() calls a C function FEED_BELT_O_A_MOTOR_OFF(). This function has to be supplied by the programmer in C. In a first design the function only stores the fact that the signal is emitted in a boolean variable. If the program halts, i.e. the complete reaction is computed, the function FEED_BELT() terminates. Now the memorized output signals are processed, e.g. the actuators are supplied with commands. This may be done with memory-mapped I/O. After this has been finished, the micro-controller is made interruptible, waits for input signals and the special timer event to run the reaction function again.

Note that the C part surrounding the Esterel program has to fulfil the perfect synchronization hypothesis, too, i.e. it has to be infinite fast w.r.t. to the environment.

After the objects of the production cell had been implemented as Esterel modules, the system can be configured in two different ways:

1. A *global configuration*. The modules are put together, compiled into one automaton. This solution has two drawbacks:

   - The designer commits very early to a nondistributed solution. It is hard to distribute a large automaton onto many micro-controllers.

   - The objects are only loosely coupled and this is not good reflected in the system's architecture.

2. A *distributed configuration*. Each module is taken as a separate development unit and compiled to an automaton of its own. This solution has the following effects:

   - For all modules the C-interface has to be programmed. This is a small drawback.

   - The distribution of the program onto one or many micro-controller is simple. With the help of a small piece of software glue, it is even possible to change the distribution at run-time. To realize different static distributions is trivial.

   - Of course the signals used for cooperation between the modules have to be propagated between the automata. In the global solution this is done by the compiler. In the distributed solution the cooperation signals are attached to C-functions. They have to be connected. Technically this is

simple: If module PRODUCER sends the signal SIG to module CONSUM-
ER, then the C interface function
PRODUCER_O_SIG() { CONSUMER_I_SIG(); }
does the job. But this is only a part of the solution. A sound integration
has to guarantee that the communicating automata show the same behaviour as the whole program. To proof this for the production cell is not difficult, but in general this is not true.

In our approach the second design alternative has been used, with one exception: The ROBOT and the PRESS are designed as Esterel modules, but not as separate objects. They are glued together into one object ROBOT_PRESS, which is a
trivial module, running both modules in parallel. Of course designing two separate
objects would be possible, but safety requirements like "the robot arm should not
be pressed" are much easier to cope with, if they can be localized within one object, as discussed before.

5.6   Experiences and Conclusion

After reading the task description of the production cell a first Esterel solution was
programmed. This helped to understand some of the more subtle parts of the task
description. This solution was thrown away. After discussions a new version was
written. The object-based partition was realized in a collection of modules.

Each module was debugged with Esterel's simulator. To use the theorem prover at this state did not help much and was too much trouble. A smoother interface
between the tools could help here. When the modules are put together, it turned
out, that Esterel's simulator was not powerful enough. The large number of input
and output signals overwhelmed the designer. This gets worse by the fact, that
most of the signals are independent from each other.

In Prolog a more powerful simulator was implemented. It can generate sequences of input events based on a small simulation language. Especially rules to
specify reactions of the *environment* can be defined, for instance "if an actuator is
triggered by an output signal, generate the corresponding input signal (to tell successful termination of the activity) within the next 10 instants, distributed equally
over the instants." The input/output events of such a run of the production cell are
stored and a pattern match language is offered to analyse the traces. This helps to
check prpositions like "motor-on and motor-off signals always alternative, and,
exclude each other in any instant."

After we were convinced, that most errors were found, AUTO was used to prove some of the safety requirements. As the realm of AUTO is Esterel, code inspections were needed to validate the interface between Esterel and the (simulated) hardware, which was based on (fortunately small) C-functions.

Writing the C function which simulates the environment and executes the Esterel program is easy. The edit-compile-test cycles are quick. Compiling all modules with the Esterel compiler, then compiling the generated C code together with the interface file and finally linking the program, takes 8 seconds CPU-time on a SPARCstation.

Although the simulation environment was available, it remains hard to write relevant test cases and to check whether the program behaves as expected or not. Proof-reading the program was more advantageous. AUTO and AUTOGRAPH as theorem prover and visualizer were not used as the ultimate proof of the correctness of the program. They are to have another perspective of the program during the process to convince ourselves, that the program is adequate. A prerequisite for both inspections and proving was a well-chosen object structure, that allowed local examinations.

With Esterel the design of the kernel of a controller application is simplified a lot, but the advantages are limited if either many kernels are loosely coupled or if the data structures used are complex. To explore this further, an object-oriented synchronous programming language is under development at GMD. One goal of that project is to retain the clear semantics of Esterel and combine it with more comfortable modularization concepts.

To give an impression of the size of the generated program: The binary code of the production cell's controller as an Unix process including a main program to simulate the environment of the production cell is about 46K. The Esterel program results in 26K, the C interface between Esterel and Unix is 2K, the main program used for simulation is 7K. A complete implementation on a micro-controller should be of the same size.

To sum up, Esterel proved to be very well suited writing a controller application for this production cell.

References

[1]   F. Boussinot and R. de Simone. *The Esterel Language. Another Look at Real Time Programming*, Proceedings of the IEEE, 79(9):1293-1304, 1991

[2]   G. Berry. The Semantics of Pure Esterel. In *Proc. Marktoberndorf Intl. Summer School on Program Design Calculi*, LNCS, Springer Verlag, 1993

[3]   V. Roy and R. de Simone. Auto and Autograph. In R. Kurshan, editor, *Proc. of Workshop in Computer Aided Verification*, New-Brunswick, June 1990

[4]   R. Budde. A Production Cell in ESTEREL - A Case Study, Technical Report, GMD, 1994.

Appendix

Esterel Code of the Modules ROBOT, PRESS, and ROBOT-PRESS

```
1     % ========================================================================
2     module ROBOT:
3
4     % ========= signals exchanged with environment =========================
5
6     % ----------------------------- % signals to actuators in the environment
7     output A_ROBOT_TURN_LEFT,      % to turn the robot                        (A7)
8         A_ROBOT_TURN_RIGHT,
9
10        A_ARM_1_TO_TABLE,          % to position arm 1                        (A2)
11        A_ARM_1_IN_PRESS,
12        A_ARM_1_RETRACT,
13
14        A_ARM_2_IN_PRESS,          % to position arm 2                        (A3)
15        A_ARM_2_TO_DEPOSIT_BELT,
16        A_ARM_2_RETRACT,
17
18        A_ARM_1_TAKE_ON,           % to pick up a plate with arm 1            (A4)
19        A_ARM_1_TAKE_OFF,
20
21        A_ARM_2_TAKE_ON,           % to pick up a plate with arm 1            (A5)
22        A_ARM_2_TAKE_OFF;
23
24    % ----------------------------- % signals from sensors in the environment
25    input S_ARM_1,                 % feedback: arm 1 movement terminated
26        S_ARM_2,                   % feedback: arm 2 movement terminated
27        S_ROBOT;                   % feedback: turning the robot terminated
28
```

```
29    % ========= signals exchanged with cooperating modules ==================
30    input ROBOT_can_TAKE_from_TABLE,
31        ROBOT_can_DROP_onto_DEPOSIT_BELT;
32    output ROBOT_has_TAKEN_from_TABLE,
33        ROBOT_has_DROPPED_onto_DEPOSIT_BELT;
34
35    % ========= signals exchanged with the press ============================
36    input PRESS_CAN_BE_LOADED,          % a plate may be put into press
37        PRESS_CAN_BE_UNLOADED;          % a plate may be taken out of the press
38    output PRESS_LOADED,                % a plate is in the press
39        PRESS_UNLOADED;                 % a plate has been taken out of the press
40
41    % ========= behavior ====================================================
42
43    signal ARM_1_HAS_A_PLATE in
44
45    [ loop % ----------------------------% table delivers and press is free
46        await tick;
47        [ emit A_ARM_1_RETRACT; await S_ARM_1
48        || emit A_ARM_2_RETRACT; await S_ARM_2
49        ];
50
51        emit A_ROBOT_TURN_LEFT; await S_ROBOT;
52        emit A_ARM_1_TO_TABLE; await S_ARM_1;
53
54        await ROBOT_can_TAKE_from_TABLE;
55        emit A_ARM_1_TAKE_ON;
56
57        await PRESS_CAN_BE_LOADED;
58        [ emit A_ROBOT_TURN_RIGHT; await S_ROBOT
59        || emit A_ARM_1_IN_PRESS; await S_ARM_1
60        ];
61
62        emit ROBOT_has_TAKEN_from_TABLE;
63
64        emit A_ARM_1_TAKE_OFF;
65        await tick;
66        emit A_ARM_1_RETRACT; await S_ARM_1;
67        emit PRESS_LOADED;
68
69        % ---------------------------------% press is occupied, table may deliver
70        trap PRESS_IN_USE in
71        loop
72        await PRESS_CAN_BE_UNLOADED;
73
74        [ emit A_ROBOT_TURN_LEFT; await S_ROBOT
75        || emit A_ARM_1_TO_TABLE; await S_ARM_1
76        || emit A_ARM_2_IN_PRESS; await S_ARM_2
```

```
77      ];
78
79      await ROBOT_can_DROP_onto_DEPOSIT_BELT;
80      emit A_ARM_2_TAKE_ON;
81      present ROBOT_can_TAKE_from_TABLE then        % both arms used concurrently
82        emit A_ARM_1_TAKE_ON
83      else                              % only arm_2 in use
84        emit A_ARM_1_TAKE_OFF
85      end present;
86
87      [ emit A_ROBOT_TURN_RIGHT; await S_ROBOT
88      || emit A_ARM_1_RETRACT; await S_ARM_1
89      || emit A_ARM_2_TO_DEPOSIT_BELT; await S_ARM_2
90      ];
91
92      emit A_ARM_2_TAKE_OFF;
93      emit PRESS_UNLOADED;
94      emit ROBOT_has_DROPPED_onto_DEPOSIT_BELT;
95        present ARM_1_HAS_A_PLATE then          % both arms used concurrently
96        emit ROBOT_has_TAKEN_from_TABLE;       % thus press remains active
97        await PRESS_CAN_BE_LOADED;
98        emit A_ARM_1_IN_PRESS; await S_ARM_1;
99        emit A_ARM_1_TAKE_OFF; await tick;
100       [ emit A_ARM_1_RETRACT; await S_ARM_1
101       || emit A_ARM_2_RETRACT; await S_ARM_2
102       ];
103       emit PRESS_LOADED
104     else                          % only arm_2 in use, thus
105       exit PRESS_IN_USE          % the press is not loaded
106     end present                  % restart robot for empty press
107     end loop
108     end trap
109
110     end loop
111
112  || loop % ----------------------------% this loop records whether ARM 1
113     await A_ARM_1_TAKE_ON;      % has taken a plate
114     do sustain ARM_1_HAS_A_PLATE watching A_ARM_1_TAKE_OFF
115     end loop
116  ]
117
118  end signal
119  end module
120
121  % ====================================================================
122  module PRESS:
123
124  % signals from sensors in the environment:
```

```
125      % - press's position: S_PRESS                                        (S 1,2,3)
126      % signals to actuators in the environment:
127       % - to move the press' lower part: A_PRESS(UP,MIDDLE,DOWN)          (A 1)
128      % signals for cooperation: with the robot:
129      % - PRESS_CAN_BE_LOADED
130      % - PRESS_CAN_BE_UNLOADED
131      % - PRESS_LOADED
132      % - PRESS_UNLOADED
133      % ================================================================
134
135      % ========= signals exchanged with environment =====================
136
137      % ---------------------------- % signals to actuators in the environment
138      output A_PRESS_UP,             % to move the press' lower part        (A1)
139          A_PRESS_MIDDLE,
140          A_PRESS_DOWN;
141
142      % ---------------------------- % signals from sensors in the environment
143       input  S_PRESS;               % the position of the press           (S1,S2,S3)
144
145      % ========= signals exchanged with cooperating modules ===============
146      output PRESS_CAN_BE_LOADED,    % a plate may be put into press
147          PRESS_CAN_BE_UNLOADED;     % a plate may be taken out of the press
148      input  PRESS_LOADED,           % a plate is in the press
149          PRESS_UNLOADED;            % a plate has been taken out of the press
150
151      % ========= behavior ==================================================
152
153      signal PRESS_MOVING in
154      [ loop
155         emit A_PRESS_MIDDLE; await S_PRESS;
156
157         do
158          sustain PRESS_CAN_BE_LOADED
159         watching PRESS_LOADED;
160
161         emit A_PRESS_UP; await S_PRESS;
162         await tick;
163         emit A_PRESS_DOWN; await S_PRESS;
164
165         do
166          sustain PRESS_CAN_BE_UNLOADED
167         watching PRESS_UNLOADED
168        end loop
169      II loop
170         emit PRESS_MOVING
171        each [ A_PRESS_UP or A_PRESS_MIDDLE or A_PRESS_DOWN ]
172      ]
```

```
173    end signal
174
175    end module
176
177    % ===================================================================
178    module ROBOT_PRESS:
179
180    % ROBOT_PRESS combines the robot and the press, signal descriptions:
181    % see the ROBOT and the PRESS module
182    % ===================================================================
183
184    % ========== signals exchanged with environment ====================
185    input  S_ARM_1, S_ARM_2, S_ROBOT;
186    input  S_PRESS;
187
188    output A_ROBOT_TURN_LEFT, A_ROBOT_TURN_RIGHT,
189        A_ARM_1_TO_TABLE, A_ARM_1_IN_PRESS, A_ARM_1_RETRACT,
190        A_ARM_2_IN_PRESS, A_ARM_2_TO_DEPOSIT_BELT, A_ARM_2_RETRACT,
191        A_ARM_1_TAKE_ON, A_ARM_1_TAKE_OFF,
192        A_ARM_2_TAKE_ON, A_ARM_2_TAKE_OFF;
193    output A_PRESS_UP, A_PRESS_MIDDLE, A_PRESS_DOWN;
194
195    % ========== signals exchanged with cooperating modules ===============
196    input  ROBOT_can_TAKE_from_TABLE,
197        ROBOT_can_DROP_onto_DEPOSIT_BELT;
198    output ROBOT_has_TAKEN_from_TABLE,
199        ROBOT_has_DROPPED_onto_DEPOSIT_BELT;
200
201    % ========== behavior, encapsulate the cooperation between robot and press ===
202
203    signal PRESS_CAN_BE_LOADED, PRESS_CAN_BE_UNLOADED,
204        PRESS_LOADED, PRESS_UNLOADED in
205
206    [ run ROBOT
207    || run PRESS
208    ]
209
210    end signal
211    end module
```

VI. LUSTRE

A Verified Production Cell Controller

Leszek Holenderski

GMD Birlinghoven

Abstract

Our aim was to fully develop (i.e. specify, program and verify) a controller for the production cell simulator. We have specified and programmed the controller in Lustre, which is a declarative language for programming synchronous reactive systems. For verification we have used a symbolic model checker, called Lesar, which allows to automatically verify those Lustre programs which use only boolean data. Since the production cell controller could be written as such a program, we were able to automatically verify all safety requirements given in the task description for this case study. Using a declarative language allowed to develop the controller in a relatively easy way, and in a relatively short time.

6.1 A Quick Introduction to LUSTRE

Lustre [1][2][3][4][5] is a declarative language for programming synchronous reactive systems. In the computational model for synchronous reactive systems time is assumed to be discrete, i.e. divided into denumerable number of non-overlapping segments called instances of time. A synchronous reactive system is a cyclic, non-terminating process. Each cycle, a so-called reaction step, is performed in one instance of time. A reaction step consists in fetching inputs, computing, and emitting outputs. All components of a system perform their reaction steps simultaneously, i.e. in the same time instance. Local communication is considered to be instantaneous.

In Lustre, a behaviour of a reactive component with input channels a, b, ... and output channels x, y, ... is specified by a set of equations of the form

$$x = expression$$

where x is an output channel, and *expression* (built from various built-in operators) may depend on both input and output channels. There must be exactly one such equation for every output channel. The equation simply means that in every instance of time the value emitted on channel x should be the value of *expression*. Since Lustre expressions may involve an operator which allows to refer to the value of an expression in the previous instance, the current value emitted on x may depend on a communication history of the reactive component.

A reactive component (called a *node* in Lustre parlance) is declared by specifying its *interface* (the input and output channels) and its *body* (the set of equations). For example, the following declaration

```
node redge (signal: bool) returns (r: bool);
let
  r = signal -> (signal and pre(not signal));
tel;
```

specifies a node which recognizes a rising edge in a supplied input signal. The signal is encoded as a stream of boolean values with *true* interpreted as *up* and *false* interpreted as *down*. Expression pre(*exp*) returns the value of *exp* in the previous instance. In the first instance, exp_1 -> exp_2 returns the value of exp_1, and in all later instances, the value of exp_2. Operator '->' is called an initialization operator since it is usually used to initialize pre(*exp*), which is not defined in the first instance.

A declarative reading of the equation for r is the following: initially r is true iff signal is up, and afterwards it is true iff signal is currently up and it was down in the previous instance.

Nodes can be used in expressions in a similar way as functions in other programming languages. For example, the following declaration

```
node fedge (signal: bool) returns (f: bool);
let
  f = redge(not signal);
tel;
```

specifies a node which recognizes a falling edge in a supplied input signal.

Equations can be mutually recursive, but recursion must be guarded by 'pre' (otherwise the equations either could not be 'solved' or would have many 'solutions'). For example, the following declaration specifies a node which sustains signal on until off comes.

```
node sustain (on, off: bool) returns (s: bool);
let
  s = on -> if on then true else if off then false else pre(s);
tel;
```

sustain(*on*, *off*) becomes true whenever *on* is up, and continues to be true until *off* is up.

Auxiliary variables can be used. For example, the following declaration

```
node after (e1, e2: bool) returns (a: bool);
  var e2_since_last_e1: bool;
let
  a = e1 and e2_since_last_e1;
  e2_since_last_e1 = e2 ->
    if e2 then true else if pre(e1) then false else pre(e2_since_last_e1);
tel;
```

specifies a node which returns true whenever event e1 happens after (or simultaneously with) event e2, for the first time since last e2. Observe that the body can be simplified, by substitution, to

```
a = e1 and sustain(e2, pre(e1));
```

Lustre can also be used as a specification language for the properties to be verified. In fact, Lustre Boolean expressions can be seen as formulas in a simple temporal logic with an immediate past operator. Using recursive equations and node declarations, one can define more complex temporal past operators (e.g. after).

In subsequent Lustre specifications, we use the following precedence of some of the Lustre built-in operators: not, and, or, ->.

Compiling a LUSTRE program

A Lustre program is a set of node declarations with one node being distinguished as a main node. As a result of compilation the Lustre compiler produces an intermediate file which contains a description of a reactive automaton, i.e. a finite state machine which implements the reactive behaviour of the program being compiled. From the intermediate file some other code can be generated, in a form suitable either for a direct implementation in hardware, or for a software simulation. In the later case, a set of C procedures implementing the automaton is generated. The procedures are then linked to a main C program whose task is to interface the controller (the automaton) with the environment it controls.

The interfacing protocol is very simple. With every input to the main Lustre node there is associated a procedure which passes to that input a value given as its parameter. After setting all inputs, the main C program should call a reaction procedure of the main Lustre node. The reaction procedure performs one reaction step during which some values are emitted on the outputs of the main Lustre node. The emission of a value on a particular output consists in calling a C procedure associated with that output. The procedure must be declared in the main C program. Its task is to pass to the environment the value given as a parameter. Setting inputs and calling the reaction procedure should be performed in a loop. Iterations of the loop define the instances of time for the Lustre main node.

6.2 Production Cell in LUSTRE

Although the Lustre program/specification for the production cell controller is relatively short, we are not able to present it in full in this short report. Instead, we present some parts of the program which illustrate Lustre in action.

6.2.1 A generic controller for a moving item

Most of the device components which constitute the production cell are items moving in two directions: the robot's base rotates left and right, the robot's arms extend and retract, the press moves up and down, and so on. In fact, only belts and magnets do not fit this schema.

A controller for a moving item generates 3 signals: start1 and start2 (which initiate movement in the respective directions), and stop (which terminates the movement). The following Lustre node describes such a generic controller.

```
node MovingItem
  (MaySafelyMove, TryToMove1, TryToMove2: bool)
returns
  (start1, start2, stop: bool);
var
  MayMove1, MayMove2: bool;
let
  MayMove1 = TryToMove1 and MaySafelyMove;
  MayMove2 = TryToMove2 and MaySafelyMove;

  start1 = redge(MayMove1 and (true -> pre(not TryToMove2)));
  start2 = redge(MayMove2 and (true -> pre(not TryToMove1)));
  stop  = fedge(MayMove1) or fedge(MayMove2);
tel;
```

The input conditions TryToMove1 and TryToMove2 are supposed to be true during the interval of time when the respective movement should last. In other words, the movement should start when redge(TryToMove), and it should stop when fedge

(TryToMove). The input condition MaySafelyMove is supposed to be true in those instances of time when the movement is not going to cause any harm to other parts. Thus, the actual movement may happen only if TryToMove and MaySafelyMove. The additional condition true -> pre(not TryToMove) in start ensures that a movement cannot be interrupted by a movement in the other direction.

Later we will formally prove two important facts about the moving item controller. First, whenever an item moves then it may safely move. Second, provided that redge(TryToMove1) and redge(TryToMove2) are always mutually exclusive then start1, start2 and stop are also mutually exclusive. The later is important, since otherwise the item could get confused by simultaneous, conflicting requests.

6.2.2 The press

The controller for the press generates signals which control movements of the press: Pup/Pdn causes the press to start moving up/down, Pstop stops the press. In addition, the controller computes the condition ArmsMayPassPress which is true whenever it is safe for the robot's arms to move in the proximity of the press. The condition is then used by the robot's controller to guard its movements.

The press controller uses the following input sensors and signals:

- Pbot, Pmid, Ptop are sensors which are true when the press is in the bottom, middle or top position, respectively.

- NearPress1 is a sensor which is true if the rotation angle of the robot's base is in the range when arm1 is dangerously close to the press and thus may be damaged by the moving press. OutPress1 is a sensor which is true if arm1 is retracted enough to safely pass the press, even if the press is moving. Similarly for NearPress2 and OutPress2.

- Rput1 and Rget2 are signals generated by the robot controller when arm1 puts a blank into the press or arm2 gets a pressed blank from the press, respectively.

The behaviour of the press is described by the following Lustre node:

```
node Press
  (Rput1, NearPress1, OutPress1,
   Rget2, NearPress2, OutPress2,
   Pbot, Pmid, Ptop: bool)
returns
  (Pup, Pdn, Pstop, ArmsMayPassPress: bool);
var
  Arm1MayPassPress, Arm2MayPassPress,
  Arm1OutOfPress, Arm2OutOfPress, ArmsOutOfPress,
```

```
Arm1CannotCollideWithPress, Arm2CannotCollideWithPress,
up, down, stopped: bool;
let
(Pup, Pdn, Pstop) = MovingItem(true, up, down);

up     = sustain(after(ArmsOutOfPress, Rput1), Ptop) or
           sustain(after(ArmsOutOfPress, Rget2), Pmid);
down = sustain(false -> Ptop and pre(Pstop), Pbot);
stopped = not sustain(Pup or Pdn, Pstop);

ArmsOutOfPress = Arm1OutOfPress and Arm2OutOfPress;
Arm1OutOfPress = not NearPress1 or OutPress1;
Arm2OutOfPress = not NearPress2 or OutPress2;

ArmsMayPassPress = Arm1MayPassPress and Arm2MayPassPress;
Arm1MayPassPress = Arm1OutOfPress or Arm1CannotCollideWithPress;
Arm2MayPassPress = Arm2OutOfPress or Arm2CannotCollideWithPress;

Arm1CannotCollideWithPress = stopped and (Pmid or Ptop) or
  sustain(Pbot, Pmid) or sustain(Pmid, Pbot);
Arm2CannotCollideWithPress = stopped and (Pbot or Ptop);
tel;
```

The definition of ArmsMayPassPress is obvious. The rest of output signals is generated by MovingItem called with suitable parameters, whose definitions follow from the observation that the press moves in a cycle consisting of the following steps:

(1) Initially, the press is in the middle position waiting for the robot to put a blank. When the blank is put the press should wait for the robot's arms to leave a danger zone and then it should start moving up. The condition after(ArmsOutOfPress, Rput1) is true precisely at that moment. When moving up the press should stop if it reaches the top position.

(2) The press should stay at the top position long enough to press the blank, and afterwards it should start moving down. Assuming that one instance of time is enough to press a blank, the condition Ptop and pre(Pstop) is true precisely at the moment when the press should start moving down. When moving down the press should stop if it reaches the bottom position.

(3) In the bottom position the press should wait for the robot to get the pressed blank. After the blank is got the press should still wait for the robot's arms to leave a danger zone and only after that it should start moving up. The condition after(ArmsOutOfPress, Rget2) is true precisely at that moment. When moving up, the press should stop if it reaches the middle position.

The condition guarding a safe movement of the press is set to true in Moving - Item(true, ...), which reflects our design decision to pass the whole responsibility for avoiding a possible collision with the press to the robot.

6.2.3 The robot

The robot controller is the most complicated part of the production cell controller. It has been divided into 3 separate nodes, which control different aspects of the robot's behaviour: Rbase controls rotation of the robot's base, Rarms controls retraction and extension of the robot's arms, and Rgrips controls switching on and off the magnets at the arms' grips.

In the analysis of the robot's behaviour, a crucial role is played by the sequence in which blanks should be fetched from the elevating table (Rget1), deposited in the press (Rput1), fetched from the press after being pressed (Rget2) and deposited on the deposit belt (Rput2). Obviously, not all sequences are correct. The correct ones must satisfy the following restrictions:

(1) projection of the sequence on Rget1 and Rput1 must be (Rget1; Rput1)*,

(2) projection of the sequence on Rget2 and Rput2 must be (Rget2; Rput2)*,

(3) projection of the sequence on Rput1 and Rget2 must be (Rput1; Rget2)*.

The most natural sequence which satisfies the restrictions is (Rget1; Rput1; Rget2; Rput2)*. However, this sequence leads to an inefficient usage of the press which has to wait for Rput1, when moved from the bottom to the middle position after Rget2. The delay is caused by the time it takes the robot to rotate forth and back to perform Rput2 and Rget1 between every pair of successive Rget2 and Rput1. By comparing the relative speeds of movement of the press and rotation of the robot's base, it turns out that the delay may be removed altogether if

(4) Rget1 does not occur between Rget2 and Rput1.

The sequence we have chosen is Rget1; (Rput1; Rget1; Rget2; Rput2)*. First, it satisfies conditions (1-4). Second, moving Rget1 between Rput1 and Rget2 does not introduce any additional delay since the time between Rput1 and Rget2 (i.e. the time it takes the press to move from middle to bottom via top) is long enough to perform Rget1 (i.e. to rotate the robot to the right, to fetch the next blank from the elevating table, and back to the left till arm2 is positioned towards the press, in order to prepare for Rget2).

In order to force the desired sequence, the robot's base must rotate in a right way. The rotation algorithm is described by the following Lustre node:

```
node Rbase
 (ArmsMayPassPress,
  Rget1, Rput1, ToTable, ToPress1,
  Rget2, Rput2, ToDBelt, ToPress2: bool)
returns
 (Rleft, Rright, Rstop: bool);
var
 left, right, BeforeFirstRput1: bool;
let
 (Rleft, Rright, Rstop) = MovingItem(ArmsMayPassPress, left, right);

 left  = sustain(Rget1, ToPress2) or
         sustain(Rput2, ToPress1) or
         sustain(Rget2, ToDBelt) or
         sustain(Rget1, ToPress1) and BeforeFirstRput1;
 right = sustain(true -> Rput1, ToTable);

 BeforeFirstRput1 = sustain((not Rput1) -> false, Rput1);
tel;
```

The inputs ToTable, ToPress1, ToDBelt and ToPress2 are sensors which recognize when arm1 is directed towards the elevating table or the press, and arm2 is directed towards the deposit belt or the press, respectively.

Condition Left is true when the base should try to rotate to the left. It is a disjunction of several conditions sustain(*start, stop*). For example, sustain (Rget1, ToPress2) means that every Rget1 should cause the base to start rotating to the left, and continue the rotation till arm2 is directed towards the press (to prepare the robot to perform the next Rget2). The last disjunct sustain (Rget1, ToPress1) and BeforeFirstRput1 is responsible for properly serving the first Rget1 in the sequence, which should be followed by Rput1 instead of Rget2. (Observe that condition BeforeFirstRput1 is true only till the first occurrence of Rput1.)

Condition Right is defined in a similar way, but with one additional trick: sustain(true -> Rput1, ToTable) means two things. First, every Rput1 should cause the robot to start rotating to the right, and continue the rotation till arm1 is directed towards the elevating table (to prepare the robot to perform the next Rget1). Second, true in true -> Rput1 means that the robot should initially rotate to the right till ToTable. This is needed to prepare the robot for the first Rget1, since initially arm2 is directed towards the press.

The rotation is guarded by the condition ArmsMayPassPress which is computed by the press controller. MovingItem(ArmsMayPassPress, ...) guarantees that whenever during the rotation ArmsMayPassPress becomes false, the rotation is stopped, and resumed when ArmsMayPassPress becomes true again.

In establishing the right sequence of Rgets and Rputs, the following node takes its part:

```
node Rgrips
  (InPress1, OverTable, ToTable, ToPress1,
  InPress2, OverDBelt, ToDBelt, ToPress2,
  Ttop, Pbot, Pmid: bool)
returns
  (Rget1, Rput1, Rget2, Rput2: bool);
let
  Rget1 = after(OverTable and ToTable and Ttop, true -> pre(Rput1));
  Rput1 = after(InPress1 and ToPress1 and Pmid, Rget1);
  Rget2 = after(InPress2 and ToPress2 and Pbot, Rput1);
  Rput2 = after(OverDBelt and ToDBelt, Rget2);
tel;
```

The input sensors InPress1, OverTable, InPress2 and OverDBelt recognize when the robot's arms are extended to the respective distances. Ttop is true when the elevating table is in the top position.

All the conditions defining the output signals have the form after(*correct position*, *previous event*) and mean that a respective signal should be generated whenever (1) some components of the production cell are in correct positions to fetch or deposit a blank and (2) the positions happen for the first time since the last occurrence of some event. The particular choice of *previous event* reflects the sequence of Rgets and Rputs we have chosen in our design. The condition true -> pre(Rput1) in Rget1 serves two purposes. First, pre(Rput1) is needed to break a causality cycle in the mutually recursive definition of Rget1 and Rput1. Second, true is chosen to initialize pre(Rput1) since the first Rget1 in the sequence is not preceded by anything.

6.3 Verification

All the safety requirements given in the *Task Description for the Case Study "Production Cell"* document can be automatically verified for our controller, by using the Lesar [5] verification tool which accompanies the Lustre compiler.

An input to Lesar is simply a pure Lustre program (i.e. one using only Boolean data) whose main node has the following structure:

```
node name (i₁, i₂, ... : bool) returns (ok: bool);
  var a₁, a₂, ... : bool;
let
  -- body
```

$a_1 = bexp_1$;
$a_2 = bexp_2$;
...
-- assumptions
assert $BEXP_1$;
assert $BEXP_2$;
...
-- conclusion
$ok = prop$;
tel;

Lines beginning with '--' are comments; i_1, i_2, ... are inputs; a_1, a_2, ... are auxiliary variables; $bexp_1$, $bexp_2$, ..., $BEXP_1$, $BEXP_2$, ..., *prop* are Lustre boolean expressions. A boolean expressions in **assert** statement is called an assumption.

The node may have many inputs but only one output. Lesar returns a *yes/no* answer: it returns *yes* iff the output is always true during any computation of the program (i.e. no matter what the input values are) and provided the assumptions are satisfied (i.e. they are always true). In other words, Lesar checks whether *prop* is invariantly true, provided the assumptions are invariantly true.

To see Lesar in action, let us consider the following node (for which Lesar returns *yes*):

```
node VerifyMovingItem
 (MaySafelyMove, TryToMove1, TryToMove2: bool)
returns
 (prop: bool);
var
 MayMove1, MayMove2: bool;
 start1, start2, stop, moving: bool;
let
 MayMove1 = TryToMove1 and MaySafelyMove;
 MayMove2 = TryToMove2 and MaySafelyMove;
 moving      = sustain(start1 or start2, stop);

 start1 = redge(MayMove1 and (true -> pre(not TryToMove2)));
 start2 = redge(MayMove2 and (true -> pre(not TryToMove1)));
 stop  = fedge(MayMove1) or fedge(MayMove2);

 assert #(redge(TryToMove1), redge(TryToMove2));
 prop = #(start1, start2, stop) and
        if moving then MaySafelyMove else true;
 tel;
```

where '#' denotes a mutual exclusion (and happens to be a built-in Lustre operator).

The node encodes the following property of a moving item: provided the input signals initiating movements in different directions never come simultaneously, (1) the output signals controlling movement are mutually exclusive and (2) whenever an item moves, it is allowed to do so.

The second property is especially important for Rbase, since it guarantees that the robot's arms never collide with the press. In order to deduce it, one needs to prove #(redge(Left), redge(Right)). In fact, this can be easily proved under the following assumptions:

(1) #(Rget1, Rput1, Rget2, Rput2) which can be proved in Rgrips by asserting #(ToTable, ToPress1, ToDBelt, ToPress2),

(2) not(Rget1 or Rget2 or Rput2) -> true which means that none of the respective signals is possible initially (and can be proved in Rgrips by asserting not Ttop -> true).

Verification of other safety requirements also depends on assumptions about the geometry of the production cell, relative speeds of various devices, and the initial positions of the devices. All the assumptions are made explicit in proofs.

6.4 Evaluation

It took us about 4 weeks to specify, program and verify the production cell controller. During that time several versions of the controller have been written, tested and partly verified.

Almost half of that time was wasted on solving some problems with interfacing the controller to the existing simulator. An initial version of the controller was obtained in 3 days (it served only one blank and was not formally verified). During the first tests it turned out that animation performed by the simulator was very slow. It seemed the controller was too slow. One week was spent on improving its efficiency, first by rewriting the original Lustre program, and second, by writing a more efficient C code generator from an intermediate code generated by a Lustre compiler. At the end it turned out that a faster controller made the animation even slower! The reason for this phenomenon was the following: since the simulator run asynchronously with the controller, the faster the controller the more requests it generated to the simulator, and as a consequence the simulator was occupied with serving the requests, instead of performing the animation. A minor change to the simulator made it synchronous and solved the problem.

Most of the remaining time was spent on verification. During verification, another phenomenon was observed: making the original program more complicated, by introducing additional conditions into equations, made the verification easier. The more information is coded into Lustre equations the less assertions are needed by the Lesar tool.

A final version of the controller is able to handle 5 blanks simultaneously. A limiting factor is the slow speed of the press which simply cannot process blanks faster. The overall throughput of the controller is optimal, since the press never waits for the arms to deposit a new blank or fetch a pressed one.

A Lustre code for the controller is relatively short: about 200 lines. Formal proofs are about 100 lines longer (since they must contain assumptions and conclusions). The controller was compiled to about 800 lines of C code.

Parts of our Lustre code may be easily reused in similar controllers, especially generic controllers for (1) an item moving in two directions and (2) a conveyor belt with a photo-element at its end.

References

[1]   N. Halbwachs, *Synchronous Programming of Reactive Systems*, Kluwer Academic Publishers, 1993, 1–175.

[2]   P. Caspi, N. Halbwachs, D. Pilaud, J. A. Plaice, *Lustre: a declarative language for programming synchronous systems*, Proc. of the 14th Symposium on Principle of Programming Languages, München, Sep. 1987, 178–188.

[3]   N. Halbwachs, P. Caspi, P. Raymond, D. Pilaud, *The Synchronous Data Flow Programming Language Lustre*, IEEE Special Issue on Real Time Programming, Proceedings of the IEEE, 79(9), Sep. 1991, 1305–1320.

[4]   N. Halbwachs, *A Tutorial of Lustre*, Lustre distribution, available by anonymous ftp from imag.imag.fr as file /ftp/pub/LUSTRE/tutorial.ps, Jan. 1993, 1–19

[5]   N. Halbwachs, F. Lagnier, C. Ratel, *Programming and Verifying Real-Time Systems by Means of the Synchronous Data-Flow Language Lustre*, IEEE Trans. on Software Eng., 18(9), Sep. 1992, 785–793.

VII. SIGNAL

The Specification of a Generic, Verified Production Cell Controller

Tochéou Pascalin Amagbegnon, Paul Le Guernic, Hervé Marchand, Éric Rutten

IRISA/INRIA, Rennes

Abstract

Our aim is to specify, verify and implement a controller for a production cell, using the Signal approach. Our contribution to this case study aims at illustrating the methodology associated with the Signal synchronous data flow language for the programming of control systems, as well as how dynamical properties can be proved using the tool Sigali. Hence, we describe the full development of the example, specifying a generic controller, safe for all scheduling scenarios. The specification is made in terms of events and boolean data, abstracting from the numerical nature of some of the sensor data; this enables the use of Sigali for the formal proof of the application's requirements satisfaction.

7.1 A Brief Introduction to Signal

7.1.1 A programming environment for real time systems

Signal [4] is a synchronous real-time language, data flow oriented (i.e., declarative) and built around a minimal kernel. It manipulates signals, which are unbounded series of typed values, with an associated clock denoting the set of instants when values are present. For instance, a signal X denotes the sequence $(x_t)_{t \in \mathbb{N}}$ of data indexed by time-index t. Signals of a special kind called event are characterized only by their clock i.e., their presence (they are given the boolean value true at each occurrence). Given a signal X, its clock is obtained by the language expression event X, resulting in the event that is present simultaneously with X. Different signals can have different clocks: one signal can be absent relatively to the clock of another signal; the absence of a signal is noted $\perp$. Constructs of the language correspond to set-theoretic operations on these clocks.

The constructs of the language can be used to specify the behavior of systems in an equational style: each equation states a relation between signals i.e., between their values and between their clocks. Systems of equations on signals are built using the composition construct. In this sense, it is a form of constraint-based programming; when equations have a deterministic solution depending on inputs, the resulting program is reactive (i.e., input-driven); in other cases, correct programs can be demand-driven or control-driven, as we will see further.

The compiler performs the analysis of the consistency of the system of equations, and determines whether the synchronization constraints among the signals are verified or not. Also, it performs transformations on the graph of clocked data dependencies, including optimizations and the synthesis of an executable control. If the program is constrained so as to compute a deterministic solution, then executable code can be produced automatically (in C or Fortran). This code basically consists of a cyclic call to a transfer function, which computes an output, and updates state variables, according to the presence and value of input signals.

The complete programming environment also features a graphical block-diagram oriented user interface, a proof system for dynamic properties of programs called Sigali (see Section 7.3), and work is in progress concerning hardware/software co-design and the synthesis of VHDL, extensions towards stochastic systems, interruptible data flow tasks and time intervals, compilation toward distributed architectures.

7.1.2   The kernel of Signal

The kernel comprises the five following features:

- **Functions** (e.g., addition, multiplication, conjunction) are defined on the types of the language. For example, the boolean negation of a signal E is not E. The signal (Y_t), defined by an instantaneous function f: $\forall\, t,\, Y_t = f(X_{1t}, X_{2t}, \ldots, X_{nt})$ is defined in signal by:

$$Y := f\{\, X1, X2, \ldots, Xn\}$$

  The signals Y, X1,..., Xn must all be present at the same time: they are constrained to have the same clock.

- **Delay** gives the previous value ZX of a signal X, as in: $ZX_t = X_{t-1}$, with initial value V_0. It is the only dynamical operator, and the only way to access past values. The corresponding instruction is

$$ZX := X\$1$$

with initialization at the declaration of ZX:

$$ZX \text{ init } V0$$

Signals X and ZX have the same clock.

- **Selection** of a signal X according to a boolean condition C is:

$$Y := X \text{ when } C$$

the operands and the result do not have identical clock. Signal Y is present at some instant t if and only if X and C are present at that instant and C has the value true. When Y is present, its value is that of X.

- **Deterministic merge** defines the union of two signals of the same type, with a priority on the first one if both are present simultaneously:

$$Z := X \text{ default } Y$$

The clock of Z is the union of that of X and that of Y. The value of Z is the value of X when X is present, or else the value of Y if Y is present and X is not.

- **Parallel composition** of processes is made by the associative and commutative operator "I", denoting the union of the underlying systems of equations. Systems communicate and interact through signals defined in one system and featured in others. For these signals, composition preserves constraints from all systems, especially temporal ones. This means that they are present if the equations systems allow it.

In Signal, for processes P_1 and P_2, composition is written:

$$(|\, P_1 \,|\, P_2 \,|)$$

The rest of the language is built upon this kernel. A structuring mechanism is proposed in the form of process schemes, defined by a name, typed parameters, input and output signals, a body, and local declarations. Occurrences of process schemes in a program are expanded by a pre-processor of the compiler. Derived processes have been defined from the primitive operators, providing programming comfort. E.g., synchro{X,Y} which constrains signals X and Y to be synchronous; when C giving the clock of occurrences of C at the value true; X cell B which memorizes values of X and outputs them also when B is true. Arrays of signals and of processes have been introduced as well.

7.1.3 Examples

An example is the process (or system of equations) STATUS (illustrated in Figure 1, obtained with the graphical interface for signal [2]). It manages a boolean state variable STATE (given at the output) in relation with two input events: ON (putting STATE to the value true), and OFF (putting STATE to the value false). As the concept of classical variable is not explicitely present in Signal, it has to be defined in terms of equations and delays. The inputs define the next value of STATE, which is called NEXT_STATE. When OFF is present it is false (i.e., the negation of OFF's value, which is true because OFF is an event). When ON is present, the value is true, which is the value of event ON. Finally, when none of them is present, it has its own previous value: STATE. STATE itself is then the delayed value of NEXT_STATE.

Figure 1   *Example of a* Signal *process: STATUS manages a boolean state variable.*

This program specifies the values of signal STATE in an unambiguous way. However, it does not define a deterministic presence clock: it only states that STATE is at least as often present as the union of ON and OFF's clocks. The clock of the output of this process, and of its internal activity (refreshing the memory) is dependent on the constraints that might exist on STATE in the context in which it is used. Figure 2 illustrates an invocation of STATUS in the context of process ACTION_CONTROL. It performs a *rendez-vous* between the two events ACTUATOR_READY and SUPPORT_READY, each of them signalling that one device is ready for an exchange (it is useful e.g., between feed belt and arm, or press and arm). For this purpose, it uses two instances of STATUS, memorizing arrival of each of the events, and computes output event DO_IT when both have occurred, resetting STATUSes at the same time. The last line of the process (featuring the synchro) specifies that the instants at wich the state variables READY and POSSIBL must be present, are the union of the clocks of the two input events.

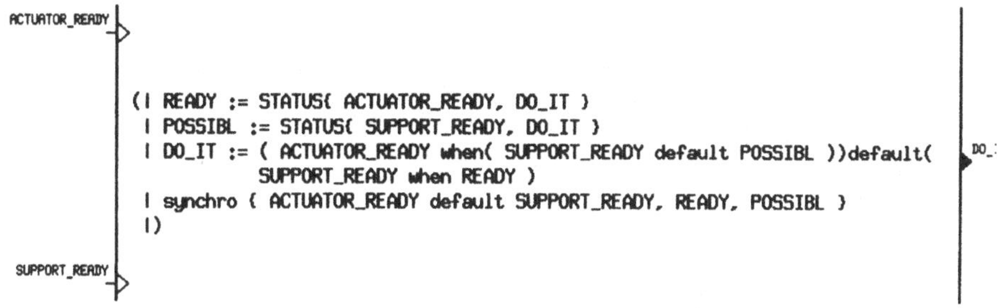

ACTUATOR_READY

```
( I READY := STATUS( ACTUATOR_READY, DO_IT )
  I POSSIBL := STATUS( SUPPORT_READY, DO_IT )
  I DO_IT := ( ACTUATOR_READY when( SUPPORT_READY default POSSIBL ))default(
                  SUPPORT_READY when READY )
  I synchro ( ACTUATOR_READY default SUPPORT_READY, READY, POSSIBL )
  I)
```
DO_

SUPPORT_READY

Figure 2   Example: CONTROL_ACTION makes a rendez-vous between two events.

7.2   Specification of the Production Cell in Signal

7.2.1   Overview of the specification

The method used for the specification consists in separating the model of the process from that of its controller. The process is described in all its possible behaviors (including "wrong" ones, for example those featuring collisions). This specification of the "natural" behavior of the components of the cell was composed with a specification of the safety rules. These rules define how to avoid dangerous situations, by sending commands to the machines. Finally, this specification of safe behaviors is composed with a scheduling, providing the production cell with an executable, safe and meaningful behavior.

The specification follows the decomposition of the cell into two groups of devices, devoted to the transportation (CTRL_TRANSPORT) or to the transformation (CTRL_ROBOT_PRESS) of the metal pieces. The controller is divided into sub-controllers for the components of the cell. Figure 3 shows how the two parts communicate: the transport part only needs to know when the table is empty (through event TABLE_EMPTY) in order to synchronize with the transformation process. The other way round, the transport process signals to the transformation process when the table is ready to be unloaded (TABLE_READY_TO_UNLOAD) (i.e., when the robot can fetch an object there), and when the deposit belt can accept a new object (event DEPOSIT_BELT_READY_TO_LOAD) without risking collisions.

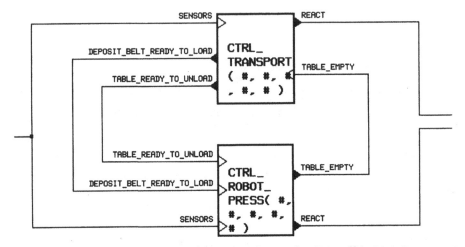

Figure 3   The overall structure of the production cell in Signal.

7.2.2   The transportation control process

The transportation control process is decomposed into the two belts, the crane, and the table (see Figure 4). Each of the processes receives sensor values as input, and outputs commands to the device it controls. Communication signals exchanged between sub-controllers are highlighted in grey. The event TABLE_EMPTY is received from the transformation controller (as seen in Figure 3), while on the other side, events TABLE_READY_TO_UNLOAD (in Figure 4 : READY_UNLOAD) and DEPOSIT_BELT_READY_TO_LOAD are sent to it (the latter is equal to BELT2_START). Event READY_LOAD informs the feed belt controller that the table is ready to accept a blank. Event DROP_ACCEPT informs the crane controller that the feed belt is ready to accept a blank. On the other side, event PICK_ACCEPT informs the crane controller that a blank can be picked up from the deposit belt. Command CRANE_TO_BELT1 notifies the deposit belt controller that it has been loaded.

Controlling the table consists in making it move either from feed belt to robot or conversely. The fact that the table is loaded is acquired when the photoelectric cell goes from true to false (which is given by TABLE_LOADED). The table controller then starts both rotation and vertical translation simultaneously. Each movement is stopped upon reception of the corresponding sensor information. Upon arrival in high position, oriented towards the robot, it sends event

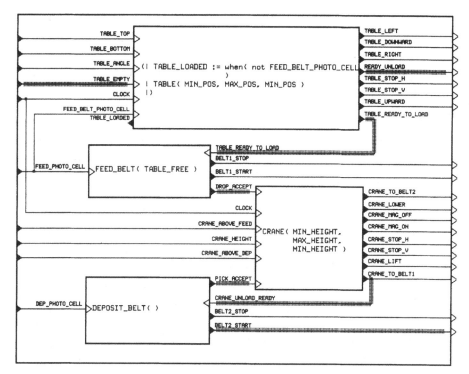

Figure 4 The TRANSPORT process, featuring feed and deposit belts, crane and table.

READY_UNLOAD. Upon reception of TABLE_EMPTY, it starts back down to the feed belt, and sends TABLE_READY_TO_LOAD.

The crane controller is decomposed along the same scheme as the table controller. Figure 4 shows that it receives the events DROP_ACCEPT and PICK_ACCEPT respectively from the feed belt (which accepts that the crane drops a blank) and the deposit belt (which is ready so that the crane can pick up a blank there). These two events contribute in triggering movements from deposit to feed belt, and conversely.

The deposit belt stops in order for the crane to deposit a blank, and then restarts when the crane sends CRANE_TO_BELT1. In the feed belt controller, the belt is stopped if the table is loaded or not in receiving position, and the photoelectric sensor detects the arrival of a new blank.

7.2.3 The transformation control process

The transformation control process (CTRL_ROBOT_PRESS in Figure 3) groups the controllers for the press and the robot. The reason for grouping them is the tight interaction needed to perform the collision avoidance between the two. Figure 5 shows how the process is decomposed into two sub-processes: TRANSFORMA-TION specifies the safe behaviors of the sub-system, while SCHEDULING specifies the specific sequencing of actions to be performed according to its use in the context of the production cell.

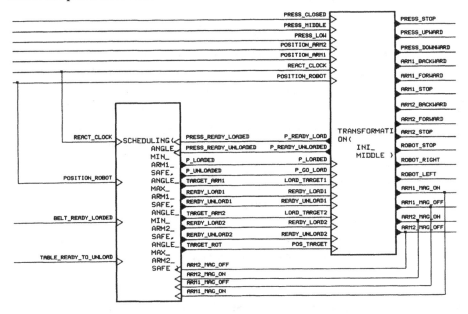

*Figure 5   Scenario and basic behaviour for
the robot and the press.*

The scheduling process

This process receives synchronization signals from the transport controller, as well as from the transformation process, informing it on the state of advancement in the production cell. Using this information, and also state information memorizing for example whether an arm is loaded or not, it schedules goals emitted towards the robot and the arms, so as to command their movements.

The transformation process

This process, illustrated in Figure 6, features the individual controllers for the robot and for the press, interconnected by a third process DEFENSE wich manages all the alarm situations.

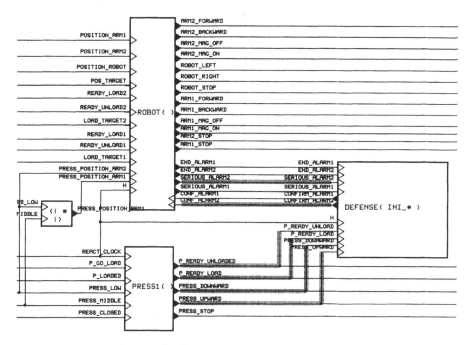

Figure 6 The Transformation process which
controls all the movements of the robot and the press

The process ROBOT: This process concerns the movements of the robot. It is decomposed into sub-processes which manage only the rotation and the retraction/ extention of the two arms according to the goals which are sent by the process SCHEDULING. Two other sub_processes indicate to the process DEFENSE when an arm enters a dangerous region (i.e., when the robot is in a position where a collision with the press can occur).

The process DEFENSE: It specifies the synchronization constraints for collision avoidance. It receives state information from the press and signals from the robot upon entering shared regions (physical space). According to them, it can control the robot by interrupting its rotation or causing the retraction of its arms during the alarm. When the dangerous situation is left, prior movements are resumed. The process DEFENSE takes input signals concerning the state of the press and the position of the robot relatively to the press.

For example, the signal SERIOUS_ALARM1 is sent by the process ROBOT when ARM1 enters a dangerous region. In this case, the process DEFENSE looks at the state of the press and sends the signal CONFIRM_ALARM1 if the press is not in safe position (for example, the press is in safe position for ARM1 if it is in bottom or middle position). When the process ROBOT receives the event CONFIRM_ALARM1, the rotation of the robot is stopped, and it has to retract Arm1. When it is done, the robot can go on with his rotation.

The process PRESS: The process controlling the press (see Figure 7) takes three sensor inputs: PRESS_CLOSED, PRESS_MIDDLE, PRESS_LOW; these are events occurring on the instants where the press arrives at the corresponding positions. Two other inputs are events received from the robot controller: P_M_LOAD and P_GO_LOAD. They respectively mean that the press is loaded with a new blank (when it has been dropped there by ARM1), and that the press is unloaded, and may go and load a new blank (when the previously transformed one has been unloaded by ARM2). The last input, H, is a base clock for the state variables.

In Figure 7, the lower sub-process encodes the state of the press. The boolean signal PRESS_MOVING_UP is true when the press moves upward, and false otherwise. It is the memorization of N_PRESS_MOVING_UP (the next value of the state). This latter signal is true when the command PRESS_UPWARD is present[1], false when PRESS_STOP is emitted, and equal to its previous value otherwise. This state variable is given the clock H (like in the case of the process STATUS in Section 7.3). The command PRESS_UPWARD is emitted when P_M_LOAD or P_GO_LOAD is present; i.e., it moves upward either when it has received a new blank (in order to press it) or when it has been unloaded (in order to go from low to middle position, and be ready to receive a new one).

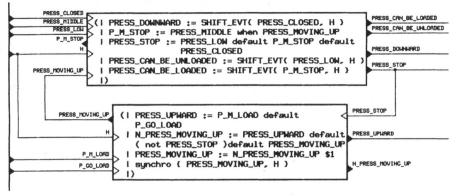

Figure 7   The press process

---

1. Recall that events are signals always true; hence, the negation of an event is a boolean signal always false, at the clock of the negated event.

The upper sub-process in Figure 7 computes the commands towards the press, and the synchronization signals towards the robot controller. The command PRESS_STOP is sent when the press arrives in lower position, or in middle position when moving upward (P_M_STOP, using the state variable described above), or when closing. The command PRESS_DOWNWARD is emitted after the press has closed, with a delay of one instant of the clock H (i.e., after it was stopped). The messages are computed as follow: PRESS_CAN_BE_UNLOADED is emitted one instant after the press has reached its lower position, and PRESS_CAN_BE_LOADED one instant after the press has been stopped in the middle position when moving upward.

7.3   Specifying and Verifying the Properties

Some of the properties required by the case study specification have been proved by using the tool Sigali, a proof system for dynamic properties of Signal programs [3].

7.3.1   Overview of the verification method

To be used by Sigali, Signal processes must be translated into systems of polynomial equations over $Z/3Z$, i.e. integers modulo 3: $\{-1,0,1\}$ [4]. The principle is to code the three possible values of a boolean signal (i.e., present and true, or present and false, or absent) by:

$$\begin{cases} \text{present} \wedge \text{true} \to 1 \\ \text{present} \wedge \text{false} \to -1 \\ \perp \to 0 \end{cases}$$

Using this encoding, the reaction events of the program are represented by a vector in $(Z/3Z)^n$. For example c:=a when b, which means "if b=1 then c=a else c=0" can be rewritten in $c=a(1-b^2)$. All the operators of the kernel of Signal can be translated into polynomial equations in the same manner as when except the delay \$, which is a dynamic operator and requires memorizing into a state variable ξ. The following table shows how all the primitive operators are translated in the case of boolean signals.

By composing the equations representing the elementary processes, any Signal specification can be translated into a set of equations called *polynomial dynamic*

$Y := \text{not } X$	$y = -x$
$Z := X \text{ and } Y$	$\begin{cases} z = xy(xy - x - y - 1 \ ) \\ x^2 = y^2 \end{cases}$
$Z := X \text{ or } Y$	$\begin{cases} z = xy(1 - x - y - xy) \\ x^2 = y^2 \end{cases}$
$Z := X \text{ default } Y$	$z = x + (1 - x \ )y$
$Z := X \text{ when } Y$	$z = x(- y - y \ )$
$Y := X\$1 \ (\text{init } y_0)$	$\begin{cases} \xi' = x + (1 - x^3)\xi \\ y = x^2\xi \\ \xi_0 = y_0 \end{cases}$

system. Formally, a polynomial dynamic system can be represented by a system of polynomial equations of the form:

$$\begin{cases} Q\,(X, Y) \ = \ 0 \\ X' \ = \ P(X, Y) \\ Q_0\,(X) \ = \ 0 \end{cases}$$

where:

- X is a set of variables, called *state variables*
- Y is a set of variables, called *event variables*
- $X' = P(X, Y)$ is the *evolution equation* of the system
- $Q(X, Y) = 0$ is the *constraints equation* of the system
- $Q_0(X) = 0$ is the *initialization equation* of the system

The polynomial vector P can be considered as a function from $(Z/3Z)^{n+m}$ to $(Z/3Z)^n$, where n and m are respectively the number of variables in X and Y. Q and Q_0 are vectorial functions.

A polynomial dynamic system can be seen as a finite transition system. The initial states of this automaton are the solutions of the equation $Q_0(X)=0$. When the system is in a state $x \in (Z/3Z)^n$, any event $y \in (Z/3Z)^m$ such that $Q(x,y)=0$ can constitute a transition. In this case, the system evolves to a state x' such that x' $=P(x,y)$.

In the case of the production cell, the polynomial dynamic system is represented into Sigali by 83 state variables and 48 event variables.

The theory of polynomial dynamical systems uses classical tools in elementary algebraic geometry: varieties, ideals and morphisms. This theory allows us to convert geometric properties, represented by sets of states or sets of events, into equivalent properties of associated ideals of polynomials. Hence, instead of manipulating explicitly and enumerating the states, this approach manipulates the polynomial functions characterizing their sets [3][6]. The tool Sigali implements the basic operators: set theoretic operators, fixpoint computation, quantifiers. With these operators, we can express CTL formulae (for example) and perform symbolic model checking. The reader interested in the theoretical foundation of our approach is refered to [3].

7.3.2   Specification of the properties

In this section, we take the press as an example and show how safety properties are proved.

Consider the following property:

"The press must not be moved downward if the press is low, and, the press must not be moved upward if the press is closed"

It is a safety property, so in order to prove that it is true on the system, it is sufficient to prove it on the press controller. This is basically due to results in automata theory: safety properties are preserved by composition (in this case: synchronous product). The proof on the whole system is much more time consuming than the proof on a subsystem, though our tool Sigali was able to handle the whole system.

In the following, we first give a methodology to achieve that kind of proof, before applying it to the example.

Method of verification

The verification of a process CONTROLLER is carried out in three steps. First, the controller to be verified is composed with two other Signal processes: an environment process (ENVIRONMENT) and a property process (PROPERTY). The environment process is an abstraction of the context. As a matter of fact, the controller might have been optimized by taking advantage of some features of the physical process. For example, physically, between the occurrences of the signals PRESS_CLOSED and PRESS_LOW, there must be an occurrence of the signal PRESS_MIDDLE. The environment process is used to model such features. The property process models the property to be checked and computes a boolean-valued signal ERROR which is true if the property is violated.

The overall process (I ENVIRONMENT I CONTROLLER I PROPERTY I) is translated by the Signal compiler into a dynamical system over $Z/3Z$. Finally, the tool Sigali is used on the resulting dynamical system to verify that the variable ERROR is never true.

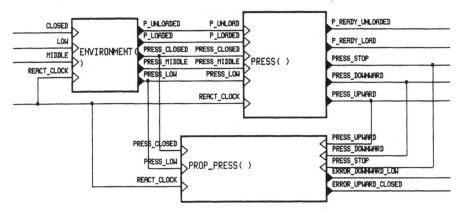

Figure 8　Verification context for the property of the press.

Example: proving the correctness of the press controller

Figure 8 shows how the verifications have been specified for the controller of the press, called PRESS. The process ENVIRONMENT models physical properties:

- *"the physical sensors PRESS_CLOSED and PRESS_LOW and PRESS_MIDDLE cannot be triggered simultaneously"*

- *"Between the occurrences of PRESS_CLOSED and PRESS_LOW, there must be an occurence of PRESS_MIDDLE"*

PROP_PRESS computes the booleans ERROR_DOWNWARD_LOW and ERROR_UPWARD_CLOSED. The former is true if the press is still moving downward after the sensor PRESS_LOW has been triggered. The latter is true if the press is still moving upward after the sensor PRESS_CLOSED has been triggered.

Let us take a close look at the computation of the signal ERROR_DOWNWARD_LOW. Figure 9 describes a process named PROP_GENERIC; the output ERROR is true if the event LIMITING_EVT is received between an occurence of STARTING_EVT and an occurence of ENDING_EVT. As shown in Figure 10 , in the case of ERROR_DOWNWARD_LOW, the input events LIMITING_EVT,

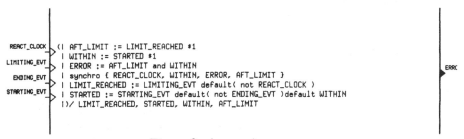

Figure 9 A generic property.

STARTING_EVT and ENDING_EVT are instantiated respectively with events PRESS_LOW, PRESS_DOWNWARD and PRESS_UPWARD default PRESS_STOP.

In the framework of Sigali, the property "ERROR_DOWNWARD_LOW *is never true*" is rephrased as "*the set of states which satisfy* ERROR_DOWNWARD_LOW = *1 (for some values of the inputs) is not accessible from the initial states*". As Sigali features the necessary primitives to compute accessible states (for example fix point computation [3]), this property is easily proved.

Most of the properties required for the production cell follow exactly the same pattern. We used our signal process PROP_GENERIC to prove properties of the elevating table and the crane [1].

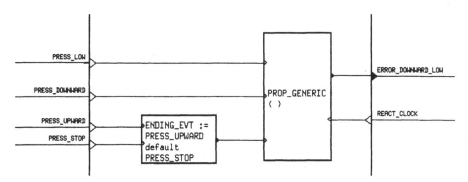

Figure 10 The property process of the press.

7.4 Evaluation

7.4.1 Work load

The total time spent on the specification, implementation, proof of the production cell in Signal, and redaction of this paper, is approximately the equivalent of four weeks for a single person. This total work load can be roughly sub-divided into the three following parts. The specification of the controller in Signal took two weeks (note that the controller was designed to be general, separating the scheduling and behavior, and not only particular to its specific use in the context of the proposed production cell). The implementation took one week (installation of the simulator, programming of the input/output interface). The specification and verification of the properties took one week (analysis, formulation in Signal processes and in Sigali operations).

7.4.2 Programming the cell

Programming the controller in an equational style proved adequate in that the declarativeness and modularity of the language facilitates the decomposition of the specification. The relative complexity of the program is the consequence of the fact that the controller was made as general as possible for each component of the production cell, and not just specific to the given configuration.

The volume of the source code, in textual Signal, is approximately 1700 lines. It must be noted that it was produced by the Signal graphical interface from which the figures in this paper are taken. The complexity of the specification is dependent upon the fact that it is designed to be general, as explained above. The duration of compilation from Signal to C is about 10 seconds; the compilation of the C code takes some seconds as well. The volume of object code in C is 56,7 Kbytes, which, once compiled, give an executable file of 90 Kbytes.

7.4.3 Proving the correctness of the controller

The properties proved could be treated on just the relevant parts of the controller; hence their specification and the duration of the computation of proofs could be kept small. The assumptions on the environment in which they hold are made explicitely. Due to the short time before the deadline for this paper, we had no time to prove all properties, but proved significant ones.

7.4.4 Flexibility and Reusability

As was explained above, the genericity of the robot controller is meant to make it reusable for different schedulings and scenarios with only very little change. The modifications can be restricted very locally to the process defining the scheduling. The structure of the specification follows the object decomposition of the production cell. The control of each device is made by a process needing only minimal synchronization communication with its environment (i.e., other devices). Hence, adding new machines is possible without changing individual controllers.

From the point of view of the properties, the fact that some of them could be proved on the local controllers, by making only small and explicit assumptions on their environments, makes them reusable if those assumptions still hold in the new context.

A research report gives a more detailed description of our treatment of this case study, from the points of view of specification as well as verification [1].

Acknowledgement. Our work on verification is supported by Electricité de France.

References

[1]   T. P. Amagbegnon, P. Le Guernic, H. Marchand, E. Rutten, *Specification, Implementation and Verification of a Production Cell with Signal.* IRISA/INRIA-Rennes, research report. (to appear)

[2]   P. Bournai, P. Le Guernic *Un environnement graphique pour le langage signal.* IRISA/ INRIA-Rennes, research report N°741, July 1993. (In French)

[3]   B. Dutertre. *Spécification et preuve de systèmes dynamiques.* Ph.D. thesis, University of Rennes 1, France, December 1992. (In French)

[4]   M. Le Borgne, B. Dutertre, A. Benveniste and P. Le Guernic. *Dynamical Systems over Galois Fields*, In Proc. of the Second European Control Conference (ECC 93), Groningen,1993.

[5]   P. Le Guernic, T. Gautier, M. Le Borgne, C. Le Maire. *Programming Real-Time Applications with signal . Another look at real-time programming,* special section of Proceedings of the IEEE, 79(9), September 1991.

[6]   H. J. Touati, H. Sajov, B. Lin, R. K. Brayton and A. Sangiovanni-Vincentelly, *Implicit state enumeration of finite state machines using BDD's*, International Conference on Computer-Aided Design, IEEE, November 1990, p.130-133.

VIII. Statecharts

*Using Graphical Specification Languages and
Symbolic Model Checking in the Verification of a
Production Cell*

Werner Damm, Hardi Hungar,
Peter Kelb, Rainer Schlör

Universität Oldenburg and OFFIS Oldenburg

Abstract

This paper discusses specification and verification of the production cell using symbolic model checking. Key features of the presented approach are the clean separation between (models of the) components of the production cell and the controller itself, the use of graphical specification techniques, and the application of methods and tools allowing compositional automatic design verification.

8.1 Introduction

The aim of this paper is to demonstrate the feasibility of automatic verification in a sense to be made precise below when restricting oneself to discrete models of systems behaviour. At the same time, it demonstrates the linkage of semiformal graphical specification mechanisms to verification based on model checking [1], [4] in a realistic case study. In our opinion, formal methods will only be accepted by designers if the verification environment allows them to express their ideas in familiar design notations. The languages chosen are a variant of the *StateChart* notation [2], and a graphical specification language for temporal logic specifications called *symbolic timing diagrams*. The StateChart language is supported by the STATEMATE system which provides detailed simulation facilities and compilers to C, ADA, and VHDL. It is widely used in industry in the development of control applications. Within the FORMAT project, part of the authors are developing an interface to symbolic model checking tools; a prototype version of the compiler described in [11] has been used in this case study. The Timing Diagram Language [10] is supported by a Timing Diagram editor and a compiler to temporal logic [7], [6] developed within the FORMAT project. Prototype versions of the tools were available at the time of writing of this paper. Given these specification mechanisms, the verification of the production cell basically comprises the following steps:

- Formalization of all requirements as timing diagram specifications
- Development of a StateChart model of the production cell
- Verification of the requirements using symbolic model checking

We developed the StateChart model and a substantial subset of the timing diagram specifications within three days. Due to the prototype nature of the tools and the limited available time, only a small, but representative number of requirements have been verified. Given more efficient versions of the tool and two more weeks, a complete verification of the design should be feasible using the compositional approach described in this paper.

In attacking the specification and verification of embedded control systems such as the production cell, a tradeoff has to be made between the degree of detail in modelling the possible behaviours of the system and the degree of automation of the proof process. Though formal specification techniques coping with continuous signals are available (c.f. e.g. [3], [5]), in general no automatic proof methods are possible because of undecidability results. In this paper we opt for abstracting continuous signals (in this case study exemplified by potentiometers) to discrete signals assuming a finite set of "critical" values. In the example, it is only important to observe whether a robot arm (say, arm 1) is directed in a position allowing for unloading the rotary table, or pointing towards the press in a position allowing for loading of the press; all intermediate positions can be collapsed into a single third value. Such abstractions can only be informally justified on the basis of our meta-knowledge about the system, a critical point we elaborate upon more in the next paragraph. The abstraction to discrete models is possible at all, because none of the requirements stipulated for the production cell needs a continuous semantical model to be expressed (in contrast e.g. to the Gas-Burner example studied in the ProCos Project [9]).

We view it as mandatory for the interpretation of a "verification result", that all abstractions performed in the modelling process are made *explicit*. Clearly all specifications of the production cell shown in this volume do make abstractions; however, in most cases, it is hardly possible to identify exactly what abstractions or implicit assumptions were made, because no clean separation between the controller on one side and the *model of the controlled system* on the other side was made. In contrast, this paper clearly separates these key components: we formalize the list of sensors and actuators as constituting the interface between the StateChart specification of the controller and the StateChart specification of the components of the production cell. The StateChart model of the components explicates all our assumptions, how the components behave in an uncontrolled world. Nothing more is assumed than those properties, which are explicitly inferable from the component models. In particular, in an unrestricted world, components may interfere in their movements and produce crashes. Only when interfaced with the controller are components expected to produce coordinated behaviour, which guarantees the functionality requirements and excludes crashes. This separation between the controller of the production cell and its components can easily be modelled in Statecharts: essentially each component and the controller correspond to concurrently active states.

The models of components and controller are described in the next section, which also recalls the key aspects from Statecharts needed to make this paper self-contained. Symbolic Timing Diagrams are introduced together with example specifications in the subsequent section. A section on verification introduces the methodology used to allow compositional model checking and gives some statistics regarding the size of the generated symbolic representations of the production cell as well as verification times.

8.2 Statecharts

Statecharts [2] enhance conventional FSMs by notions of hierarchy (abstraction) and orthogonality (parallelism) to factor out common transition (sub-)structures and to avoid the problem of state explosion in the graphical representation.

A statechart consists of a set of states, graphically represented as rectangular boxes, and a set of transitions between them, drawn as arcs. States may have substates and are either of type *AND* or *OR* to express parallel composition and sequential composition, resp.; to be in an *AND*-state — where to be in a state is the same as to say a state is active — implies to be in all of its substates and to be in an *OR*-state implies to be in exactly one of them. The resulting overall structure of a statechart is an *AND/OR*-tree, where the leaves are called *basic* states. One substate of each *OR*-state is distinguished as *default* and is entered whenever its father but none of its siblings is entered.

Transitions can be labelled with a label of the form [*Cond*], where *Cond* is a boolean expression over states. A transition becomes enabled when the expression holds, i.e., if the parallel (sub-) statecharts fulfill the expression; then the source of the transition is exited and the target is entered. One transition will in general interconnect states on different levels in the hierarchy. Thus, with each transition we have to associate a *scope*, representing the lowest state that is left unaffected by the transition, formally given as the lowest common strict *OR*-ancestor of source and target of the transition.

In general, several transitions, i.e. those that affect only orthogonal components, are taken simultaneously, constituting a *step* of the system. *Maximal progress* requires that steps be always *maximal* in the sense that all transitions are included as long as they are currently enabled and not in *conflict* with another transition already in the step, i.e., both have an *AND*-state as least common ancestor of their respective scopes. If several steps can be built this way at the same time, the system chooses nondeterministically between them.

A step takes a system from one *state configuration* to the next; a state configuration is a maximal set of states that can be active at the same time, i.e., it contains the root, with each *AND*-state all of its children and with each *OR*-state exactly one of them.

8.2.1   General modelling concepts

For each machine component of the production cell two (sub-)statecharts have to be specified. One models the part of the controller "responsible" for this component, the other describes the component itself. To model the reality adequately both substatecharts are only allowed to communicate via the given sensors or actuators, i.e., the controller can only act based on observations provided by the sensors of the machine components and its own state while the component only changes its state based on the setting of the actuators by the controller. The actuators and sensors are modelled via (statechart) states. A state representing an actuator is active means that the actuator is set; analogously if the component is in a state which represents a sensor it means that this sensor is set. The reaction of the controller (resp. component) to the component (resp. controller) is modelled by transitions which are labelled by boolean expressions over the states representing the sensors (resp. actuators). For example if a controller should leave a state S_1 in order to enter state S_2 if a sensor s_1 or s_2 of a component C is set, a transition from S_1 to S_2 is labelled with [C.s$_1$ ∨ C.s$_2$]. The model of the production cell is the parallel composition of the implementation of the controller, and of the models of the machine components. Both sub-statecharts are again parallel compositions of sub-statecharts; each of these sub-statecharts represent one component of the production cell like the robot, the press or the crane.

The modelling of sensors and actuators by states induces one problem. Both the controller and the machine components have to take into account that a reaction on a change of a sensor or an actuator takes at least one step. If a component enters a "sensor" state, the controller can react on this one step later. Another step later the component "can be sure" that the controller was able to react on the sensor. This leads to an internal splitting of the sensor states into two states in order to model that there exists the possibility that a sensor is set and the controller has not yet reacted. This is explained in more detail in the discussion of the following example.

Due to the absence of fairness in statecharts, it is difficult to model that a state is eventually entered. Since the specification of the production cell can not deal with quantitative time expressions, one has to model explicitly that the feed belt can transport a blank from the start position to the end position at earliest in three steps, but at latest in five steps, for example. We cope with this problem by introducing a timer component. This timer has a fixed number of states which are stepped cyclically. Other components of the production cell can act depending on whether a special state of the timer is reached. Since the other components do not know which internal state the timer has when they are entering a state which can only be left by ending of the timer a non-deterministic waiting will be performed. In the best case the timer ends at the next step and the component waiting for it can continue. The worst case occurs if the timer is at start state: in this case the component has to wait as many steps as the timer needs to reach the *time_out* state.

8.2.2 An example: the rotary table

The statechart specification of the rotary table is used to explain our specification technique. Consider a part of the model of the rotary table depicted in Figure 1.

The model of the machine component consists of three parallel sub-statecharts. Figure 1 depicts two of them. The *Rotate* component is an *OR*-state consisting of four substates. The task of the state is to model the rotation movement of the rotary table. We can distinguish four states. The position of the table is correct for the feed belt or it is correct for the first arm of the robot or it is in none of these two position but between. Besides these allowed positions the table can rotate exceed its safe

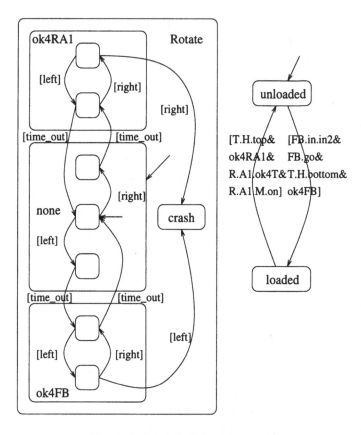

Figure 1   Model of the rotary table

movement. In this case an error state is entered represented by *crash*. The system starts in the middle *none* position. If the controller is in state *right*, i.e., the actuator *right* is set, the table begins to rotate clockwise, by entering the upper none state. At this position the sub-statechart waits until the timer generates a time-out by entering the state *time_out*. After this the table is in the correct position for the first robot arm. If the controller still sets the actuator *right* the environment does not enter its error state since the controller must have a chance to react to the sensor indicating that the table is in its correct position *ok4RA1*. After one step the controller is able to leave its state *right*, i.e., does not set the actuator *right* any longer, and the rotation of the table stops. But if the controller does not change the actuator immediately the component enters the *crash* state. This situation describes the restriction on movement of the table. To leave the *ok4RA1* state the controller must set the *left* actuator by entering the corresponding state.

The vertical movement of the rotary table is implemented by a similar component. Besides these two components the *load/unload* states describes whether a blank is situated on the table or not. The system starts with an empty table. The table becomes loaded when the feed belt is in its (last) *in* position, the controller says *go* to the feed belt, the table is at the bottom and its position is correct for the feed belt. The table becomes unloaded when it is up, the first robot arm is in its correct position, its magnet is on and the table has rotated towards the arm.

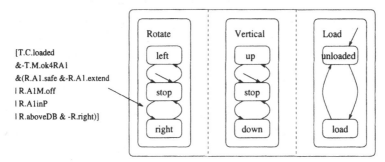

Figure 2   Model of the rotary table

For the specification of the controller part of the table consider Figure 2[1]. The controller of the table consists of three parallel components. One controls the rotation, one the vertical movement and the last keeps track of whether the table is loaded or not, since this information will also be used by other controller components. The system starts with the state configuration *stop, stop, unloaded*. The controller enters the *left* state if the table is unloaded (which is true in the start configuration) and the table is not in the correct position for the feed belt indicated by the sensor *ok4FB* which also holds at the beginning. The movement to left will be stopped by entering the *stop* state when the environment indicates that the correct position for the feed belt is reached by setting the sensor *ok4FB*. The controller sets the *right* actuator when

- the *unload* state has changed to *load,* and
- the table is not in the correct position for the first robot arm, and
- the first robot arm is in a safe position, i.e.,
  - the arm is in its safe position and the controller does not generate the extend actuator for extending the arm, or
  - the arm carries no blank indicated by the switched off magnet, or

---

1. Since the conditions of the transitions are very complex they are not part of the picture.

— the arm is in the press, or

— it is above the deposit belt and the controller does not move the robot
to the right.

The rotation of the table stops as soon as this safety condition for the first robot
arm fails or the position *ok4RA1* is reached by the component. The controller for
the vertical movement is quite similar.

Since the controller can only communicate with the components via the sensors and actuators, it cannot use the information whether the table is loaded or not
by using the corresponding states of the components. These are internal to the
model of the component, and not communicated via sensors. Thus it has to store
and compute this information itself. A transition from the unload to the load state
will be performed when a blank is in the *in* position of the feed belt and the controller sets the *go* actuator for the feed belt. It becomes unloaded again when the
magnet of the first robot arm is on and the arm is in the correct position for the table
and the table is in the correct position for the robot arm.

8.3   Requirement Specifications Using Symbolic Timing Diagrams

In order to specify the requirements of the production cell system as stated in [8]
we used a graphical formalism called Symbolic Timing Diagrams (STDs), which
is derived from the notion of classical timing diagrams as used by hardware designers. In contrast to these diagrams, which are often used informally or with an
ad-hoc semantics in mind, STDs have a precise semantics, which is defined by
translation into a characterizing temporal logic formula (as introduced in [7]).
STDs are intended to be used for a declarative specification of the behaviour of individual system components as observable at their interface. While in the realm of
hardware design specifications often take the form of protocols with concrete timing constraints, STDs are directed towards the early steps of a complex design (i.e.
the system level or an even more abstract one), where requirements are abstract
system properties wherein only the *order* of events and their (under given premises) guaranteed occurrence matters.

STDs are designed to be able to specify these types of constraints in a visual
form which is easily comprehensible, thus inviting their use by the designer who
is not familiar with formal methods such as temporal logic. In the following we
give a short introduction to STDs; a more detailed exposition can be found in [10]

and [6]. A STD consists of a number of *symbolic waveforms*. Each waveform has several *regions* which are separated by *edges*. The regions are labelled by predicates (denoted by boolean expressions over the variables observable at the components interface). Between any two edges on different waveforms there may be *constraints* (denoted be respective arrow shapes) used to express a required temporal ordering of these edges (Figure 3).

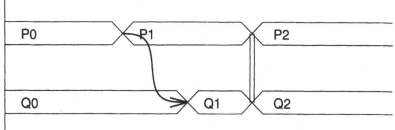

Figure 3   Symbolic waveforms with constraints

Figure 4 show different forms of constraints. A causality constraint expresses that (1) an event matched by edge *e1* must occur before an event matched by edge *e2* and (2) if the event matched by edge *e1* occurs, then an event matched by edge *e2* will occur eventually. A precedence constraint is weaker than a causality constraint and expresses only condition (1). A synchrony constraint expresses that an event matched by edge *e1* and an event matched by edge *e2* must occur simultaneously.

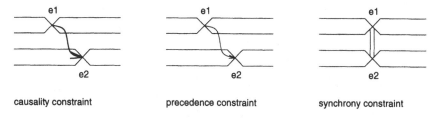

Figure 4   Basic types of constraints and their graphical denotation

A STD can be regarded as a dynamic pattern called into effect whenever the values of the interface variables conform to the conjunction of the initial predicates on each symbolic waveform of the diagram. In the following part of the run of the system, the constraints have to be obeyed. Each constraint arc has also a dashed (weak) form. The weak form expresses a *premise*, not a requirement: If the premise is violated, the STD becomes deactivated, it does not restrict the system further. The property expressed by the diagram can also be formulated in temporal logic.

We used the temporal logic translation of STDs as input to the model checker, see section 8.4.

Example 1 (safety requirement). In a situation where the press is in its top position (*P_S.top*) and the robot is rotated such that arm1 points towards the press (*R_S_R.A1_in_Press*) and *arm1* is in an unsafe position (i.e. in a position where either a collision with the press or the table might occur), then the arm must be retracted to a safe position (*R_S_A1.safe*), before the press may leave its top position (which means that it will start to move towards the middle position, which is dangerous for the robot arm), unless the robot rotates *arm1* out of the dangerous position (Figure 5). The dashed arrow is an example for a precedence premise, whereas the solid arrow is a precedence requirement. The following formula is a translation.

$$\textbf{always } (P\_S.top \wedge \neg R\_S\_A1.safe \wedge R\_S\_R.A1\_in\_Press \rightarrow$$
$$(P\_S.top \wedge \neg R\_S\_A1.safe \wedge R\_S\_R.A1\_in\_Press \textbf{ unless}$$
$$R\_S\_A1.safe \vee P\_S.top \wedge \neg R\_S\_A1.safe \wedge \neg R\_S\_R.A1\_in\_Press))$$

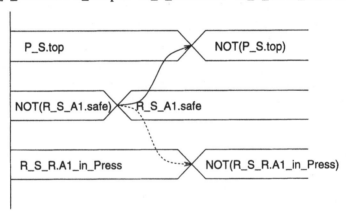

Figure 5   A STD expressing a (conditional) safety requirement

Example 2 (progress requirement). We assume that the following facts have been established: (1) The magnet of the gripper of the travelling crane is *on* iff the crane carries a metal plate. (2) The magnet may switch from *on* to *off* only if the crane is in a correct position above the feed belt and no blank lies underneath. Then in a situation where the crane is in a position ready to drop a plate onto the feed belt and the magnet of the gripper is on (*C_A_M.on*) while there is no plate at the end of the feed belt, a change of the magnet state from *on* to *off* is required to lead to a situation where a plate reaches the end of the feed belt, unless the position of the crane changes earlier or another part already carried on the feed belt reaches the end of the feed belt earlier. Note that from the diagram it can be inferred that given the crane position is kept stable sufficiently long, then a change of the magnet from *on* to *off* will lead to a situation where (some) part arrives at the end of the feed belt (Figure 6).

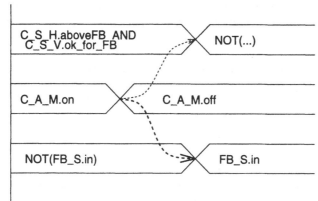

Figure 6   A STD expressing a (conditional) progress requirement

8.4   Verification by Symbolic Model Checking

The behaviour of a statechart can be described by a finite transition system. The states of such a transition system — in the following denoted transition states — represent the state configurations of the statechart. Two transition states of the transition system are connected by a transition if and only if a step from one state configuration represented by the first transition state to another state configuration represented by the other transition state is possible. Due to the nondeterminism of statecharts the behaviour is characterized by a next-step relation over the state configurations.

For finite transition systems, the validity of several modal logics can be checked automatically. This is called *model checking*. In particular, CTL (computation tree logic) can be checked with space and time complexity linear in both the size of the system and the size of the formula [1]. But in practice, even linear complexity is prohibitive: The estimated number of transition states of our finite model of the production cell (including the controller) is $8 \cdot 10^{19}$, far beyond the possibilities of a model checker which enumerates the state set. Those programs can handle approximately up to 10^7 states. We describe in short the essentials of *symbolic model checking*. More complete treatments can be found in the literature, e.g. [4]. What is new is the application to an example like the production cell.

8.4.1   Principles of symbolic model checking

The trick to overcome the limitations mentioned above is to avoid *explicit enumeration* of the transition state set and to represent the system *symbolically*, in a way such that the standard operations a model checker has to perform can easily be done. These operations are:

- Computing the set of states satisfying an atomic formula.

- Computing the backward image of the next step relation of the system and a given set of states, i.e. those states which can reach one of a given set in one step.

- Comparing two sets of states.

Thus, we have to represent sets of states and the next step relation of the system. The first step is to define a binary encoding of the state set. Given the encoding, any set of states can be described by a boolean formula with as many variables as the bits of the encoding. For the next step relation we need a formula with twice as many variables. It is true of $(\hat{x}, \hat{x}')$, if the state encoded in $\hat{x}'$ is one of the successors of the state in $\hat{x}$. All the operations listed above can be expressed by boolean operators. The only drawback is that boolean formulas tend to get very large and testing the equivalence of two formulas (for comparison of two sets) is an NP-problem. But part of these difficulties can be overcome by using so-called ROBDDs (reduced ordered binary decision diagrams) as a compact way for representing formulas.

An ROBDD is very similar to a minimal automaton accepting the set of minterms of the formula. (Again, for a more complete treatment, we refer the reader to the literature, e.g. [12]). The size of the ROBDD is highly sensitive to the order of the variables in the minterms. Very often, if the order is chosen carefully, the ROBDD is very small. Since ROBDDs are canonic representations, checking the equivalence of two ROBDDs amounts to testing whether they are isomorphic, which can be done very fast. Boolean operations have complexities linear in the size of the arguments and the result, and the result of an operation has at most the size of the product of the inputs.

This means, that all operations needed for model checking can be done in time linear in the size of the involved ROBDDs, and one operation most often will not radically increase the size of the ROBDD. But this does not mean that exponential growth does not occur: In the worst case, n operations with ROBDDs of size k might produce an ROBDD of size k^n.

Experience shows that many problems which could not be handled with explicit model checking are easy when applying symbolic techniques. Approximately 25 bytes are needed to store one node of an ROBDD. Using a workstation with more then 100 Mbyte of core memory, ROBDDs of sizes in the range of 10^6 to 10^7 nodes can be handled. But all too often, even symbolic model checking does not succeed immediately, the production cell considered here being one example.

8.4.2 Application to the production cell

The binary encoding we use for the statechart language is described in [11]. For the production cell, 88 bits are needed for the encoding of one configuration of the system.

After experimenting with the variable order, we had no problems generating the next step relation for the controlled production cell. In the shortest representation we found, the ROBDD has 73199 nodes. In general, a good order must keep the distance between variables whose values are closely related very small. Otherwise, the automaton (viewing the ROBDD as an automaton accepting minterms) would have to remember the values of the variables which occur first for a long time, namely until it is told the values of the related variables. If the related values appear soon, the automaton could decide early whether the relation is satisfied and either reject the minterm immediately, or go on with examining the rest, now only keeping the information that the relation was satisfied.

The consequences for the production cell are the following. The statecharts for the controller and the machine components have a very similar structure. They exhibit essentially the same parallelism. Many of the states of the machine model are mirrored in the controller (e.g. the controller of the table has a substate *Load* which corresponds to a similar substate of the table model). Their values are of course highly correlated, most of them are indeed reached and left either synchronously or one lags at most one step behind the other. It would be impossible to generate an ROBDD for the next step if the blocks of variables for the machine components and the controller were ordered one after the other, i.e. each environment variable would precede each controller variable. Instead, we have merged the variables for corresponding parallel components as much as possible. The result were, as already said, 73199 nodes, and we hoped to be able to check relevant properties in the global system representation.

This, however, proved to be wrong. The ROBDDs for intermediately computed sets grew enormously. And whenever the size of the ROBDDs which are operated on exceed the available core memory, the performance of the system drops drastically. Most of the time is spent on paging rather than on computing. Only very rarely was the model checker able to compute a result after exhausting the memory. So we had to resort to *compositional* techniques to cope with the example.

8.4.3   Compositional model checking

The properties we want to verify have the form:

"All runs of the system satisfy P"

This includes liveness properties: $\Diamond$ 'Q means that Q eventually holds on every run. Thus, if we can show that P holds for a system which has more possible runs, we have verified P for the system under consideration, too. This principle can be applied in two ways.

First, consider the safety property from Figure 5. Obviously, if it holds, it will already be guaranteed by the robot and the press alone, the rest of the system does not matter at all. Thus, we can *abstract* from the concrete behaviour of the rest of the statechart and allow it to behave arbitrarily. This is a rather typical situation: The safety requirements can be localized to a few components and checked in isolation.

With liveness properties, the picture is a little different. The fact that a blank will infinitely often arrive at the end of the deposit belt and will infinitely often be removed, formalized in $p =_{df} \Box \Diamond \text{DB\_S.in} \land \Box \Diamond \neg \text{DB\_S.in}$, needs the whole system for its verification, although only the deposit belt is mentioned explicitly. To verify it without being able to do a global model checking, we have to find local *liveness properties* whose conjunction implies p. Note that p would be valid for the production cell, if it were true for the deposit belt. It only happens to be the case that p, as all other interesting liveness properties, is not as easily localized as the safety properties. Nevertheless, the liveness properties can also be localized to a conjunction of properties, each valid for a few components in isolation.

Now we have to explain how the model of the production cell can be reduced to a few components. We cannot simply take the statechart models of, say, the robot and the press, and the corresponding controller subsystems. The press only interacts with the robot, but the behaviour of the robot also depends on the table and the deposit belt. The conditions labelling the transitions mention states of those other parts of the statechart. Therefore, we have to provide additionally a statechart which exhibits at least all possible behaviours of the complete statechart for those components.

We realized this in the following way. For each component, we programmed an abstract statechart visiting at random all states of the controller and machine part any other controlled component might refer to, thus having at least all behaviours of the detailed statechart. Any statechart containing one version of each component with its controller, either the detailed or the abstract one, could be used for verification. The size of the ROBDD for the next step relation could in some cases be drastically reduced, enabling us to carry out the model checking. The next section contains detailed information on space and time requirements for some of the examples we tried.

The method explained above is a little bit ad hoc and does not produce optimal results. For example, the press and the robot need only very little of the other components. Feed belt and crane are not referenced at all, and it also does not matter whether there is a blank at the end of the deposit belt. We chose the method because it is an easy way to generate abstractions of arbitrary parts of the cell *manually*. A program doing abstractions automatically would not be hard to write, and a more elaborate verification system should include one. But we have performed this case study with prototype tools only. And this was one of the deficiencies we had to cope with.

8.4.4 Experiences

For each type of requirements, we checked some examples. Apart from the lack of time, the prototype nature of our tools was responsible for the fact that we did not check most or all of the requirements. Nearly every step of the verification process needed some user interaction. We used a timing diagram editor, but we had to translate its output manually to the input format of the model checker. Because the structure of the statechart fixes the variable order the ROBDD compiler chooses, changing the variable order required a restructuring of the statechart, which in some cases was tedious. Abstractions had to be performed manually. Additionally, some optimizations of the model checker became necessary, including a (rudimentary) garbage collection. All these obstacles slowed down the verification process which, including the formalization of the requirements and the development of the statechart model of the system and its controller, should be a matter of at most two or three weeks.[2]

Table 1 gives sizes of the ROBDDs for the next step relation, and the CPU time needed to generate them. Only information concerning the smallest representation

---

2. Provided that all requirements pass the model checking after sufficient localization.

we have found are given. For comparison, the estimated number of configurations of the statechart is given, too. In Table 2, the CPU time needed for model checking some of the requirements is presented. We used a SUN SPARC 10 with 95 MByte of core memory. We experienced that a timing diagram describing a run in a step-by-step fashion is usually verified rather fast, whereas a property relating two events which require many intermediate steps takes considerably longer. The temporal logic formula expressing that whenever the magnet of the crane is on *C_A_M_on*, eventually a blank will appear at the end of the feed belt *FB_S_in* exemplifies this. The other formula is again a safety requirement.

Represented Components	Number of Nodes	Number of Variables	CPU Time [sec]	Number of States
full model	73,199	88	22.3	$8.4 \cdot 10^{19}$
crane, user, feed belt, table	7,878	70	11.3	$2.3 \cdot 10^{17}$
robot, press	20,445	73	14.6	$3.5 \cdot 10^{17}$
table, robot	25,174	74	15.2	$5.3 \cdot 10^{17}$
crane, user, feed belt	4,210	66	9.4	$5.9 \cdot 10^{16}$

Table 1   Space and time requirements for the representation of next step relations

Represented Components	Property	CPU [min:sec]
full model	STD of Figure 5	1:27
full model	STD of Figure 6	1:52
full model	$\Box\,(C\_A\_M\_on \rightarrow \Diamond\, FB\_S\_in)$	204:39[a]
crane, user, feed belt, table	$\Box\,(C\_A\_M\_on \rightarrow \Diamond\, FB\_S\_in)$	1:42
crane, user, feed belt, table	$\Box\,(FB\_S\_in \rightarrow FB\_S\_in$ **unless** $(T\_S\_V\_bottom \wedge T\_S\_R\_ok4FB)\,)$	5:36
robot, press	STD of Figure 5	0:20

Table 2   Running time of the model checker

Represented Components	Property	CPU [min:sec]
crane, user, feed belt	STD of Figure 6	0:17

Table 2   Running time of the model checker

a. To verify this property a special ROBDD-operation was necessary. Instead of computing the precessor-set of a given ROBDD w.r.t. the next-step relation by computing first the product of both ROBDDs and then to quantify the primed variables existentially, a new ROBDD-operation *and_exist* performs these two operations in one step. This leads to much more efficient memory usage.

8.5   Conclusion

The selection of discrete models constitutes an abstraction — analog signals like the potentiometers in the case study are only observed at discrete values. While the justification of such abstractions is outside the scope of the formal methods employed in this paper, it constitutes but one example that the designer has to exercise extreme care in setting up his model of real system behaviour. We have emphasized the importance of making such modelling assumptions explicit and indeed done so in our specification by cleanly separating controller and controlled components.

In general modelling decision may well have an impact on the feasibility of automatic verification methods. As an example, our StateChart specification needs some abstract notion of the relative timing-properties of system components in order to ensure that the setting of actuators in critical states (like the metal piece having reached the feed belt position monitored by the sensor) affects the component sufficiently quickly (e.g. before the metal piece falls on the floor). Our model handles this problem by essentially distinguishing *almost* critical states from critical states; another option possibly simplifying verification would ensure that the controller and the controlled components alternate in taking steps. Another such modelling decision relates to avoidance of crash states. In our approach, avoidance of *crash* states constitutes part of the (timing-diagram) specification of the system, which sometimes yield complex formulae; alternatively, one may construct the model of components in such a way, that all crash situations are expressed as states and thus reduce the proof of crash-freedom to a pure reachability property. Though clearly models become more complex, the verification task itself may well be substantially simpler.

The work reported shows the principal feasibility to employ symbolic model-checking of designs of the complexity of the production cell. The case study was the first test case for our BDD-compiler and the symbolic model checker supporting StateChart Verification. Yet we were able to verify formula covering all requirement classes, ranging from simple safety properties excluding crashes to liveness properties ensuring that the metal blank indeed is moving properly from component to component. We did not have the time to deduce the maximal number of metal blanks handled by the controller; a safe guess is, that certainly five or perhaps more blanks could be handled concurrently. To make movement of, say, the crane more efficient, it would in some cases be better to have potentiometers rather than switches as sensors. In case of the crane it would be unsafe not to first withdraw the gripper before moving it horizontally.

Acknowledgements

We thank Thomas Lindner for his invaluable support in clarifying and developing our understanding of the production cell.

This research was supported by the German Ministry of Technology within the KORSO project and ESPRIT within the FORMAT project.

References

[1] E.M. Clarke, E.A. Emerson and A.P. Sistla. Automatic Verification of Finite State Concurrent Systems Using Temporal Logic Specifications: A Practical Approach. In *Proceedings of the 10th ACM Symposium on Principles of Programming Languages*, 1983, pp. 117-126.

[2] D. Harel. Statecharts: A Visual Formalism for Complex Systems. Science of Computer Programming, vol. 8, 1987, pp. 231-274.

[3] Z. Manna and A. Pnueli. Models for reactivity. Acta Informatica, 1993, to appear.

[4] J. R. Burch, E.M. Clarke, K.L. McMillan, D.L. Dill and J. Hwang. Symbolic Model Checking: 10^{20} States and Beyond. In *Proceedings of the Fifth Annual Conference on Logic in Computer Science*, Jun 1990, pp. 428-439.

[5] Z. Manna and A. Pnueli. Verifying Hybrid Systems. In *Proc. Workshop on Hybrid Systems, LNCS, Springer Verlag*, 1993.

[6]   Werner Damm, Bernhard Josko and Rainer Schlör. System-Level Verification of VHDL-based Hardware Designs. In *Specification and Validation Methods for Programming Languages and Systems*. Oxford University Press, 1993, to appear.

[7]   Rainer Schlör and Werner Damm. Specification and verification of system-level hardware designs using timing diagrams. In *The European Conference on Design Automation with the European Event in ASIC Design*, 1993, pp. 518-524.

[8]   Th. Lindner. *Case Study Production Cell: Task Definition*. Forschungszentrum Informatik, Haid-und-Neu-Strasse 10-14, 76131 Karlsruhe, Germany, 1993.

[9]   M. Hansen and E.-R. Olderog. *Constructing Circuits from Decidable Duration Calculus*. Technical report, University of Oldenburg, 26111 Oldenburg, Germany, 1993.

[10]   J. Helbig, R. Schlör, W. Damm, G. Doehmen and P. Kelb. VHDL/S - integrating statecharts, timing diagrams, and VHDL. Microprocessing and Microprogramming 38, 1993, pp. 571-580.

[11]   J. Helbig and P. Kelb. An OBDD-Representation of Statecharts. EDAC, 1994, to appear.

[12]   R.E. Bryant. Graph-based algorithms for boolean function manipulation. In *IEEE Transactions on Computer, C-35(8)*, 1986.

IX. TLT

Distributed Specification and Verification with TLT
A Case Study

Jorge Cuellar, Martin Huber

Siemens AG, Corporate Research and Development

Abstract

The Temporal Language of Transitions (TLT) is a framework for the design and verification of distributed systems being developed at Siemens. Similar to UNITY and TLA, it is a formalism to model systems and specify their properties. It further includes methods for refinement and composition, used repeatedly in this case study. We were able to specify, verify and simulate a distributed controller that can handle up to eight plates. The correctness proofs for both safety and liveness properties were done automatically using a model checking tool based on BDDs.

9.1 Introduction

The Temporal Language of Transitions (TLT) framework contains a first order logic based, state/transition oriented language to model systems at different levels of abstraction (the user's view of TLT), a simple but expressive logic which can be used for giving meaning to both TLT programs and their properties (the logic view of TLT), and an operational semantics corresponding to the TLT logic (the semantic view of TLT). Currently, tool support based on verification techniques like higher-order theorem proving, tableau methods, and model checking is being developed.

Two important formal approaches which laid the foundations for TLT are Chandy and Misra's UNITY [2] and Lamport's Temporal Logic of Actions (TLA)

[7]. The UNITY design methodology consists of a language for describing parallel programs, a temporal logic for expressing specifications, and a calculus with rules for proving that programs meet their specifications.

TLT combines the simplicity of UNITY with the flexibility of TLA and it offers the possibility of structuring distributed systems into modules using techniques as information hiding, synchronization between modules and explicit modelling of the environment of a module.

We worked out this case study in a series of refinement steps, starting with an abstract centralized controller and ending with implementable distributed controllers The total effort was of approximately 30 hours. It is impossible to separate design and verification efforts: each design step was automatically checked (in almost no time) by a TLT tool based on the symbolic model checker SVE[6]. This helped us to detect errors immediately. We believe that this distributed design without our verification tools would have been incorrect or would have needed much more time for validation.

The structure of the paper is as follows: in section 9.2, we start presenting an abstract centralized solution describing coarsely the behaviour of the whole system. Next, we formulate and prove freedom of deadlock and some liveness properties. In section 9.3, we complete the abstract solution by adding the environment, that is, the handling of sensors and actuators, to our controller. In doing so, the movements of the components get refined in a way not disturbing the liveness properties. The abstract solution then is distributed in section 9.4. It turns out that distributing also is a property preserving refinement. Thus the properties proved in section 9.2 continue to hold. The distributed controllers then get further refined to fulfil all safety properties as shown exemplary for press and robot in section 9.5. In the last two sections we present the simulation and the conclusions, respectively.

9.2   An Abstract Centralized Controller

In this section we present a very abstract controller for the production cell is developed. We also introduce some features of the TLT programming language. The abstract controller describes the behaviour of the devices necessary for handling the plates in the correct order. This order is not purely sequential: we want the robot to be flexible enough as to chose or change its strategy depending on the states of the table and the press. In the subsequent sections, this centralized controller will be distributed and refined into communicating components. The components of the system will synchronize, but between two synchronization points, these devices will also interact via asynchronous message-passing.

Programs start with a *head* listing name and parameters of the program:

PROGRAM   abstract_centralized_ProdCell(number_of_plates : nat)

The only parameter of the production cell is the number of plates that are to be handled. Initially, they are stored outside the cell and may be put on the feedbelt whenever the belt is not occupied by a plate that has not yet reached the photoelectric cell. (As suggested in the task description, the belts are capable of carrying at most two plates).

The *declaration* section contains *variable* declarations. Each variable has a given finite[1] data type as well as an environment class (GLOBAL, LOCAL, READ or WRITE), restricting the scope of a variable with respect to the environment. In this very abstract solution we do not consider an environment. All variables are thus declared to be LOCAL (i.e., they are read and changed only by the program).

```
DECLARATIONS
    LOCAL   new_plates       : integer_range(0,number_of_plates);
    LOCAL   plate_at_cell_FB, plate_at_head_FB : BOOLEAN;
    LOCAL   phase_T          : {collect, moving_to_R, deliver, moving_to_FB };
    LOCAL   plate_at_T       : BOOLEAN;
    LOCAL   phase_R          : { collect, moving_arm1_to_P, fill_press,
                                 moving_arm2_to_P, empty_press,
                                 moving_arm2_to_DB, deliver,moving_arm1_to_T } ;
    LOCAL   plate_at_arm1, plate_at_arm2   : BOOLEAN;
    LOCAL   phase_P          : { collect, forging, unloading, deliver };
    LOCAL   plate_at_P       : BOOLEAN;
    LOCAL   plate_at_cell_DB, plate_at_head_DB : BOOLEAN;
    LOCAL   phase_C          : { collect, moving_to_FB, deliver, moving_to_DB };
    LOCAL   plate_at_C       : BOOLEAN
```

Here the variable phase_X keeps track of the phase entered by device X, it is the "program counter" of device X. The variables plate_at_X indicate whether there is a plate at device X. They will be used for proving progress.

The *always section* allows to define auxiliary variables. To simplify notation, we define Boolean variables for each phase of the devices. For the robot, for example, we define

---

1. In general, TLT includes first-order data types. For model checking, however, only the finite datatypes booleans, integer intervals, enumerations as well as arrays and structures built up out of them are supported in the tools.

ALWAYS
```
    collect_R              := phase_R = collect ;
    deliver_R             := phase_R = deliver ;
    empty_press_R         := phase_R = empty_press;
    fill_press_R          := phase_R = fill_press;
    moving_Rarm1_to_P     := phase_R = moving_arm1_to_P;
    moving_Rarm2_to_P     := phase_R = moving_arm2_to_P;
    moving_Rarm1_to_T     := phase_R = moving_arm1_to_T;
    moving_Rarm2_to_DB    := phase_R = moving_arm2_to_DB;
```

and similar for the other components. Once defined, each further appearance of the left hand side of one of the definitions in the program will be replaced by the corresponding right hand side when the program gets translated into a (temporal) logic formula.

Next follows the *initially section* of the program. The initial states are described as a predicate. In our example, there is exactly one initial state: Most devices are in phase collect (except robot and crane that move to phase collect) and do not carry plates.

INITIALLY
```
    new_plates=number_of_plates ∧ ¬ plate_at_head_FB ∧ ¬ plate_at_cell_FB ∧
    collect_T ∧ ¬ plate_at_T ∧ moving_Rarm1_to_T ∧ ¬ plate_at_arm1 ∧
    ¬plate_at_arm2 ∧ collect_P ∧ ¬ plate_at_P ∧ ¬ plate_at_head_DB ∧
    ¬ plate_at_cell_DB ∧ moving_C_to_DB ∧ ¬ plate_at_C
```

The *instruction section* contains a finite number of TLT instructions. It is indeed possible to write down correct programs with an infinite number of (parametrized) instructions, but this is not the case here. TLT instructions are written as guarded-commands, and a TLT program is executed by non-deterministically selecting an instruction from the set of instructions whose guards evaluate to true and executing it. To accommodate refinement, the behaviour of a TLT program is allowed to contain *stutter steps* (i.e., steps which do not modify any program variable), but an implicit *progress assumption* assures that only finitely many stutter steps occur as long as at least one program instruction can be executed.

In this case study only a restricted class of instructions of type

$$[l]\ g \to \alpha_1 \parallel ... \parallel \alpha_n$$

is used. l is called the *label* of the instruction, g is called the *guard* and the α_i are called *actions*. The label is used to refer to the instruction. The guard determines if an instruction is *enabled*. Formally, the guard is a *state predicate*, i.e., a first order logic predicate containing only program variables. For each program variable x, a fresh variable x' (x prime) is introduced, creating a new set of so called *primed* program variables. *Transition predicates* are first order logic predicates naming both unprimed and primed program variables. Actions are transition predicates with primed variables referring to the values of the program variables in the next step of the execution sequence. *Parallel composition* $\alpha_1\|\alpha_2$ is used to execute actions simultaneously.

INSTRUCTIONS

[FB1]   new_plates>0 $\wedge \neg$ plate_at_head_FB $\rightarrow$
        new_plates'=new_plates-1 || plate_at_head_FB'

[CFB]   ¬plate_at_head_FB $\wedge$ deliver_C $\rightarrow$
        ¬plate_at_C' || moving_C_to_DB' || plate_at_head_FB'

[FB2]   plate_at_head_FB $\wedge$ ¬plate_at_cell_FB $\rightarrow$
        plate_at_cell_FB' || ¬plate_at_head_FB'

[FBT]   plate_at_cell_FB $\wedge$ collect_T $\rightarrow$
        plate_at_T' || moving_T_to_R' || ¬plate_at_cell_FB'

[T1]   moving_T_to_R $\rightarrow$ deliver_T'

[T2]   moving_T_to_FB $\rightarrow$ collect_T'

[TR]   deliver_T $\wedge$ collect_R $\rightarrow$
        moving_T_to_FB' || ¬plate_at_T' || plate_at_arm1' ||
        (plate_at_P $\wedge$ ¬plate_at_arm2 $\Rightarrow$ moving_Rarm2_to_P') $\wedge$
        (¬plate_at_P $\wedge$ ¬plate_at_arm2 $\Rightarrow$ moving_Rarm1_to_P') $\wedge$
        (plate_at_arm2 $\Rightarrow$ moving_Rarm2_to_DB')

[RP1]   fill_press_R $\wedge$ collect_P $\rightarrow$
        forging_P' || plate_at_P' || ¬plate_at_arm1' ||
        (plate_at_T $\Rightarrow$ moving_Rarm1_to_T') $\wedge$
        (¬plate_at_T $\Rightarrow$ moving_Rarm2_to_P')

[RP2]   empty_press_R $\wedge$ deliver_P $\rightarrow$
        unloading_P' || ¬plate_at_P' || plate_at_arm2' ||
        (plate_at_T $\wedge$ ¬plate_at_arm1 $\Rightarrow$ moving_Rarm1_to_T') $\wedge$
        (¬(plate_at_T $\wedge$ ¬plate_at_arm1) $\Rightarrow$ moving_Rarm2_to_DB')

[RDB]   deliver_R $\wedge$ ¬plate_at_head_DB $\rightarrow$
        plate_at_head_DB' || ¬plate_at_arm2' ||
        (¬plate_at_P $\wedge$ plate_at_arm1 $\Rightarrow$ moving_Rarm1_to_P') $\wedge$
        (plate_at_P $\wedge$ plate_at_arm1 $\Rightarrow$ moving_Rarm2_to_P') $\wedge$
        (¬plate_at_arm1 $\Rightarrow$ moving_Rarm1_to_T')

[RT]   deliver_R $\wedge$ plate_at_head_DB $\wedge$ plate_at_T $\wedge$ ¬plate_at_arm1 $\rightarrow$
        moving_Rarm1_to_T'

[RP]   deliver_R $\wedge$ plate_at_head_DB $\wedge$ ¬plate_at_P $\wedge$ plate_at_arm1 $\rightarrow$
        moving_Rarm1_to_P'

[R5]   moving_Rarm1_to_P $\rightarrow$ fill_press_R'

[R6]   moving_Rarm2_to_P $\rightarrow$ empty_press_R'

[R7]   moving_Rarm1_to_T $\rightarrow$ collect_R'

[R8]   moving_Rarm2_to_DB $\rightarrow$ deliver_R'

[P2]   forging_P $\rightarrow$ deliver_P'

[P3]   unloading_P $\rightarrow$ collect_P'

[DB2]   plate_at_head_DB $\wedge$ ¬plate_at_cell_DB $\rightarrow$
        plate_at_cell_DB' || ¬plate_at_head_DB'

[DBC]   plate_at_cell_DB $\wedge$ collect_C $\rightarrow$
        plate_at_C' || moving_C_to_FB' || ¬plate_at_cell_DB'

[C1]   moving_C_to_FB $\rightarrow$ deliver_C'

[C2]   moving_C_to_DB $\rightarrow$ collect_C'

That completes the first version of the controller. Some of the instructions refer to only one component (e.g. [FB1],[FB2]) whereas others describe the passing of a plate from one device to the next one. For example, a plate may be passed from the table to the robot (see guard of instruction [TR]) only if the table is in phase deliver and the robot in phase collect. The action then describes the relation between the current and the next state. In the next state the table will be moving towards the feed belt, there will be a plate at arm1 but no one at the table any more. Depending on the current situation, the robot will start to move either to unload the press, to fill the press or to place a plate at the deposit belt. This flexible reaction of the robot allows maximum utilisation of the robot arms and the press. To achieve maximum utilisation, we also had to add instructions [RT] and [RP], which were omitted in a first specification. They are used to avoid deadlock when 7 or 8 plates are to be handled (see the following section). If fewer than 7 plates are to be handled, there is no deadlock even if these instructions are not included.

As mentioned before, the instructions are translated into logical formulas. Instruction $g \rightarrow \alpha_1 \parallel ... \parallel \alpha_n$ corresponds to the transition predicate

$$t(\mathbf{x},\mathbf{x}') := g(\mathbf{x}) \wedge \alpha_1(\mathbf{x},\mathbf{x}') \wedge ... \wedge \alpha_n(\mathbf{x},\mathbf{x}') \wedge \textit{Frame}$$

where $\mathbf{x}$ is the vector of all unprimed program variables, $\mathbf{x}'$ its primed copy, and *Frame* is defined as $x_1' = x_1 \wedge ... \wedge x_k' = x_k$ for all program variables x_i not mentioned in instruction t. This frame predicate roughly says: "... and nothing else changes". Instruction [TR], for example, becomes the transition predicate

$$
\begin{aligned}
t_{TR} := \quad & \text{phase\_T=deliver} \wedge \text{phase\_R=collect} \wedge \\
& \text{phase\_T' =moving\_to\_FB} \wedge \neg \text{ plate\_at\_T'} \wedge \text{plate\_at\_arm1'} \wedge \\
& (\text{ plate\_at\_P} \wedge \neg \text{ plate\_at\_arm2} \quad \Rightarrow \text{phase\_R' = moving\_arm2\_to\_P}) \wedge \\
& (\neg\text{plate\_at\_P} \wedge \neg \text{ plate\_at\_arm2} \quad \Rightarrow \text{phase\_R'=moving\_arm1\_to\_P}) \wedge \\
& (\text{ plate\_at\_arm2} \qquad\qquad\qquad \Rightarrow \text{phase\_R' = moving\_arm2\_to\_DB}) \wedge \\
& \text{new\_plates' = new\_plates} \wedge \text{plate\_at\_cell\_FB' = plate\_at\_cell\_FB} \wedge ...
\end{aligned}
$$

We also require that the instructions are consistent, in the sense that if an instruction is enabled in state $\mathbf{x}$ (the guard is true) then it is also executable, that is, there is a choice of values for $\mathbf{x}'$ such that $(\mathbf{x}, \mathbf{x}')$ satisfies the transition predicate. In other words, we require $g(\mathbf{x}) \Rightarrow \exists_{\mathbf{x}'} t(\mathbf{x},\mathbf{x}')$ to be logically true. In this example all instructions are trivially consistent. The whole instruction section now simply consists of the disjunction of all transition predicates.

For more complete descriptions of the translation of TLT programs to logic we refer the reader to [3], [4] and [5].

9.2.1 How to handle eight plates

In this section we demonstrate how to prove liveness and freedom of deadlock for up to eight plates in the abstract program presented above.[1] The key idea to solve both the problem of liveness (every plate that is put on the feed belt will eventually be picked up by the crane) and freedom of deadlock (as long as there are no more than eight plates to be handled there is always at least one instruction enabled) is quite simple. Consider the following 9 places that can hold a plate: two on each belt, two at the robot, one on the table, the press and the crane. These places form a cycle with two places being neighbours if a plate can be handed over from one to the other.

Initially, we can put a plate on the feed belt ([FB1]). Afterwards there are at least one and at most eight plates in the cycle. Since there are 9 places there has to be one occupied place followed by an empty one. This is the place where progress always can (freedom of deadlock) and will (liveness) take place. More formally, one has to prove 9 formulas all being of type

plate_at_X1 ∧ ¬ plate_at_X2   UNTIL   ¬ plate_at_X1 ∧ plate_at_X2

to be able to conclude that there is no deadlock and that the plates are processed in the right order. p UNTIL q holds iff on any execution sequence it is true that whenever p holds then it continues to hold up to a point where q holds.

In the deadlock situation mentioned in the last section, the robot is in phase deliver after having picked up a plate from the press and having moved to the deposit belt. If arm1 does not carry a plate, this sequence is only possible if the table is not occupied when the robot unloads the press (compare instruction [RP2]). If there are meanwhile 7 plates in the cycle, all places besides arm1 and press now carry plates, that is, a plate should be passed from table to arm1. But since the robot waits at the deposit belt,

plate_at_T ∧ ¬ plate_at_arm1   UNTIL   ¬ plate_at_T ∧ plate_at_arm1

does not hold. Indeed, we realized this mistake in our first specification only through looking at the debugging sequence delivered by the model checker SVE for this property. As already pointed out, the deadlock gets resolved by the introduction of instruction [RT], that allows the robot to move back from the deposit belt to the table. Similar considerations led to the introduction of instruction [RP].

---

1. In all infinite fair execution sequences there will at the end be 8 plates in the cycle, since every time a plate reaches the photoelectric cell of the feed belt nondeterministically either a new plate is put on the belt ([FB1]) or (if possible) the crane delivers a plate ([CFB]).

Freedom of deadlock as already shown indirectly above (using the cyclic setting of the production cell) can also be expressed in a more general way. The formula

ALWAYS-TRUE $\exists_{instruction\ i}\ g_i$

expresses the fact that on any execution sequence (starting in any initial state) there always has to be at least one instruction enabled. We have proven this formula to be correct with the model checking tool in 31 seconds (cpu time on a SUN Sparc 10) for number_of_plates set to 8 (with 9 plates we got a debugging sequence that leads to a deadlock after 88 steps).

The UNTIL-properties took between 3.1 (for plate_at_head_FB $\wedge$ $\neg$ plate_at_cell_FB UNTIL $\neg$ plate_at_head_FB $\wedge$ plate_at_cell_FB) and 80 seconds (for plate_at_cell_DB $\wedge$ $\neg$ plate_at_C UNTIL $\neg$ plate_at_cell_DB $\wedge$ plate_at_C). These 80 seconds were improved to 36 seconds by precalculating the set of reachable states, which is one of (compared to theorem proving) few possibilities to influence the fully automatic calculations of the model checker. This precalculation reduced the state space from 2^{27} to 163730 states[1].

9.3   The Environment of the Controller: Dealing with Sensors and Actuators

Up to now the program only models the elementary phases of the devices and the places where there are plates. As a next step the environment of the controller will be modelled. The sensor values of the simulation will be 'read' by assuming that there are some additional variables set by the environment that can only be read by the controller. This is indicated by the keyword READ instead of LOCAL.

READ   angle : integer_range(-90,50)

Vice versa we use PRIVATE to indicate that a variable used in a program may be read but not changed by the environment:

PRIVATE   motor : { robot_left, robot_stop, robot_right }

---

1. The 27 bits making up the total state space are composed of 22 bits to encode the program variables and 5 bits to encode the instructions. These extra bits are necessary to handle fairness. The number of reachable states depends, for example, on the number of plates, but also on the algorithm used for the robot. With number_of_plates set to 6, there are 88134 states reachable. Using a fixed order for the robot (moving from the table to fill the press, then to empty the press, then to deliver the plate, then back to the table, etc.) and with 6 plates (which is maximal for this simple algorithm), there are only 14522 states reachable.

GLOBAL variables finally may be read and changed by both a program and its environment. We only have one global variable

GLOBAL disable_env_X : BOOLEAN

for each device X. It is used to enforce that the controllers keep pace with the environment (i.e., the simulation). The environment is only enabled to do a nonstuttering step[1], if all disable_env_X are set to false. As long as there is a at least one device X with disable_env_X set to true, the environment is disabled (blocked). Whenever the environment does one step, it immediately blocks itself by setting all disable_env_X to true. The controller now does at least one step in each device X and eventually sets disable_env_X to false. In real life however, the environment will not be disabled. This is not necessary under certain additional constraints, for example the assumption that the controllers react very fast (before the environment changes twice) and that the environment indeed reacts to the controllers as expected (i.e., the motor does not move if set to stop). These assumptions may be weakened to more realistic timing conditions.

Generally it will be impossible to prove any property of a program if its environment is allowed to change arbitrarily some or all the programs variables. Therefore, the behaviour of the environment has to be restricted. This is done in the *link-constraints-section*. Collectively, they define *the most general environment* of the program. Each other module that manipulates the program variables has to be a refinement of the most general environment, or, in other words, has to respect all link-constraints.

For example, the robot requires the following link-constraints:

LINK-CONSTRAINTS
[robot1]	¬disable_env_R
[robot2]	angle'≠angle ∨ extension1'≠extension1 ∨ extension2'≠extension2
	⇒disable_env_R'
[left]	angle>arm1_at_P ∧ robot_left ⇒ angle'=angle-1
[right]	angle<at_T ∧ robot_right ⇒ angle'=angle+1
[stop]	robot_stop ⇒ angle'=angle
[ext_arm1]	...

Whereas the instructions form a disjunction (either instruction 1 is executed, **or** instruction 2,...), the link-constraints have to be read as a conjunction (all of them must be fulfilled by the environment).

---

1. Similar as for programs, we must ensure that the environment does not stutter all the time.

In the case study, the environment may only execute a step when it is not waiting for the robot controller [robot1]. If one of the sensors relevant for the robot are changed, then the environment gets disabled [robot2] to allow the controller to react to the change. If the robot has not yet reached its leftmost (arm1_at_P is a macro defined to equal -90) position and the motor is set to robot_left then the sensor value will be decremented by one [left1].

Typically, the environment of a TLT program consists of other TLT programs. If some programs are composed to form one system, it has to be proven that they respect each others link-constraints. Both for the explicitly named constraints in the link-constraints-section as well as the implicit ones due to the environment classes of the variables, this typically results in numerous but simple (often even syntactic) proof obligations. In this example the environment is not given as a TLT program, and it is not possible to prove that the simulation fulfils the link-constraints of the controller. Therefore, one tries to rely on a minimal set of properties. For example, we do not restrict the behaviour of the environment if the angle of the robot is less than -90 and the motor is set to robot_left. But this is not a problem, since the controller makes sure that this situation never occurs.[1]

The instructions now are extended to also consider sensor values and set the actuators properly. Instruction [TR], for example, becomes:

[TR]   deliver_T ∧ collect_R   →
        moving_T_to_FB' ‖ ¬plate_at_T' ‖ plate_at_arm1' ‖
        (plate_at_P ∧ ¬plate_at_arm2 ⇒ moving_Rarm2_to_P') ∧
        (¬plate_at_P ∧ ¬plate_at_arm2 ⇒ moving_Rarm1_to_P') ∧
        (plate_at_arm2   ⇒ moving_Rarm2_to_DB')   ‖
        arm1_mag_on' ‖ ¬**disable_env_R'** ‖ ¬**disable_env_T'**

If it is executed, both the components table and robot now enable the next step of the environment. The only further change consists of turning on the magnet of arm1[2].

The most significant changes concern the movement of the robot. [R5] for example now becomes

---

1. Given the link-constraints [left], [right] and [stop], it is sufficient for the controller to stop rotating left whenever angle=arm1_at_P is reached. For simplicity, we assume here (see [left]), that the value of angle is decremented exactly by one. Really essential is only that we know an upper bound for the change of the angle within one step of the environment.
2. As plate_at_arm1⇔ arm1_mag_on is an invariant of the new program, the introduction of the extra variable arm1_mag_on is not really necessary.

[R5] moving_Rarm1_to_P →
    ¬disable_env_R' ||
    (¬angle=arm1_at_P ⇒ robot_left') ∧
    (angle=arm1_at_P ⇒ robot_stop') ∧
    ... extension1= ...   ∧
    (angle=arm1_at_P ∧ extension1=extended_arm1 ⇒ fill_press_R') ∧
    (¬(angle=arm1_at_P ∧ extension1=extended_arm1) ⇒ moving_arm1_to_P')

compared to the original

[R5] moving_Rarm1_to_P → fill_press_R' .

In essence there are two independent movements: the robot moves to its left-most position and extends its first arm appropriately. The phase is changed to fill_press only when reaching this position. Until then, the phase remains un-changed (i.e. moving_Rarm1_to_P). The kind of refinement used here is called *atomicity refinement*: one step in the original program is broken up into a series of 'small' steps. The property

plate_at_arm1 ∧ ¬ plate_at_P UNTIL ¬ plate_at_arm1 ∧ plate_at_P

is preserved by this refinement.

Since the environment (i.e., the simulation) is waiting for all components of the production cell to reset disable_env_X, it has to be taken care that no component disables it because it is disabled itself. Initially, for example, the robot is waiting for the first plate to be delivered by the table. As long as no plate arrives, all guards concerning the robot are false. For this special case we now have to introduce one new instruction per component, which enables the environment:

[RG]   ¬ (deliver_T ∧ collect_R)∧
            ...   ∧
        ¬ (moving_Rarm1_to_P) ∧
            ...   → ¬ disable_env_R'

9.4   Distributing the Solution

In this section the program will be distributed. There are several reasons that jus-tify the distribution of the solution:

— The verification task gets too large.

    At least for the liveness properties with fairness assumptions, problems of the size of the production cell cannot be handled by our model checker without concessions to the specification (as, for example, restricting the number of plates or using simpler algorithms).

— The distributed solution is closer to reality.

A production cell with different components typically is controlled by one processor local to each component. This simplifies the replacement of a component or a rearrangement of existing components.

Most instructions of the program map completely to one of the components, that is they only name variables that unambiguously are local to one component (e.g. [R5]). These instructions can be executed independently (which in our logic means arbitrarily interleaved) from the other controllers. Whenever plates are transferred from one place to the next the situation gets more complicated. Consider instruction [TR]. At first glance one might be tempted to divide it as follows:

```
[TR.T]   deliver_T   →   ¬ disable_env_T' || ¬plate_at_T'   || moving_T_to_FB'
[TR.R]   collect_R   →   ¬ disable_env_R' || plate_at_arm1' || arm1_mag_on' ||
                         ( plate_at_P              ⇒ moving_Rarm2_to_P' ) ∧
                         ( ¬plate_at_P ∧ ¬plate_at_arm2 ⇒ moving_Rarm1_to_P') ∧
                         ( plate_at_arm2           ⇒ moving_Rarm2_to_DB' )
```

This looks natural and easy, but when executed asynchronously, it is difficult to guarantee that the two controllers execute their instruction within a time period short enough for the plate to be passed safely from table to robot. One alternative consists in writing a 'protocol' that restricts the order of some finer grained instructions in such a way that it blocks the first device until the second arrives, switches on the magnet of arm1 and afterwards changes the states of the devices. Although not difficult, such protocols decrease the readability and increase the verification complexity; we regard them as implementation details that should not be considered yet.

In order to cope with this sort of situations, the TLT semantics extends the usual model of interleaved execution of components: it is also allowed to force components to execute instructions simultaneously. Syntactically this is indicated by adding the component name to the labels of the instructions (above [TR] is decomposed to [TR.T] in the table controller and [TR.R] in the robot controller). The semantic model of the decomposed program is defined to be identical to the original centralized program. Therefore, the decomposed program will have exactly the same execution sequences and properties.

Instead of distributing a correct program, as we have done, it is also possible to compose and synchronize independent modules. If we had started with such an explicit distributed solution, it would now be necessary to show that the synchronization required is indeed implementable[1]. There are several types of synchronization, depending on the number of programs involved in the synchronization, whether the programs wait at the synchronization points or not, the conditions under which these synchronization points are reached, if value passing is needed, etc. The "synchronization point" is nothing else but the guard of the instruction to be synchronized, and, as such, it is a state predicate. A program "waits in a synchronization point" means this state predicate remains true unless the synchronization takes place. In general, this does not imply that the program is blocked, but the enabled transitions may not change the predicate to false. In this case study we only need synchronization between two programs without value passing, where both programs wait at the synchronization point. In order to guarantee to be able to implement such a synchronization we require *weak synchronization fairness* for both programs. This is:

— Once enabled, each of the instructions to be synchronized remains enabled (as long as they are not executed).

— Once one process has reached its synchronization point, the second one will eventually reach the same synchronization point too.

— If both instructions to be synchronized are enabled and remain enabled indefinitely long, then they will eventually be executed simultaneously.

The first property is easily checked by proving

$$g_i(\mathbf{x}) \wedge t_j(\mathbf{x},\mathbf{x}') \Rightarrow g_i(\mathbf{x}')$$

for all program instructions $j \neq i$ and

$$g_i(\mathbf{x}) \wedge \text{env}(\mathbf{x},\mathbf{x}') \Rightarrow g_i(\mathbf{x}')$$

where env is the transition predicate corresponding to the step of the most general environment. In our example, the components not only wait at the synchronization points, they are also blocked there (no other instruction is enabled).

One case in which the second property is often trivial is when there is only one synchronization point in the programs. As soon as there is more than one synchronization point in a program it is also necessary to show that the programs do not deadlock by waiting at different synchronization points (e.g., the robot is in phase fill_press and the press is waiting to get unloaded). In our case, the second property is obtained by construction: we distributed a correct program in a correct way. A deadlock is impossible: in the original program the corresponding state is unreachable.

---

1. More general, the synchronization needs to be implementable after the last refinement. It is thus possible to postpone some or all proof obligations until, for example, some nondeterminism in the programs gets refined.

The third property finally is the fairness assumption itself. It guarantees that the execution of the synchronized instructions is weakly fair and creates no further proof obligations.

9.5   Refining the Program: Safe Movements

In the first abstract solution, the movement of the robot was modelled by one 'magic' step taking the robot to its goal position. By considering sensors and actuators, we refined this movement into a series of 'smaller' steps directed towards its goal position. Moving strictly closer towards the goal simplified the proof that the small steps implement the magic one. In reality, however, the movement of the robot is more complex since collisions between press and robot have to be prevented.

The behaviour of the press is straightforward: Whenever the press enters the phase collect (deliver), it sets a PRIVATE flag waiting_at_middle (waiting_at_bottom). After synchronizing with the robot, the press enters a newly added phase waiting_to_forge (waiting_to_unload). When the robot sets a flag arms_safe_out, the press resets waiting_at_middle (waiting_at_bottom) and starts moving by finally entering phase forging (unloading):

[RP1]	col_P	$\rightarrow$	waiting_to_forge_P' II plate_at_P' II ¬disable_env_P'
[P1]	waiting_to_forging_P ∧ arms_safe_out →		forging_P' II ¬waiting_at_mid'
[P2]	forging_P	$\rightarrow$	¬disable_env_P' II

(history_P=middle ∧ at_middle   ⇒ press_upward' ∧ history_P'=middle) ∧
(at_top   ⇒ press_downward' ∧ history_P'=top) ∧
(history_P=top   ∧ at_middle   ⇒ press_downward' ∧ history_P'=top) ∧
(at_bottom   ⇒ press_stop' ∧ history_P'=bottom ∧ waiting_at_bottom'∧ del_P') ∧
(¬at_bottom   ⇒ waiting_at_bottom'=waiting_at_bottom ∧ forging_P')

One point worth to be mentioned about the movement of the press is the use of the variable history_P for storing the last sensor value. This is necessary to decide in middle position whether to move upwards or downwards.

From the task description follows that robot and press can not collide if the following invariant holds:

[INV]	ALWAYS-TRUE	(arm1_out ∨ waiting_at_middle ∨ waiting_at_bottom) ∧
		(arm2_out ∨ waiting_at_bottom)

with arm1_out and arm2_out being defined as

arm1_out	:=	angle ≥ -70 ∨ extension1 ≤ safely_retracted1 and
arm2_out	:=	angle ≤ 15 ∨ extension2 ≤ safely_retracted2 .

For the press controller we proved the stronger invariant

ALWAYS-TRUE   (arms_out ∨ waiting_at_middle ∨ waiting_at_bottom)

in less than 1.5 seconds (with arms_out := arm1_out ∧ arm2_out).

To do the proofs locally in the press controller, we needed two additional link-constraints:

[l_P1]   ¬waiting_at_middle ∧ ¬waiting_at_bottom ∧ arms_out ⇒ arms_out'
[l_P2]   arms_safe_out'   ⇒   arms_out'

The first link-constraint forces the environment of the press to keep the robot arms out of the press whenever the press is moving. The second one is necessary since the press controller does not know the actual definitions of arms_safe_out and arms_out in the ALWAYS section of the robot controller. It reflects the invariant

arms_safe_out ⇒ arms_out

holding by definition since arms_safe_out is

arms_safe_out   :=   arm1_safe_out ∧ arm2_safe_out

with

arm1_safe_out   :=   angle > -70 ∨ extension1< safely_retracted1   and
arm2_safe_out   :=   angle < 15 ∨ extension2 < safely_retracted2 .[1]

Looking closely to the definitions of arms_safe_out and arms_out already reveals the strategy we use for the robot: The robot controller sets the actuators to move towards its goal position only if after the following step of its environment [INV] still holds. It is therefore necessary for the robot to look one step ahead: only if

safe_next_move   :=   (arm1_safe_out ∨ waiting_at_middle ∨ waiting_at_bottom) ∧
                                        (arm2_safe_out ∨ waiting_at_bottom)

holds the movement can be continued directly. If safe_next_move does not hold, then (as the simplest possible strategy) both arms get retracted in order to get into a state where safe_next_move again holds. Instruction [R5], for example, becomes

[R5] moving_Rarm1_to_P → ¬disable_env_R' ||
(**safe_next_move**   ⇒
    (¬angle=arm1_at_P ⇒ robot_left') ∧
    (angle=arm1_at_P ⇒ robot_stop') ∧
    ... extension1= ...   ∧
    (angle=arm1_at_P ∧ extension1=extended_arm1 ⇒ fill_press_R') ∧

---

1. For the sake of a uniform description, we assume the existence of 2 different constants completely_retracted2 < safely_retracted2. In the real simulation arm2 is safe only if it is retracted completely.

$(\neg (\text{angle=arm1\_at\_P} \wedge \text{extension1=extended\_arm1}) \Rightarrow \text{moving\_arm1\_to\_P'})$
$) \wedge$
$(\neg$ **safe_next_move** $\Rightarrow$
    robot_stop' $\wedge$ **moving_arm1_to_P'** $\wedge$
    (**extension1>completely_retracted1** $\Rightarrow$ **arm1_backwards'**) $\wedge$
    (**extension1=completely_retracted1** $\Rightarrow$ **arm1_stop'**) $\wedge$
    (**extension2>completely_retracted2** $\Rightarrow$ **arm2_backwards'**) $\wedge$
    (**extension2=completely_retracted2** $\Rightarrow$ **arm2_stop'**)
$)$

The following additional link-constraints were necessary to prove [INV] locally in the robot:

[I_R1]   ¬arms_safe_out ∧ waiting_at_middle ⇒   waiting_at_middle'
[I_R2]   ¬arms_safe_out ∧ waiting_at_bottom ⇒   waiting_at_bottom'

This time we have to show that

    $t \Rightarrow$ [I_R1]   and $t \Rightarrow$ [I_R2]

hold for all instructions t in the press controller.[1] Together these proofs take less than 1.5 seconds. The proof of [INV] in the robot controller takes 22.7 seconds.

Again we claim that the transformations done in this section make up a proper refinement and therefore all properties proved earlier still hold in the refined program. Since we lack tool support to confirm this claim we also have proven

[U1]   plate_at_arm1 ∧ ¬ plate_at_P UNTIL   ¬ plate_at_arm1 ∧ plate_at_P
[U2]   plate_at_P ∧ ¬ plate_at_arm2 UNTIL   ¬ plate_at_P ∧ plate_at_arm2
[U3]   plate_at_arm2 ∧ ¬ plate_at_DB UNTIL   ¬ plate_at_arm2 ∧ plate_at_DB

in the most concrete distributed robot controller. To do the proofs locally, we extended the assumption-commitment style of reasoning used above to prove [INV], to also handle liveness properties.

Safety properties typically are proved by showing that they hold initially and that neither any program instruction nor the step of the most general environment can violate them. For liveness properties, however, it is not sufficient to consider only one step: typically a series of steps is necessary to do all the progress necessary. This also holds for the environment and leads to fairness assumptions required from the most general environment. In our case we had to assume that whenever the robot is moving arm1 towards the press with the arms outside the press, then eventually the press will be waiting in bottom position:

    WF(moving_Rarm1_to_P ∧ arms_out ,waiting_at_middle)

---

1. Since waiting_at_middle and waiting_at_bottom are PRIVATE variables of the press controller, the press is the only component that might violate [I_R1] or [I_R2].

This assumption and its dual, WF(moving_Rarm2_to_P ∧ arms_out ,waiting_at_bottom), make up the *fairness section* of the robot controller.

Using these assumptions, the proof for [U1], [U2] and [U3] took between 13 and 36 seconds after precalculating the set of reachable states in 96 seconds and after generating a model that also includes the fairness assumptions in 70 seconds. These precalculations also reduced the 22.7 seconds necessary to prove [INV] to about 0.4 seconds.

9.6 Running the Simulation

Currently, there does not exist tool support to automatically compile TLT programs into executable code. However, we are able to interactively simulate TLT programs by executing them stepwise with the model checking tool. Therefore, we just had to select all steps of the environment and instead of executing them we sent the values of the program variables to the simulation and built the next state of the controller using the new sensor values from the simulation. The necessary changes were done in one afternoon and resulted in two extra pages PROLOG code. Connecting PROLOG with the simulation and the feedback pipe then took 2 further days,...

The only changes in the original 'tksim_9' file were

- changing max_blanks from 5 to 8 and
- calling do_get_status without checking the status_flag

As expected, the performance of the resulting controller is quite poor.

9.7 Conclusion

The main focus of our contribution was on writing distributed controllers and on the verification task. Since TLT has no compiler yet, we simulated the controllers at the expense of much overhead.

We developed a complete distributed solution in a series of refinement steps and we verified all relevant properties. Most of the required properties are trivial; the interesting ones were proven automatically. The most difficult proof (to guarantee that there are no collisions between press and robot) is presented in the text. Freedom of deadlock and liveness were proven in the centralized abstract version. Since refinement, and decomposition in particular, are property preserving, these properties also hold in the distributed version. Independently, we also did compositional proofs for the most concrete distributed controllers based on assumption-commitment reasoning.

TLT as programming language can easily be used on different abstraction levels. It also increases reusability. On the most abstract level table, press and crane are distinguished only by the different names given to their phases (the belts even are the same on all abstraction levels). Later, devices can be changed or replaced by other devices given the link-constraints of all other components are still fulfilled.

We intend to do more work on this case study. We have already started writing the specification using interfaces in a spirit similar to [1], but we use transition predicates and history variables to describe the relation between the controllers. We plan to recast all compositional proofs in this setting and develop tools to support this type of reasoning. We expect that this will increase the readability of the specification and also make the assumption-commitment proofs clearer. One immediate achievement is the possibility to disguise the variables disable_env_X in the controllers. They are used only to force alternating execution of simulation and controllers but add nothing to the actual controlling algorithms.

We plan to extend the specification to increase the robustness and allow exception handling. For example, the robot should also react to situations where the press does not behave correctly (somebody might have removed a plate, etc.). We will use timers to supervise all movements. These issues will be investigated as part of the second author's Ph.D. thesis.

Acknowledgements

The tools used in this study were developed in collaboration with Dieter Barnard and Gerd Gouverneur. Klaus Winkelmann, who specified the production cell in CSL [8], carefully read an earlier draft of this paper.

References

[1]   R. Allen, D. Garlan, *Formal Connectors*, technical report CMU-CS-94-115, Carnegie-Mellon-University, 1994

[2]   K.M. Chandy, J. Misra, *Parallel Program Design - A Foundation*, Addison-Wesley Publishing Company, 1988

[3]   D. Barnard, J. Cuellar, *A Tutorial Introduction to TLT - Part I: The Design of Distributed Systems*, Siemens ZFE BT SE 11, 1994[1]

[4]   D. Barnard, J. Cuellar, M. Huber, *A Tutorial Introduction to TLT - Part II: The Verification of Distributed Systems*, Siemens ZFE BT SE 11, 1994

[5]   J. Cuellar, I. Wildgruber, D. Barnard, *Combining the Design of Industrial Systems with Effective Verification Techniques*, FME '94, Formal Methods Europe 1994, to appear[1]

[6]   T. Filkorn, H.-A. Schneider, A. Scholz, A. Strasser, P. Warkentin, *SVE System Verification Environment*, Siemens ZFE BT SE 11, to appear

[7]   L. Lamport, *The Temporal Logic of Actions*, digital systems Research Center, 1991

[8]   K. Nökel, K. Winkelmann, *Controller Synthesis and Verification with CSL*, Siemens ZFE BT SE 15, in this volume

---

1. available by e-mail: martin.huber@zfe.siemens.de or jorge.cuellar@zfe.siemens.de

X. SDL

The Specification and Description Language
Applied with the SDT Support Tool

Stefan Heinkel, Thomas Lindner

Forschungszentrum Informatik, Karlsruhe

Abstract

We describe the application of SDL, a method for the construction of reactive systems, to an example of the field of production control. The control program for a production cell system is constructed, fulfilling safety constraints imposed on the system.

The effectiveness of standard validation techniques for SDL for guaranteeing the safety properties is investigated. It turns out that usual techniques like interactive and automatic simulation are helpful, but not sufficient for safety-critical systems. Possible alternatives are discussed.

10.1 Aim

The aim of this case study is to investigate whether SDL [1], a well-known method for constructing reactive systems, developed and used in telecommunication applications, is suitable for the systematic development of control software for a safety critical system. Our main focus is on whether SDL and the supporting toolkit used in this case study, SDT [5], provides the developer with adequate means to construct software fulfilling certain safety requirements, necessary to avoid injury of people. Additionally, we try to figure out how flexible the approach is: was the effort spent for verifying and validating the software in vain, when changes in the requirements force changes in the software?

10.2 SDL

SDL is a language standardized by CCITT [2]. The most recent version is from 1992; by default we refer to SDL '92 when writing SDL. The former version is referred as SDL '88 in this paper.

In SDL, the behavior of systems is described by means of processes. Processes communicate with each other by exchanging signals (which may carry parameters for data exchange) via channels. Processes describe finite state machines: a process is almost always in one of a finite set of states. It changes this state only, when a signal arrives. In response to the signal the process may sent other signals and update local variables, and eventually enters the same or another state. Examples of process diagrams are shown in Figure 5 and Figure 7.

Processes may be grouped into blocks. Blocks are the most important structuring construct in SDL. A block may contain other blocks, connected by several channels, or it may contain a set of processes, as well connected by channels. The system is finally composed of several blocks.

An SDL system communicates with the environment by sending and receiving signals to and from the environment. These signals are transferred via channels like any internal communication. An example of a system diagram is shown in Figure 1. Arrows connecting blocks with the border of the system box are channels for communication with the environment.

SDL provides a set of object-oriented constructs:

- Instantiation: All processes, blocks, procedures are defined by process types, block types, and procedure types, from which multiple instances of the same type may be created. These instances are distinguishable via unique identifiers (e.g. process identifiers, PIDs).

- Inheritance: All types may be specialized by inheriting all features from a supertype. On some language levels (e.g. processes) one may redefine some components (e.g. transitions or procedures), if they are declared to be virtual in the supertype.

- Genericity: Some types may have generic parameters. This is for instance useful for reusing a block by renaming the channels connecting the block with its environment.

- Encapsulation: The only way to change both state and local variables of a process is to send a message according to the protocol the process understands.

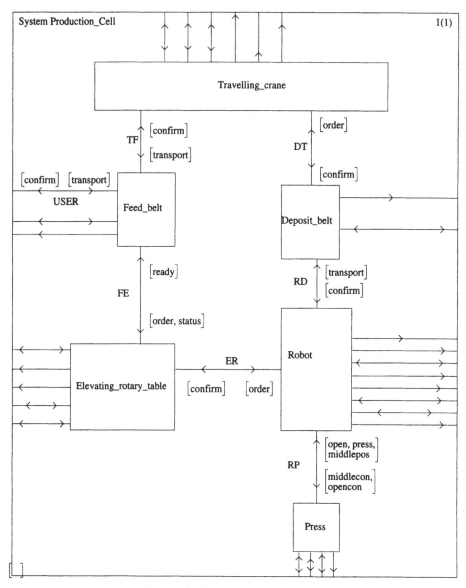

Figure 1 System Production _Cell

Communication in SDL is asynchronous. With each process a message queue is associated which buffers the messages sent to the process. Thus, even if a process is currently changing its state, it can still receive messages and process them immediately after having finished its transition. Which signals a process understands depends on the state the process currently is in. Undefined signals are discharged.

A means for validating SDL specifications are so-called Message Sequence Charts (MSC). Using MSCs, one can formulate requirements of an SDL specification, and afterwards validate it using the SDT Validator. An example is shown in Figure 2: we observe the exchange of messages between the environment and the production cell system. Vertical bars represent the time, horizontal arrows stand for messages. A *transport* signal is sent from the environment to the system each time a blank is put into the system at the feed belt. If a *transport* signal is sent back from the system to the environment, we know that a blank has arrived at the travelling crane. The SDT Validator is only able to validate MSCs which observe *all* messages exchanged between the participants, therefore we have to include the two *confirm* messages into the MSC of Figure 2. Thus, this MSC expresses kind of a liveness condition of the system: if two blanks[1] are inserted into the system, one of them will eventually arrive at the travelling crane.

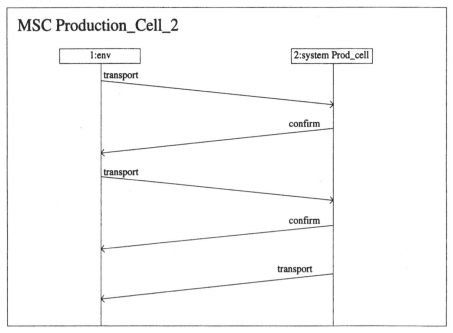

Figure 2   An example of a message sequence chart (MSC)

---

1. It is necessary to insert *two* blanks, as we choose to model the system in a way that it only makes the robot to move a forged piece out of the press if a new blank is available for putting it into the press on the elevating rotary table.

10.3 Production Cell in SDL

10.3.1 The system level

Sensors and actuators are viewed as part of the environment. As a consequence, each sensor and each actuator is represented by a channel connecting the corresponding block with the surrounding box representing the environment. The names of these channels were removed in Figure 1 for the sake of simplicity.

In a first step, the controller specification at system level was divided into six blocks representing the components of the production cell (cf. Fig. 1). Where communication is necessary between two blocks, a channel connects them (e.g. TF^1). Via the *USER* channel it is possible to start the controller and to simulate the insertion of a metal blank into the production machinery.

10.3.2 The component level

Each controller of a component consists of two different parts. The first part contains the description of the special actions, which must be performed in this particular production cell. The name of this block has the suffix *manager* in a block/process name. The other part describes the general actions of a component (e.g. to take or put a metal blank by a robot arm).

Figure 4 shows an example, the block *Elevating_Rotary_Table*. The block *table_manager* is responsible for the cooperation with other parts of the configuration of *this* production cell. Therefore, it is connected via channels with other blocks, for instance the block *Robot* via *ER*. *Table* (an instance of the generic block type *General_elevating_rotary_table*) manages elementary actions, for instance moving the table in upper position. This block is connected with actuators (*A7, A8*) and sensors (*S8, S9*). Obviously, *Table* can be re-used when designing a controller for another production cell, which contains an elevating rotary table. Only the first, configuration-specific part must be replaced.

There are other possibilities of re-use when specifying the controllers for the different components. The robot and the elevating-rotary-table, for instance, are both machines, which have to rotate. By grouping the corresponding actions of the controller in a separate block type it is possible to use this block in the definition of both controllers. All such so-called *part-of relations* are shown in Figure 3.

---

1. All channels connecting two blocks are named consistently: Their names consist of two letters that are the first letters of the names of the two blocks which are connected.

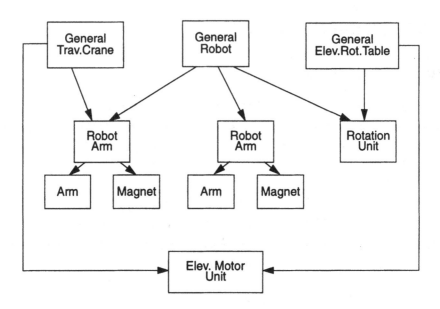

Figure 3 Part-of-relationships

Specialization can be applied if an existing controller can be reused by just *adding* new features. As an example, consider the two conveyor belts. The deposit belt has to transport a metal blank from its beginning to its end, where from the travelling crane picks it up. Additionally, the feed belt has to put the metal blank on the elevating rotary table. The common part of the belts' behaviour is modelled in *GeneralBelt*.

A complete example: the block elevating_rotary_table

According to the proceeding section the block contains two subblocks, the *Table_manager* and an instance of the generic block type *General_elevating_rotary_table*, named *Table* (Figure 4). This block type is divided into two processes: one instance of the process type *rotation_unit* (Fig. 5) to perform the rotation of the table, and one instance of the generic process type *elev_mot_unit* for the vertical movement. Both types can be re-used, for instance in the controller of the travelling crane to handle the horizontal movement of the crane. Only the signal names are different.

The process type *rotation_unit* (Figure 5). It is assumed that the motor performing the rotation is initially switched off. The state of this motor is stored in the variable *motor_on*. If the process reaches the state *Wait* an input is expected. The two possible signals accepted in this state are *rot_left* and *rot_right* (i.e. rotate counter-

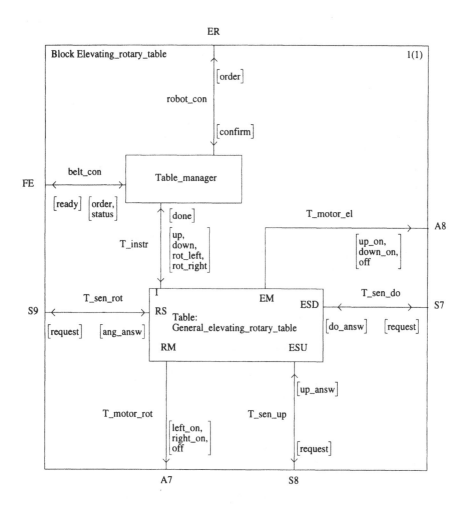

Figure 4   Block Elevating_Rotary_Table

clockwise or clockwise, respectively). The parameter *angle* determines how far to rotate.

After receiving an input signal the current position has to be determined and is stored in the variable *value*. Depending on the content of *angle* and *value* different actions are performed in order to reach the requested position. After having reached it, the signal *done* is returned to the sender of the input signal, in order to confirm that the requested position is reached and all actions are finished.

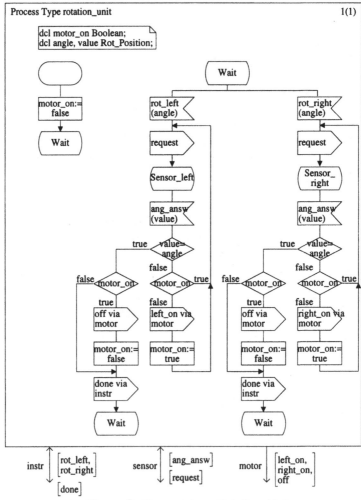

Figure 5   Process type Rotation_Unit

The process *table_manager*. In order to get a more abstract level of specification the rotation positions are denoted by some literals. These literals are collected in the enumeration type *Rot_Pos*. In the case of the elevating rotary table, the literal *O* denotes the position of the table needed to receive a metal blank from the feed belt, *N* denotes the position where the robot can pick up the metal blank from the table (cf. Fig. 6). At the beginning it is assumed that the position *O* can be reached by rotating left from the current position.

Figure 6   Rotation of the table

When the process (Fig. 7) is started, two signals are sent in order to reach the position required for getting a metal blank from the feed belt. After receiving the confirmation that this action is finished, the process is waiting for the signal *status*. This signal is sent by the process *feed_belt_manager* each time a metal blank has arrived at the end of the feed belt. Only if *table_manager* is in the state *DownO* the elevating rotary table is in the position required to receive the metal blank, therefore a *ready* signal is sent to the *feed_belt_manager*. Now, an *order* signal is awaited indicating that the metal blank was put on the table. Then the table can be moved to the position required by the robot in order to pick up the metal blank. After reaching it, an *order* signal is sent to the *robot_manager* and a *confirm* signal is awaited. This signal is sent by the *robot_manager* after the robot has finished picking up the metal blank.

Due to the fact that *table_manager* and *feed_belt_manager* are concurrent processes, a *status* signal can be sent at any time. In our modelling, *status* signals can only be consumed in the state *DownO*. According to the semantics of SDL, these signals would be discharged in any other state. Therefore, it is necessary to save this signal in every state in which it is not possible to consume it. This saving is depicted by a parallelogram box. In the state *Ready* saving is not necessary, as the *feed_belt_manager* sends the signals *status* and *order* alternatingly.

The SDL '88 version of the specification

After the design of the specification we wanted to prove some properties using a tool for SDL '92. Unfortunately, at this time only a prototype version of this tool is available, which does not support validation of specifications. Therefore, the SDL '92 specification was transformed into a specification containing only SDL '88 constructs. This was done by copying and renaming blocks and processes instead of using type instances. Obviously, the length of the specification was increased. This step is straightforward, as the additional constructs of SDL '92 are completely defined using SDL '88.

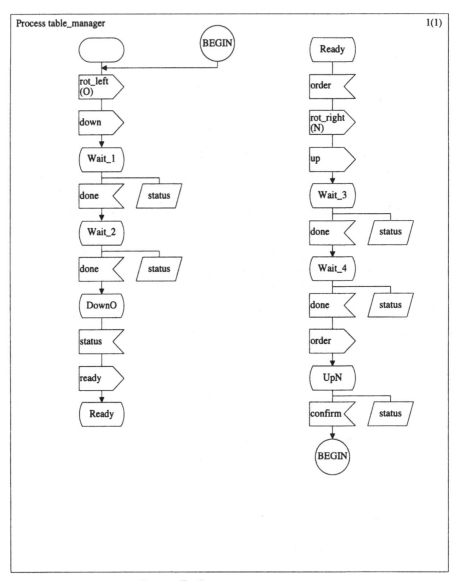

Figure 7   Process Table_Manager

10.4 Evaluation

Proved properties

The message sequence chart shown in Figure 2 was proved to hold using the SDT

Validator, which performs an exhaustive search on the state space of the finite state machine representing the SDL specification. Safety properties could have been proved this way, too, but we found it too complicated to express these properties using MSCs. It would be necessary to construct huge MSCs which are difficult to understand. Some properties are not expressible at all using MSCs.

The state space of the production cell is very large for automatic verification on the system level. It comprises in the order of magnitude of about 10^{12} states. Only very simple MSCs can probably be validated for the whole model. Our modelling of the environment (see below) is quite raw, a more detailed model of the environment would probably make the validation tool fail to prove anything — due to state space explosion.

Assumptions about the behaviour of the cell

In the original SDL specification, sensors and actuators were modelled as part of the system's environment. When simulating the cell, it was the user's task to provide the sensor values in a "reasonable" way.

This proved to be an impractical way for modelling the behavior of the real world for two reasons: First, it is impossible to perform automatic simulations (exhaustive state space exploration), as the system will very quickly run into a state where it simply waits for input signals to occur in every process. Secondly, modelling the environment just in the head of the specifier can lead to odd situations: the specifier does a wrong specification due to wrong assumptions about the environment, and validates his specification using the same, wrong assumptions. These assumptions are never made explicit and exist just in the specifier's mind.

For these reasons, the behavior of the environment was documented by modelling the sensors in SDL itself, grouping them together with their corresponding actuators. Still, this model is by no means complete: if one sends this process the signal *up_on* and a bit later *off*, the sensor values remain unchanged, as no request for a sensor value has occurred. But, as opposed to the former approach, the assumptions are now explicit and validation can be performed automatically.

Measurements

The specification was written, validated and corrected within one month. The SDL '92 is about 1800 lines in length, the SDL '88 version is more then 2000 lines in length. Both figures are derived from the textual syntax of SDL, which can automatically be generated from the diagrams.

It took about another month to learn the usage of the toolkit, to analyse and validate the specification and prove the liveness property mentioned above. For this proof, the validation tool ran not more then 10 minutes on a SPARCstation 10.

Flexibility

Our modelling is very flexible with respect to two different kinds of changes:

If the production cell would be expanded by adding a new machine into the production circuit, we would only have to change the control managers dealing with the machines directly cooperating with the new device. In almost all cases this would only require to change the names of channels and signals.

It is not possible to reuse the proof of the liveness property, but due to the fact that this proof can be run automatically in short time, this does not matter.

Efficiency

It is beyond the scope of SDL to prove that the constructed controller has maximal efficiency, i.e. that it achieves maximal throughput. Additionally, questions like maximum usage of single machines, or guaranteeing a minimum time after which the control program has moved one plate through the whole cycle, are not formally treatable inside SDL. These properties can only be tested. Unfortunately, the additional libraries for SDT, necessary for generating control software from the SDL model, were not available, as not affordable, for us. A careful analysis of our model convinced us that the production cell is able to handle five blanks simultaneously, achieving good throughput.

10.5 Conclusion

SDL proved to be a suggestive visualization of the behavior of the cell's components. Two important characteristics of this special case study are very well expressed by the graphical language features of SDL: first, one has to capture quickly how communication flow works in the system: SDL blocks describe that nicely. Secondly, the control flow is well depicted by process diagrams. Thus, it was easy to write the SDL specification.

From the formal methods peoples' point of view, the means for verification and validation in SDL are quite limited. MSCs proved to be not expressible enough to be very useful in this case study. The verification tools can only do tests, though in a very systematic way. The formal semantics of SDL allow for a mathematically more precise style of verification, but this way is currently not well supported by tools: a theory for symbolic model checking of SDL specifications would be helpful. Unfortunately, the asynchronous nature of the language and queuing of messages will probably significantly enlarge the binary decision diagram encoding the state transition relation necessary for model checking.

In this context, the results of the SPECS project (part of the RACE Esprit project) seem to be interesting [3]. There, a translation of a significant part of SDL into a process algebra, called CRL, was done. Tools for model checking and proving have been developed for CRL. Future work will check the capabilities of this approach.

Whether to recommend or not SDL for the usage in similar problems depends on the needs of the user: if he or she really needs the formal proof of properties for instance required by any national or international norm, one will probably spent a lot of money for this verification, if possible at all. Formal proofs are currently not straightforward in SDL. If one is satisfied with simulating and testing as verification or validation means, SDL may be one of the right choices.

Graphical languages are from our experience useful in those cases, where a lot of people need to communicate about the specification. If the transition from the informal to the formal specification is likely to fail or be erroneous, a graphical language, easily understandable for the customer, can be extremely helpful. But our experience shows as well that single users tend to use a textual syntax after a short time, as the textual syntax is more compact and easier to change. As SDL is a language having both textual and graphical syntax, this question reduces to the presence or absence of tools supporting the automatic transformation from one representation into the other.

References

[1]   R. Brok, Ø. Hangen: *Engineering Real-Time Systems,* Prentice Hall, 1993.

[2]   ITU: *Z.100 ITU Specification and Description Language (SDL)*, Geneva, 1993.

[3]   Bo Bichel Nøbaek, *Final Methods and Tools for the Handling of SDL Specifications,* deliverable D4.15 of the ESPRIT Project SPECS, CEC identifier: 46/SPE/WP4/DS/A/015/b1

[4]   S. Heinkel: Verifikation in SDL, diploma thesis, Universität Karlsruhe, Germany (in German language), 1994, to be finished.

[5]   TeleLOGIC Malmö AB. *SDT user manual,* 1993.

[6]   TeleLOGIC Malmö AB. *OSDT user manual,* 1993.

XI. Focus

Formal Development of a Production Cell in Focus
A Case Study

Max Fuchs, Jan Philipps

Technische Universität München

Abstract

A specification for a production cell is developed using the design method Focus. The specification comprises both the components of the production cell and the corresponding control programs. This work investigates the suitability of a rule system for the stepwise refinement of distributed systems in an assumption/commitment style. As a running example we consider the elevating rotary table of a production cell. Besides descriptive and constructive specifications for this component an executable simulation program in a functional language is presented. All design steps except the final transformation to an executable program are proven correct.

11.1 Introduction

In the development of distributed systems, methodical issues and the formalism used are of great importance. This is because distributed systems are often very complex and there are strict requirements regarding correctness and reliability. The formalism used to specify the production cell is Focus [1]. The specification comprises both the components and the control programs of the cell.

As our main interest in specifying the production cell was the methodical aspect of stepwise refinement, we did not prove all the safety requirements of the task description. We did show, however, that all refinement steps are correct, and that the cell fulfills a basic liveness property: every part that enters the factory leaves it again.

The rest of this chapter is structured as follows:

- In section 12.2 we give an overview about Focus and introduce the formalism used for our specifications.

- In section 12.3 we develop specifications and a (functional) control program for the elevating rotary table of the production cell.

- In section 12.4 we discuss some shortcomings and extension of our approach, and describe how general safety properties could have been included and verified.

11.2 Method

Focus is a design method for distributed reactive systems. Focus provides models, formalism and verification principles for stepwise specification and development. The development process is split into three different abstraction levels. These levels are *requirement specification*, *design specification*, and *implementation*. Since all development steps take place on a formal basis, the overall correctness can be proven, if required.

The most abstract specification, which is called requirement specification, is formulated either using trace logic or stream processing functions. A trace specification gives an overview over the global system behaviour by describing the allowed sequences of actions in the system. A specification on the design level requires an interface description — for the production cell we do have an interface description and therefore we start at the design level.

At the level of design specifications, the system is modeled by a network of components that work concurrently and communicate over unbounded FIFO channels. Focus follows the tradition of Kahn [4] by describing the components as functions that map sequences of input messages to sequences of output messages (stream processing functions). System components have a clearly defined interface, and they can be developed in a modular way. Design decisions can be checked at the point they are taken, components can be refined independently, and already developed components can be reused in other systems.

It is at this level, where most of the developement in the Focus framework takes place. For this reason, Focus was recently extended with a rule system [7] similar to Hoare's well-known rules for sequential systems. The rules allow refinement of specifications formulated in the assumption/commitment-style.

The implementation level finally gives executable programs for the system's components. There are a number of possible target languages for Focus specification, ranging from abstract, not yet implemented languages like AL and PL [1], over other specification languages like SDL [8] to clocked hardware [10]. For the production cell we used Concurrent ML, a version of Standard ML enriched with threads and communication primitives.

11.2.1 Basic concepts

In this section we describe the formalism used at the design level of Focus.

For function application we write *f.x* instead of f(x) to save brackets. The dot is left-associative, i.e. $f.g.x = (f.g).x$.

The most important data structure in Focus is the stream. Streams are used to describe the communication history of the channels between the various components of a distributed system. A stream is a finite or infinite sequence of data elements called *messages*. Given a data set *D* we write:

- $D*$ for the set of finite streams over *D*,

- D^∞ for the set of infinite streams over *D* and

- D^ω for the set of all streams over D ($D^\omega = D* \cup D^\infty$).

For the empty stream we write $\langle\rangle$ and for a finite stream of *n* elements over D we write $\langle d_1, d_2, ..., d_n \rangle$.

We define the following operators on streams. For $s, t \in D^\omega$ and $A \subseteq D$:

- *ft.s* denotes the first element of *s*, if *s* is non-empty.

- *rt.s* denotes *s* without its first element.

- #*s* denotes the number of elements in the stream *s*, or ∞ if *s* is infinite.

- *A©s* denotes the stream *s* with all elements not in *A* removed. For singleton sets $A = \{ a \}$ we write *a©s*.

- $s \circ t$ denotes the concatenation of *s* and *t*.

- s^n denotes the result of concatenating *s* with itself *n* times ($s^0 = \langle\rangle$).

We also define a partial order ("prefix order") $.\angle.$ on streams via

$$s \angle t \equiv \exists u \in D^\omega : s \circ u = t$$

The set of all streams over a given data set D together with the prefix order forms a complete partial order with the empty stream as the least element, and the elements of D^{∞} to guarantee the existence of least upper bounds.

We model components of distributed systems as stream-processing functions (SPF). A SPF is a function from tuples of input streams to tuples of output streams, that is monotonic and continuous w.r.t. the prefix order $\angle$. Monotonicity implies that a component cannot undo any output that has already been emitted. Because of continuity, a component's behaviour is fully described by its behaviour for finite inputs. Therefore components can work in parallel.

A simple SPF is the function *map*, which applies a given function f to every element of its input stream:

$$map.f.(\langle d_1 \rangle \circ s) = \langle f.d_1 \rangle \circ map.f.s$$

For composing SPF there are the three classical operators, namely sequential composition $(f \, ; \, g).x = g(f(x))$, parallel composition $(f \parallel g).(x,y) = (f.x, \, g.y)$, and a feedback operation μ.

Each composition yields again a SPF.

11.2.2 Specifications

We specify components by formulating relations between input and output of their SPFs. Generally, we don't specify the behaviour for all possible input streams, but only for those that fulfil certain expectations of the component. This style of specification is usually called assumption/commitment-style.

We write specifications as a pair $[A, C]$ where A (the assumption) is a predicate with the input streams as free variables, and C (the commitment) is a predicate with the input and output streams of the agent as free variables. The denotation of such a specification is the set

$$\{ f \in I^{\omega} \to O^{\omega} \mid \forall i \in I^{\omega} : A(i) \Rightarrow C(i, f.i) \}$$

of all functions f such that when the input i of f fulfils the assumption of the component, the output o of f is related to i according to the component's commitment. In this case, we considered a component that consumes input data of a set I and produces output of a set O, but it is easy to generalize this definition to arbitrary sets and tuples of input streams.

For example, we can specify a component that applies a function f to its input messages, where no particular properties of the input are required, as

$$[\textbf{true}, o = map.f.i]$$

A specification $[A_2, C_2]$ is called a *refinement* of a specification $[A_1, C_1]$, if the denotation of $[A_2, C_2]$ is a subset of the denotation of $[A_1, C_1]$. We then write

$$[A_1, C_1] \Rightarrow [A_2, C_2]$$

For each composition form of SPFs there is a corresponding decomposition rule in Focus. As an example, we introduce the composition form of a master/slave-system (Figure 1). Master/slave-systems can be used to describe the interaction of two components, where one controls the other. For instance, a control program sends commands to the controlled machine, which returns acknowledgment messages. Given two functions, f and g, the output of a master/slave-system $o = (f \oplus g).i$ for a given input i is defined by the least fixpoint of the following set of equations:

$$(o, y) = f(i, x)$$

$$z = g.y$$

$$x = z$$

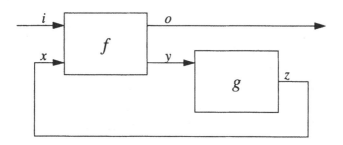

Figure 1   Master/Slave-System

Since the set of all streams forms a complete partial order, and the functions f and g are monotonic, least fixpoints always exist (they may be infinite). Moreover, because of the continuity of the functions, the least fixpoint can be computed by the stepwise communication between components.

The corresponding decomposition rule is shown in Figure 2. It allows us to refine a specification $[A, C]$ of the function $f \oplus g$ to specifications $[A_1, C_1]$ for the master and $[A_2, C_2]$ for the slave function. The rule is closely related to the WHILE-rule in Hoare's calculus. In our rule, the assumption predicate A_1 of the master specification serves as the invariant.

A_1 *is admissible*

$A(i) \Rightarrow A_1(i, \langle \rangle)$

$A(i) \wedge A_1(i, x) \wedge C_1(i,x,o,y) \Rightarrow A_2(y)$

$A(i) \wedge A_1(i, z) \wedge C_1(i,z,o,y) \wedge A_2(y) \wedge C_2(y,z) \Rightarrow C(i,o)$

$A(i) \wedge A_1(i, x) \wedge C_1(i,x,o,y) \wedge A_2(y) \wedge C_2(y,z) \Rightarrow A_1(i, z)$

$[A,C] \Rrightarrow [A_1,C_1] \oplus [A_2,C_2]$

Figure 2   Master/slave refinement rule

The idea behind the rule is as follows. Assume the stream i fulfils the assumption predicate A of the original component. Initially the feedback stream x is empty, and the invariant A_1 is valid (second premise). Therefore the master produces output in the streams o and y that fulfils its commitment C_1. Because of the third premise, the assumption A_2 of the slave component is also fulfilled. The slave's output z fulfils the commitment predicate C_2, and because of the fifth premise the assumption A_1 stays valid, when the slave's output is fed back. This process goes on, until a stable situation, the fixpoint, is reached. Then the streams x and z are identical, and the fourth premise implies the commitment of our original component. The fixpoint, however, may be infinite. The purpose of the first premise is to ensure that the iteration process is valid even for infinite computations. A predicate over streams is *admissible*, if it holds for a stream whenever it holds for all finite prefixes of it.

11.3 Application of the Method

In our approach we model each component of the production cell as a stream-processing function. The streams between the components describe the flow of metal blanks and control signals.

As an example for the use of Focus and its refinement rule system, we consider the elevating rotary table (ERT) of the production cell. We make a further simplification: our table moves only vertically. It is quite easy, however, to add rotation to our specifications. From [5] we take the following informal description:

- The ERT receives a metal blank.
- The ERT moves to its upper position.

- The ERT informs the robot that a blank is available.
- The ERT receives an acknowledgement that the robot removed the blank.
- The ERT moves to its lower position and awaits the next blank.

First we need to consider the messages that the ERT consumes and emits. We need a message to express that a new metal blank has arrived at the table, that the table should go up or down, and that the robot has removed a blank from the table. We use three sets of messages:

- $B = \{ Blank \}$ to model the flow of metal blanks.
- $C = \{ Up, Dn \}$ to model the command signal flow from the table's controller to its motor.
- $A = \{ Ok \}$ to model acknowledgements from the robot back to the table.

Whenever the ERT's motor has executed a command, it will send an acknowledgement of this command back to the controller. For the acknowledgement of a message m we write $ack(m)$. We extend this notation to sets of messages by:

$$ack(M) = \{ \, ack(m) \mid m \in M \, \}$$

11.3.1 Descriptive specifications

A decomposition of the production cell (we skip the details here) yields the following specification of the table (as well as similar specifications for the other components):

$$ERT = [\, \#a \leq \#i \leq \#a + 1, o = i \,]$$

The streams i, a, and o are elements of B^{ω}, A^{ω} and B^{ω}, respectively. The table simply passes on every metal blank received on its input channel i (which is connected to the feed belt) to the output channel o (which is connected to the robot). The assumption of the specification states that the table is only required to pass on the blanks, if enough acknowledgements from the robot arrive via channel a (one for each blank, or possibly one for each except the most recent one).

The next refinement step decomposes the specification *ERT* into a control component specification *CTRL* and a component specification *MOT* that simulates the abstract behaviour of the ERT's motor:

$$ERT \Rightarrow CRTL \oplus MOT$$

The result of the refinement is shown below. The streams *i*, *a*, and *o* are as above; *y* is an element of C^ω, and *x* and *z* are elements of $ack(C)^\omega$.

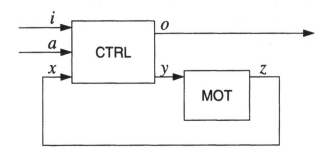

$CTRL = [A_{CTRL}, C_{CTRL}]$

$A_{CTRL}(i,a,x) \equiv \#a \le \#i \le \#a + 1 \wedge$
$\qquad \#ack(Up)\copyright x \le \#i \wedge \#ack(Dn)\copyright x \le \#a \wedge$
$\qquad x \angle \langle ack(Up), ack(Dn) \rangle^\infty$

$C_{CTRL}(i,a,x,o,y) \equiv \#o = \#ack(Up)\copyright x \wedge$
$\qquad \#Up\copyright y = min(\#i, 1 + \#ack(Dn)\copyright x) \wedge$
$\qquad \#Dn\copyright y = min(\#a, \#ack(Up)\copyright x) \wedge$
$\qquad y \angle \langle Up, Dn \rangle^\infty$

$MOT = [\textbf{true},$
$\qquad z = map.\{ Up \to ack(Up), Dn \to ack(Dn) \}.y]$

The motor component *MOT* simply acknowledges the commands to go up and down. The commitment of the control component C_{CTRL} describes the operation of the controller:

- Whenever the motor component signals that the table reached its upper position, the blank is offered to the robot (via channel *o*). Since *o* only consists of messages *Blank*, it is sufficient to specify the length of the stream.

- A command is sent to raise the table for each blank that arrives, provided that the table is in its lower position (i.e., the motor returned a sufficient number of *ack(Dn)* messages).

- A command is sent to lower the table again whenever the robot signalled via channel *a* that it removed a blank, and the table is in its upper position.

- The controller alternately sends *Up* and *Dn* and starts with the message *Up*. This expresses an assumption for the cell: initially the table is in its lower position.

The controller's assumption predicate A_{CTRL} contains the assumption of the specification *ERT* (that there are enough acknowledgments from the robot) as well as some restrictions on the feedback from the motor:

- There are no more acknowledgements for the command *Up* than blanks arrive at the table.

- There are at most as many acknowledgements for the command *Dn* as blanks were removed by the robot.

- The acknowledgements arrive in the same order as the commands are sent by the controller.

This decomposition can easily be proven correct with our refinement rule for master/slave-systems.

11.3.2 Constructive specifications

So far we have an abstract specification: what we have basically described are the contents and length of the ERT's input and output streams. In the next refinement step we aim at a more operational characterisation of the table. In functional programming languages higher-order functions are often used to get shorter and clearer programs. Focus also allows higher-order functions, and indeed the function *map* we have introduced before allows us to concisely express the behaviour of the control component: incoming messages are converted to different messages and output again. Using two *map*-functions in parallel, we can handle the two output channels of *CTRL*, but what about the three input channels? We solve this problem with a separate SPF that merges the input streams into one. We call this function *zip*, as it zips its three input channels together. While in general this leads to some interesting problems (see [3] for a discussion), in our case it is easy because we know the order in which the messages arrive: first a blank, then the acknowledgement that the table has reached its upper position, then the signal that the robot has removed the blank and finally that the table is again in its lower position. We pass this information to the function *zip* in form of a sequence of numbers. This is the sequence of the channels to read the next message from: 0 is channel *i*, 1 is channel *a*, 2 is channel *x*. We skip the definition of this function and only show how it can be used for the constructive specification of the ERT's controller:

$$CRTL_{CONS} = [\,A_{CONS},\ C_{CONS}\,]$$

$$A_{CONS} = A_{CTRL}$$

$$C_{CONS}(i,a,x,o,y) \equiv \exists h \in (B \cup A \cup ack(C))^{\omega} :$$
$$h = zip.\langle 0, 2, 1, 2\rangle.(i,a,x) \wedge$$
$$o = map.\{\ ack(Up) \rightarrow Blank\ \}.(ack(Up)\copyright h) \wedge$$
$$y = map.\{\ Blank \rightarrow Up,\ Ok \rightarrow Dn\}.(\{\ Blank,\ Ok\ \}\copyright h)$$

The assumption predicate of the controller is the same as above. Only the commitment has been changed. Again, the validity of the refinement

$$CTRL \Rrightarrow CRTL_{CONS}$$

can be shown using a refinement rule from [7] and predicate calculus.

11.3.3 Executable programs

We now have a constructive specification for the ERT. How can we transform this specification to an executable program? Up to now, Focus offers two target languages, an applicative language for abstract programs (AL), and a procedural language for concrete programs (PL). However, so far neither is implemented. We therefore chose to use Concurrent ML as the target language for the executable programs. Concurrent ML is an extension of ML with primitives for synchronous communication, but it is easy to implement asynchronous communication with unbounded buffers and the functions *map* and *zip*. We implement the function *map* using associative lists (the operator .-+->. constructs an association) so that elements not in the domain of the function are always removed from the input stream. Then the simulation program for the ERT closely resembles the commitment of the constructive specification:

```
fun ertmot i = mapper ["Up" -+-> ack "Up", "Dn" -+-> ack "Dn"] i;
fun ertzip (i1,i2,i3) = zip [0,2,1,2] [i1,i2,i3];
fun ertctrl i =
      (mapper [ack "Up" -+-> "Blank"] ||
      mapper ["Blank" -+-> "Up", "Ok" -+-> "Dn"]) i;
fun ert (i1,i2) = let
      val h1 = channel()
      val h2 = ertzip(i1,i2,h1)
      val (h3,h4) = ertctrl h2
      val h5 =ertmot h4
    in
      h5 ->- h1;
      h3
    end;
```

The infix function ->- connects two channels. Here we used strings as messages, because it simplifies the code to generate traces of the system. Of course, Concurrent ML supports the full set of ML data types for communication.

Since so far Concurrent ML has no denotational semantics, we cannot prove the correctness of our implementation, but we believe that the informal semantics of the language are close enough to our formalism to justify this step. True control programs for the ERT can be developed by further refinement of the specification *MOT* that yields a second master/slave-system with a low-level control component as master and the table's actuators and sensors as slave component.

Functional programming languages are not yet of industrial interest, but simple programs such as this one can be transformed into procedural languages using transformations rules, as explained in [1].

11.4 Conclusion

The remaining components can be refined in a similar way, yielding specifications and programs for the complete system. In our modeling we made some assumptions about the behavior of the cell to simplify the specifications. Most of them only concern initial positions of the machines; two more restrictive assumptions are the following:

- The press is initially non-empty. This allows an easier description of the robot's control cycle.

- The feed belt knows the time it takes for a blank to reach the ERT, and only then sends the message *Blank* to the ERT. We could specify this behavior by extending Focus to deal with real-time aspects. This area, however, is still part of ongoing investigations.

How well does our system fulfill the requirements of the informal task description? We guarantee that the production cell is alive, by demanding that the complete cell behaves basically like the identity function: everything that goes in, must come out again. This property is maintained throughout all refinement steps.

The task description contains two kinds of safety requirements. Local requirements are concerned with the sequence of control commands sent by the controllers, and of the acknowledgements sent back by the controlled hardware. We specify and verify these requirements whenever we introduce a new component. Further refinement steps leave them intact.

Because of the assumption/commitment-style of our specifications, in many cases it is possible to simplify the specifications of slave components, because they can rely on their master to send commands in the proper sequence. The second kind of safety requirements are global requirements about components that are not directly connected. For example, a robot arm may not pass the closed press, when it is extracted. For global safety requirements (and also for more general liveness properties than the simple deadlock-freedom in our work) it is necessary to begin development at the level of trace specifications, where only safe system runs are specified. The SPECTRUM specification in [11] is an example of such a trace specification. To verify that a design specification fulfills a trace specification, and therefore also the specified safety properties, we have to show that the set of traces generated by stream processing functions is a subset of the traces specified by the trace specification. While in general it is somewhat difficult to prove this inclusion, in many cases it is possible to derive a first design specification from a trace specification by syntactic transformations, as shown in [9].

We believe that the main advantage of Focus is its wide range of applications. It supports the complete development process starting with a very abstract specification and ending up with an implementation. A further advantage is the modularity of Focus. The specification can easily be adapted for other production cells using the same or other similar machines. In the decomposition of the specification for a new cell only the material flow has to be considered. Since the individual components of a distributed system are only loosely coupled, the internal operations of each machine are irrelevant for the decomposition.

The final specification of the production cell has a length of about 80 lines. We are sure that even if the restrictions mentioned above were removed, the specification would remain compact. This is due to the use of the assumption/commitment-style as well as higher-order functions in specification.

References

[1]   M. Broy, F. Dederichs, C. Dendorfer, M. Fuchs, T. F. Gritzner, and R. Weber. The design of distributed systems - an introduction to Focus.Technical report SFB 342/3/92 A, Technical University Munich, 1992

[2]   M. Broy, F. Dederichs, C. Dendorfer, M. Fuchs, T. F. Gritzner, and R. Weber. Summary of case studies in Focus - a design method for distributed systems.Technical report SFB 342/3/92 A, Technical University Munich, 1992

[3]   M. Broy. Functional specification of time sensitive communication systems. In J. W. de Bakker, W. P. de Roever, and G. Rozenberg, editors, *Stepwise Refinement of Distributed Systems: Models, Formalism, Correctness. Lecture Notes in Computer Science 430*, pages 153-179. Springer, 1990.

[4]   G. Kahn, The semantics of a simple language for parallel programming. In J.L. Rosenfeld, editor, *Information processing 74*, pages 471-475. North-Holland, 1974

[5]   T. Lindner, A. Rüping, and E. Sekerinski. Aufgabenstellung für die Fallstudie "Fertigungszelle". Internes Arbeitspapier, Forschungszentrum Informatik, Karlsruhe, 1992.

[6]   J. Philipps. Spezifikation einer Fertigungszelle - Eine Fallstudie in Focus. Diploma Thesis, Technical University Munich, 1993.

[7]   K. Stølen, F. Dederichs, and R. Weber. Assumption/commitment rules for networks of asynchronously communicating agents. Technical report SFB 342/2/93 A, Technical University Munich, 1993

[8]   M. Broy. Towards a Formal Foundation of the Specification and Description Language SDL. Formal Aspects of Computing, Vol. 3, p. 21-57, 1991.

[9]   C. Dendorfer and R. Weber. From service specification to protocol entity implementation - an exercise in FOCUS. Technical Report SFB 342/4/92 A, Technical University Munich, 1992.

[10]   M. Fuchs. Technologieabhängigkeit von Spezifikationen digitaler Hardware. Technical Report SFB 342/14/94 A, Technical University Munich, 1994, PhD-Thesis.

[11]   C. Lewerentz, T. Lindner. Case Study Production Cell. LNCS, Springer, 1994.

XII. SPECTRUM

Describing Traces in an Algebraic Specification Language Abstractly by Predicates and more Concretely by CSP-like Programming Constructs

Dimitris Dranidis, Stefan Gastinger

Ludwig-Maximilian-Universität München

Abstract

This work aims at describing the behaviour of the production cell on two levels of abstraction. First, the production cell is modelled using a sublanguage of the process description language CSP. Informal description methods, like diagrams, are used for the development of the first model as well. A pseudo-interpreter for the CSP-sublanguage is given, which allows to validate the model against the informal description. To formulate and verify critical properties, an abstract trace specification of the production cell is given, that accepts further, especially more efficient models of the production cell. This abstract specification introduces several views to the production cell, each of which separates a different aspect such as data flow or performed actions.

12.1 Introduction

It is always difficult to justify or even argue for choosing a certain specification method or language to describe a problem. Does an appropriate method exist at all for each problem and how should one choose it? In our case we did not employ ourselves too long with these questions. We were given an informal (textual) description of the production cell [6] and some figures of it, but we haven't seen the model in real. So our first goals were

- to understand the problem and its requirements,
- to design a computational model,
- to specify the designed model (in a formal way),
- to "play" with the specified model (i.e. prototype).

After achieving these goals and getting a deeper understanding of the problem, we could then employ ourselves with questions of correctness, verification etc. So we started by reading carefully the informal description and noticed that the production cell is decomposed into physical subsystems, which can be described each independently. These subsystems operate concurrently and interact with each other as well as with their environment.

This first insight motivated us to use CSP for specifying a first model of the production cell. In CSP systems are described as process terms. Larger systems can be build from smaller systems by composing their terms with the concurrent operator. Interaction is achieved through common events. However, the development of the production cell description was not limited to CSP. We also used informal design methods, such as event diagrams, which helped us understanding tedious terms. Furthermore, we wrote a small pseudo-interpreter for the CSP terms in order to observe traces of the system and to validate our description against the informal description.

To be able to formulate and verify critical properties of the CSP description, we tried to reach an abstract trace specification of the production cell. Trace specifications are used to describe runs of distributed systems as possibly infinite sequences of observable actions. This kind of specification suits well for the description of distributed systems on a very abstract level. It allows to formalize properties of the traces given by the semantics of the CSP description. Verification of the CSP description against the trace specification may be done by using appropriate calculi. The structure of the trace specification follows the modular structure of the CSP description and additionally introduces several views to the production cell which separate different aspects of the distributed system.

A future goal of our work is to formalize the semantical relation between both, the CSP description and the trace specification, and to prove the CSP description to be an implementation of the trace specification. To do so, it has to be shown that its observable actions can be interpreted as observable actions of the trace specification, and that any possible trace in its interpretation fulfills the trace specification. SPECTRUM is used as an common framework for the formalization of both, the CSP description and the trace specification: the sublanguage of CSP, used for the description of the production cell, is modelled in SPECTRUM and the trace specification is modelled by stream based specification techniques in SPECTRUM. This common framework helps to formalize the semantical relation be-

tween these two specifications.Figure 1 illustrates in an overall picture how the
parts of the work reported here relate to each other.

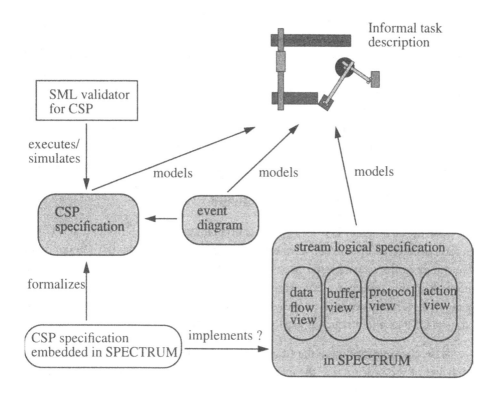

*Figure 1   Structure and relationships between different parts of SPEC-
TRUM specifications*

12.2 A CSP Modelling

In the following sections we will shortly represent the methods we applied in order
to provide a first formal model of the production cell. At first we will describe the
language we used for the formalisation, the diagrams we used for the visualisation
of the model and the program we used for the validation of the model. We will then
illustrate these methods by providing the description of a subsystem of the production cell. Finally, we will discuss the evaluation of our methodology.

12.2.1 CSP

CSP (Communicating Sequential Processes) [5] is a language for the description of systems operating in parallel (concurrently). Processes are used to describe the behaviour of concurrent systems. A process defines the behaviour of an object, insofar this can be done in terms of a finite set of events, selected as its alphabet. The actual occurrence of an event in the life of an object should be regarded as an instantaneous or an atomic action without duration; if we wish to express duration we must introduce two events to represent the start and the end of an event. The set of events which are considered relevant for a particular description of an object is called its alphabet. The alphabet is a permanent predefined property of an object. It is logically impossible for an object to engage in an event outside its alphabet.

CSP viewed as a specification language

One can argue whether CSP is a specification language or not. Some claim that it is too concrete and constructive to serve as a specification language. On the other hand it is not a conventional programming language since it is not executable (though in this report we will present a pseudo-interpreter for a part of the language). So it really depends on how one approaches the language. In this report we viewed CSP as a (constructive) specification language.

A process description in the (trace[1]) semantics is a tuple (A, T) where A is the alphabet of the process, i.e. the events in which the process can engage, and T is the set of traces (subset of $A*$) of the process, i.e. the set of all sequences of events in which the process can actually participate. The semantics of a process term P is also a tuple of its alphabet aP and the set of its traces.which is defined inductively for each operator involved in the definition.

The difference between the two descriptions is the following:

- predicates *restrict* the set $A*$ by specifying its allowed members;

- terms *construct* the subset of $A*$ which represents the allowable behaviour of the process.

One usually uses predicates over traces as specifications and CSP terms as products (or product designs). Hoare [5] gives a calculus for proving that a product satisfies a specification.

---

1. We don't involve non-determinism in the description of the production cell. So the simple trace semantics suffices for the semantics of process terms.

We find that sometimes it is more convenient and comprehensible to write down a term instead of a predicate, when we can express the same thing in both ways. We will give an example to illustrate what we mean:

$$Sp \equiv (\{a, b\}, \{ t : t \in \{a, b\}^* \text{ and } \forall s \subseteq t . 0 \leq (\#(a@s) - \#(b@s)) \leq 1 \})$$

$$P = a \rightarrow b \rightarrow P$$

where A^* is the set of traces over the set of events A, $\subseteq$ is the trace prefix relation and $\#(a@s)$ is the number of occurrences of the event a in the trace s. It is easy to show that P and Sp denote the same alphabet and traces tuple.

In our description we used CSP terms to describe the acceptable behaviour of the system. The disadvantage is that we cannot describe the permitted behaviour of the system, since the language is constructive.

Nevertheless the description is abstract concerning the event selection. Our events represent gross observations of the system behaviour that need to be refined (to motor actions and sensor signals) in a future implementation.

The sublanguage

For the needs of the product line specification we did not make use of the whole CSP language.

We did not involve nondeterminism in our description. That is guaranteed since (a) we did not use internal choice and event hiding, (b) by the use of external choice[1] we always choose between different events and (c) all recursive definitions are guarded. That guarantees that the simple trace semantics suffices for the description of terms.

In the description we used the following CSP operators:

1. **STOP$_A$** is a constant process that never actually engages in any event. It usually denotes a deadlock situation. The alphabet of STOP$_A$ is A.

2. $a \rightarrow \mathbf{P}$ is a process that at first engages in the event a and then behaves exactly as the process P. The event a must already belong to the alphabet of P.

3. **P$\Box$Q** is a process that behaves either like the process P or like the process Q. The environment controls which of the two processes will be chosen by means of their very first event. P, Q and P$\Box$Q (must) have all the same alphabet.

---

1. External choice was actually used only in one term.

4. **P** $\|$ **Q** is a process that behaves like the compound system with the processes P and Q running concurrently. All those events, that are in both, the alphabet of P and the alphabet of Q, require simultaneous participation of both P and Q. The alphabet of P $\|$ Q is the union of the alphabets of P and Q.

5. **rec$_A$** f is the process P which satisfies the recursive definition $P = f(P)$, where f is built from the above CSP operators. The usual notation for a recursive process is for example: $P = a \rightarrow P$. The alphabet of rec$_A f$ is A.

All the events are elementary and atomic. We did not use channel communication and for the synchronization we only used elementary atomic actions (no variable assignment).

12.2.2 Event diagrams

We used diagrams to design individual terms and to illustrate the synchronisation between the terms. There is a major distinction between the diagrams which we used (from now on called event diagrams) and the labelled transition diagrams (cf. [5]). In event diagrams we use the events to label the nodes instead of the edges. By doing so we can include in the diagrams edges that illustrate the synchronization. Connection diagrams were used in [5] for this purpose, but there one could not see the internal structure of the terms.

Designing diagrams

It is easy to construct event diagrams from state transition diagrams and vice versa. The former is done in the following way: (a) add a node to each edge and label this node with the label of the edge, (b) let the edges flow through the nodes of the initial diagrams and (c) finally erase the nodes (see Figure 2).

A process consisting of a sequence of events is drawn vertically (imagine an event flow from top to bottom). A process consisting of the parallel execution of two processes is drawn with the processes side by side. Common events of the two processes are joined with a dashed line. Process diagrams are then stretched in order to keep the synchronization lines horizontal (a good drawing program can do that automatically).

Note that these are not strict rules for the drawing since sometimes it is impossible to satisfy them. However if they can be satisfied it is a good tactic to follow.

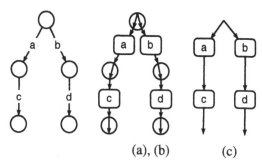

(a), (b)   (c)

*Figure 2   The construction of an event diagram from a state transition
diagram (see text for explanation of steps (a),(b) and (c)).*

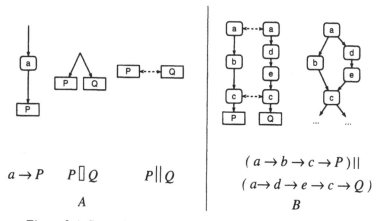

$a \rightarrow P \qquad P \Box Q \qquad\qquad P \| Q$

A

$$(a \rightarrow b \rightarrow c \rightarrow P)\|$$
$$(a \rightarrow d \rightarrow e \rightarrow c \rightarrow Q)$$

B

*Figure 3 A. Some diagrams and their corresponding terms
B. Example of a term, its diagram, and the
action causality graph.*

Understanding diagrams

Event diagrams illustrate an event flow (from top to bottom and again to top in case
of recursive definitions). When the flow reaches an event which is connected with
a synchronisation line (dashed arrows) it must wait until all the other flows have
reached this line. Then they all proceed simultaneously. This can be better understood if we view these diagrams as graphs for the causality relation between actions (see Figure 3).

12.2.3 A Pseudo-interpreter for CSP-terms in SML

In the following we will shortly describe the program and sketch the way it works. The program we wrote in order to observe the behaviour of our specified system is based on the small example programs found in [5]. We used the functional programming language SML.

Processes are declared as datatypes generated by the used CSP operators and events, where events are declared as an equality datatype. Processes are applied to (or engage in) events via the apply function which has the following functionality:

apply : Process x Event -> Process;

The definition of the apply function is actually based on the operational semantic of process terms as defined via a labelled transition system $(\mathbf{P}, \rightarrow)$ with $\mathbf{P}$ the class of processes and $.\rightarrow.$ a relation, where $p \overset{a}{\rightarrow} q$ means that the process $p \in \mathbf{P}$ can engage in the event a and after it engages, it reaches a state where the behaviour is described by $q \in \mathbf{P}$. So apply$(p, a) = q$. However apply is a function and no relation so we cannot model nondeterminism, i.e. the case where a process p can evolve either to a process q, or to r after engaging in the event a $(p \overset{a}{\rightarrow} q, p \overset{a}{\rightarrow} r)$.

The program can work in two modes: interactive and non-interactive. In the interactive mode, in each interaction the user chooses an event from the menu of initial events (the set of events in which a process can engage in its first step) and the process engages in this event (via the apply function). Then the user chooses another event and so on. When all events are refused from the process then the menu is empty. This resembles the operation of the reactive ready trace machine described in [4], where the user is provided with menu-lights showing the initial actions and buttons for the execution of an action. In the non-interactive mode the program makes a choice from the initial menu and proceeds without waiting for interaction from the user. The operation can be interrupted or it can stagnate in case of deadlock by showing the user that a deadlock has occurred. This is the behaviour of the completed traces machine as described in [4].

In both cases we can detect a deadlock. We can then search for the cause of the deadlock and change the process definition in the program. This change also propagates to the CSP-term definition and diagrams.

12.2.4 Example

We will illustrate the development methods described earlier with one example taken from the production cell. The entire descriptions can be found in [1].

We have choosen a subsystem of the production cell whose development is interesting and mirrors at some degree the development of the whole cell without bearing its complexity. A good candidate is the elevating rotary table (from now on ER-Table or ERT).

In Figure 4 we present the informal description of the elevating rotary table as it is found in [6].

Description of the Production Cell	Actuators
...	The control program can ...
1. The feed belt conveys the metal plate to the elevating rotary table.	6. rotate ERT (electric motor);
	7. move ERT vertically (el. motor);
2. The elevating rotary table is moved to a position adequate for unloading by the first robot arm.	**Sensors**
	The control program receives information from the sensors as follows: ...
3. The first robot arm picks up the metal plate.	7. Is the ERT in its lower position? (switch)
...	8. Is the ERT in its upper position? (switch)
Elevating Rotary Table	
The task of the ERT is to rotate the plates by about 45 degrees and to lift them to a level where they can be picked up by the first robot arm..	9. How far has the table rotated? (potentiometer)
	Restrictions on mobility
...	The ERT is not moved down below the feed belt. The table must not rotate when it is set at this kind of low level.

Figure 4   Informal description of the elevating rotary table

Action set

As already mentioned we will abstract in the level of actions. That means, our action set will not include actions that are as fine as noted in the informal description. This is not a mistake since we can refine these actions in a future implementation.

In the sequel we will use the following abbreviations for the subsystems of the elevating rotary table (ERT): EM for the elevating motor, RM for the rotating motor, LS for the lower position sensor, US for the upper position sensor, P for the potentiometer, T for the table. We define the following actions:

EM :	up_{EM}	(the elevating motor starts moving up)
	$down_{EM}$	(the elevating motor starts moving down)
	$stop_{EM}$	(the elevating motor stops)
RM :	$left_{RM}$, $right_{RM}$, $stop_{RM}$	(analogous to EM)
LS :	on_{LS}	(the lower sensor returns on)
	off_{LS}	(the lower sensor returns off)
US :	on_{US}, off_{US}	(analogous to US)
P :	$belt_P$	(the potentiometer has shown that the table faces the feed belt)
	$robot_P$	(the potentiometer has shown that the table faces the robot)
T :	$loaded_T$	(a blank is being thrown on the table)
	$unloaded_T$	(a blank is being removed from the table)

One should always have in mind that each action is instantaneous, without any duration. Note that all the defined events, except those for the table (T), are also motor actions or sensor information. However one should not regard them as commands from the controller or signals from the sensors, but consider them as observations. Also note that the events of the potentiometer (P) introduce an event abstraction; we do not use any angle information as an implementation would do.

CSP terms

First, we define the terms for all the subsystems. Then, for each pair A, B of terms we define an interaction term A-B which serves for the synchronisation of terms when they are combined with the parallel operator; the term A-B includes only events from the alphabet of process term A merged with process term B.

For the description of the elevating rotary table (ERT) we must assume an initial position: the table is in its lower position and faced to the feed belt; so it ready to receive a blank.

The process terms for the subsystems are very simple since no concurrency exists at this level.

$$EM = up_{EM} \rightarrow stop_{EM} \rightarrow down_{EM} \rightarrow stop_{EM} \rightarrow EM$$

$$RM = left_{RM} \rightarrow stop_{RM} \rightarrow right_{RM} \rightarrow stop_{RM} \rightarrow RM$$

$$LS = on_{LS} \rightarrow off_{LS} \rightarrow LS$$

$$US = off_{US} \rightarrow on_{US} \rightarrow US$$

$$P \quad = belt_P \rightarrow robot_P \rightarrow P$$

$$T \quad = loaded_T \rightarrow unloaded_T \rightarrow T$$

The next step is the definition of the interactions between the subsystems. As in the description of the whole production cell we begin by defining terms for each interaction between two subsystems. Each term is a requirement for the system. We define which actions should happen before another one occurs.

$$EM\text{-}T = loaded_T \rightarrow up_{EM} \rightarrow stop_{EM} \rightarrow unloaded_T \rightarrow down_{EM} \rightarrow stop_{EM} \rightarrow$$
$$EM\text{-}T$$

$$RM\text{-}T = loaded_T \rightarrow left_{RM} \rightarrow stop_{RM} \rightarrow unloaded_T \rightarrow right_{RM} \rightarrow stop_{RM} \rightarrow$$
$$RM\text{-}T$$

$$LS\text{-}T \; = on_{LS} \rightarrow loaded_T \rightarrow LS\text{-}T$$

$$US\text{-}T \; = on_{US} \rightarrow unloaded_T \rightarrow US\text{-}T$$

$$P\text{-}T \; = belt_P \rightarrow loaded_T \rightarrow robot_P \rightarrow unloaded_T \rightarrow P\text{-}T$$

$$RM\text{-}P = belt_P \rightarrow stop_{RM} \rightarrow robot_P \rightarrow stop_{RM} \rightarrow RM\text{-}P$$

The above terms are quite comprehensible and it is relatively easy to understand the intention behind them. For the next ones we will need some additional explanations.

$$EM\text{-}US \; = EM\text{-}US_1 \; \| \; EM\text{-}US_2 \; \text{where}$$

$$EM\text{-}US_1 = stop_{EM} \rightarrow on_{US} \rightarrow stop_{EM} \rightarrow EM\text{-}US_1$$

$$EM\text{-}US_2 = off_{US} \rightarrow up_{EM} \rightarrow down_{EM} \rightarrow EM\text{-}US_2$$

So the upper sensor (US) is initially off, and becomes on when the machine has stopped (after it has moved up). We can only require that something can happen before something else. We can not express that something must happen right after something else has happened. And we surely do not want an implementation in which the elevating motor (EM) keeps moving up (up_{EM}) after it reached the upper position (on_{US}). We only observe (and we want to observe only) that the elevating motor (EM) has stopped and it has stopped at the right position (examine the P-T term too). Similar to EM-US is the EM-LS term:

$$EM\text{-}LS \; = EM\text{-}LS_1 \; \| \; EM\text{-}LS_2 \; \text{where}$$

$$EM\text{-}LS_1 = on_{LS} \rightarrow stop_{EM} \rightarrow stop_{EM} \rightarrow EM\text{-}LS_1$$

$$EM\text{-}LS_2 = up_{EM} \rightarrow off_{LS} \rightarrow down_{EM} \rightarrow EM\text{-}LS_2$$

At last we formulate the critical property that the table must not rotate in its lower position:

$$EM\text{-}RM = up_{EM} \rightarrow left_{RM} \rightarrow EM\text{-}RM$$

There are no other interaction constraints between the subsystems. For example the sensors (US and LS) and the potentiometer (P) are pairwise independent subsystems (actually their independencies are expressed implicitly through the interaction with the rest of the subsystems).

The description of the elevating rotary table (ERT) is defined as the parallel operation of all the subsystems and their interactions.

$$
\begin{aligned}
ERT = \; & EM \parallel RM \parallel LS \parallel US \parallel P \parallel T \parallel \\
& EM\text{-}RM \parallel EM\text{-}LS \parallel EM\text{-}US \parallel EM\text{-}T \parallel \\
& RM\text{-}P \parallel RM\text{-}T \parallel \\
& LS\text{-}T \parallel US\text{-}T \parallel P\text{-}T
\end{aligned}
$$

Diagrams

The above example shows, that single terms are quite easily understood separately. But what happens with the complex ERT term? Does it describe what we had in mind? We believe that event diagrams can help the developer getting a clearer insight into the structure of the term and a better understanding of the synchronisation involved. We present the diagram of the ERT term in Figure 5.

By examining the structure of the diagram we recognize two subsystems:

1. The rotating motor (RM) connected with the potentiometer (P) and the table (T) and

2. the elevating motor (EM) connected with the lower sensor (LS), the upper sensor (US), and the table (T),

which interact via their common subsystem, the table (T). Furthermore by examining the event flow in these subsystems we realize that the descriptions are rather redundant. That is the result of describing subsystems and their dual interactions independently. Actually the terms in Figure 6 would do the same job (see also diagrams in same Figure).

This does not mean that we have to throw away our first description. One of the advantages of the first description is its modularity. The finer grained terms could be reused in a different setting. The second description is the collapse of all these terms; it can be easily understood but does not provide us insight in the subsystems. After all this is a small subsystem of the whole production cell. One

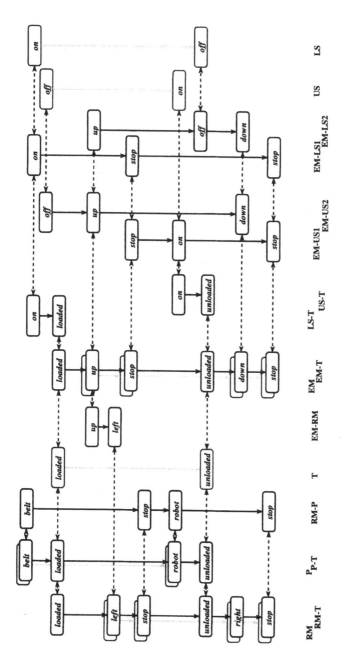

Figure 5   The event diagram of the process ERT.

$$RM\text{-}P\text{-}T = belt_P \rightarrow loaded_T \rightarrow left_{RM} \rightarrow stop_{RM} \rightarrow$$
$$robot_P \rightarrow unloaded_T \rightarrow right_{RM} \rightarrow$$
$$stop_{RM} \rightarrow RM\text{-}P\text{-}T$$

$$EM\text{-}LS\text{-}US\text{-}T = EM\text{-}LS\text{-}US\text{-}T_1 \parallel EM\text{-}LS\text{-}US\text{-}T_2$$

where

$$EM\text{-}LS\text{-}US\text{-}T_1 = on_{LS} \rightarrow loaded_T \rightarrow up_{EM} \rightarrow st\text{-}$$
$$op_{EM} \rightarrow on_{US} \rightarrow unloaded_T \rightarrow down_{EM} \rightarrow$$
$$stop_{EM} \rightarrow EM\text{-}LS\text{-}US\text{-}T_1$$

and

$$EM\text{-}LS\text{-}US\text{-}T_2 = off_{US} \rightarrow up_{EM} \rightarrow off_{LS} \rightarrow$$
$$down_{EM} \rightarrow EM\text{-}LS\text{-}US\text{-}T_2$$

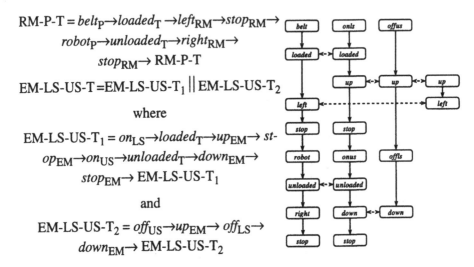

Figure 6   Collapsed form of terms.

should work the same way by defining terms for the subsystems and their interaction. These can be easily re-used, re-examined, and changed. The second form of description does not have these advantages.

12.3 A Stream Based Specification

The aim of this work is to establish a formal specification of the production cell, which allows to describe its behaviour by axiomatizing its possible runs as (infinite) sequences of observable actions. This kind of specification, also called trace specification, suits well for the description of distributed systems on a very abstract level. The resulting specification is used to formalize the safety properties given by the informal description [6] and in the future it will be used to verify the CSP description from section 12.2 against it. Observable actions are introduced using algebraic specification techniques and traces, i.e. sequences of events, which are labelled by actions, are modelled by the data type of streams in SPECTRUM. Predicates over these streams are used to describe interesting properties of the traces and they are formalized in the standard SPECTRUM logic. We will refer to this kind of specification as stream logical specification. The development process follows three main objectives, that have to be met by the resulting specification.

- **Abstraction**: The stream logical specification has to be more abstract than the CSP description: the latter one has to be an implementation of the former one and additionally the former one allows further, especially more efficient implementations, too.

- **Modularity**: The modular structure of the CSP description has to be taken over by the stream based specification: every separate component of the production cell and every interface between interacting components has to be described within a separate specification module.

- **Separation of different aspects**: In contrast to the CSP description, the stream logical specification has to separate different aspects of the system description, such as data flow, communication protocol and actions performed, by introducing different views of the production cell.

In the following, the basics of stream logical specification techniques are briefly presented and the application of these techniques to the development process of our stream logical specification is discussed. A small part of the production cell, i.e. the feed belt and the elevating rotary table, will be used as an example to illustrate the method followed. For a detailed elaboration of this work see [1].

12.3.1 Traces are modelled by streams

This section shortly presents the notion of streams in SPECTRUM (cf. also [1], [1]) which are used to model traces. Streams are possibly infinite sequences of events, where each event of a stream is unique and is labelled by an action. Given a type *Action* of actions, the type of all finite and infinite streams over actions from *Action* is called *Stream(Action)*.

STREAM = {

    enriches NAT, SET

    data *Stream* α = • & • : *Elem* × *Stream* α → *Stream* α **strict in 1**

    • ^ • : *Stream* α × *Stream* α **to** *Stream* α **strict in 1**

    #• : *Stream* α → *Nat*

    #. • : *Set* α × *Stream* α → *Nat*   **strict in 1**

    • @• : *Set* α × *Stream* α → *Stream* α **strict in 1**

    • **prefix** • : *Stream* α × *Stream* α **to** *Bool*

    • **'causal.** • : *Set* α × *Stream* α × *Nat* × *Set* α **to Bool**   **strict in 1,3,4**

    axioms $\forall^{\perp}$s,t∈ *Stream* α . $\forall$A,B∈ *Set* α . $\forall$a,b∈ α . $\forall$k∈ *Nat* .

        $\#\perp = 0$ $\land$ #(a&s) = 1+#s $\land$ $\#_A(s) = \#(A@s)$ $\land$

        $A@\perp = \perp$ $\land$ A@(a&s) = if a∈ A then a&(A@s) else A@s fi $\land$

        $\perp^\land s = s$ $\land$ (a&t)^s = a&(t^s) $\land$

        $\perp$ **prefix** s $\land$ (a=b $\land$ t **prefix** s $\Rightarrow$ (a&t) **prefix** (b&t)) $\land$

        A ^{s}**causal**$_k$ B $\Leftrightarrow$$\forall$t∈ *Stream* α . t **prefix** s $\Rightarrow$0 <= $\#_A(s) - \#_B(s)$ <= k

}

Hereby we assume NAT and SET to be a predefined specification of natural numbers and finite sets. Streams are constructed using the undefined stream $\perp$ and the & constructor which appends an event labelled by a certain action to a given stream. Streams may be concatenated using the $\wedge$ operator. # counts the number of elements in a stream, $\#_A$ counts the number of elements, which occur in the set A of actions. A filter operation @ is introduced, which yields the substream of s, that consists only of elements occuring in the set A. The **prefix** relation on stream is the usual (pointwise) prefix relation on streams. The predicate **causal** will be the most important one. It asserts, that the action a is causal for the action b, i.e. in any prefix of s, b must occur at most as often as a does. Additionally, the constant number k gives a limitation to the occurances of a: a must never occur more than k times the occurrences of b in a prefix of s. In the case of singleton sets we omit the brackets and write a $^s\mathbf{causal}_k$ b instead of {a} $^s\mathbf{causal}_k$ {b}. We use the usual short notation a $^s\mathbf{causal}_k$ b $^s\mathbf{causal}_k$ c instead of writing a $^s\mathbf{causal}_k$ b $\wedge$ b $^s\mathbf{causal}_k$ c. The **causal** predicate allows to describe cyclic runs of parts of the production cell. In section 13.3.5 we give examples, i.e. the cycle of dataflow, where further cycles may start before the previous one is finished, and a protocol cycle, where a new cycle must not start before the previous one has finished.

12.3.2 Stream logical specification

As mentioned above, the resulting specification has to meet three requirements: it has to be abstract, modular and it has to separate different aspects in separate views. Stream logic specification techniques are used to describe runs of the production cell. During the development process it has been regarded that the CSP description is a model of the stream logical specification w.r.t. its possible traces. Whenever sensible the stream logical specification is kept more abstract than the CSP description in the sense that it fixes less design decisions. For example the capacity of the feed belt is exactly one in the CSP description whereas it is allowed to be a constant but unlimited natural number in the stream logical specification. This is realized by assuming an additional sensor at the beginning of the belt, which does not have to be realized by hardware but also may be implemented by a piece of software: we will introduce a signal indicating that a metal plate has been dropped onto the feed belt by turning the magnet of the travelling crane off after it has reached the right position. The program must be able to recognize this signal as well as the signal indicating whether the motor of the feed belt is on or off. With these two signals given, a program may delay some constant amount of time while the motor is on, allowing the feed belt to remove the metal plate from the critical zone at the front end of the feed belt. After this period of time, the program produces a virtual sensor signal, indicating that the front end of the belt is now clear again.

12.3.3 Views separate different aspects of the system

The resulting target specification introduces four separate views of the production cell, each of which describes a particular aspect. A *data flow oriented view* divides the whole system into subsystems –called components–, describes the kind of data that are exchanged, and declares between which components these data are exchanged. A *buffer oriented view* refines this description of data exchange by explicitly modelling internal buffers of the components. A *protocol oriented view* formally describes the procedure of interaction by introducing signals indicating, which components play an active role during the data exchange. An *action oriented view* describes the physical actions performed by each component in order to implement these protocols. The data flow as well as the protocol oriented view are introduced in order to explicitly describe certain properties of the system, which are only implicit in the CSP description. In contrast, the buffer and the action oriented views describe aspects of the system, which also are described explicitly within the CSP description: internal states, representing buffers as well as physical actions performed by the actions and sensors of the production cell.

12.3.4 Modular structure of the specification

For any of these views there exists a modular description of the whole system in the sense that it contains a separate specification module for any component of the production cell. Based on these component descriptions, there also exist separate interface descriptions for the synchronization aspects between interacting components. As the only exception the data flow view is hold monolithic, since it contains the components as objects.

In contrary to the modular structure of the description within each view, which can be seen as a kind of horizontal structuring, the specification of the production cell in different views yields a vertical structuring: each view has its own modular specification and between these there exist semantical refinement relations. For example, the data flow view is an abstraction of the buffer oriented view. In general, these refinement relations have a more complex structure and are to be discussed in detail. Here, we only sketch these semantical relations. For a detailed formal explanation see [1].

The modular specification of each view follows the same procedure: in a first step algebraic specification techniques are used to describe the set of observable actions used within this view. These specifications are called VIEW_ACTION and introduce a sort *ViewAction* of observable actions. In a second step, stream logical specifications describe the set of possible traces of the production cell by a predicate PC_{VIEW} over the given set of observable actions. Due to the modular structure of the description, there exists a separate specification module for each component which is called COMPONENT_VIEW (e.g. BELT_BUFFER). Since the definition of the predicate PC_{VIEW} has to be split into several modules, it is defined by putting together several predicates PC^i_{VIEW} for each component and each interface description, where each PC^i_{VIEW} is defined in the corresponding module.

We call a given model (A, T_A) a model of the stream logical specification of a view, if its set of actions A is a subset of the set of possible actions *ViewAction* of the stream logical specification and the predicate T_A, describing all possible traces of the model, is an implementation of the predicate PC_{VIEW}, i.e each trace of the model also is a trace of the stream logical specification.

12.3.5 Examples from the stream logical specification

This section presents some example modules from the stream logical specification. The modules specify the feed belt and the elevating rotary table. We give examples for each particular view.

The data flow view introduces the components of the system, the data objects, the production cell processes on (i.e. metal plates), and the flow of these data objects through the system. On this level of abstraction it does not make sense to give a modular specification. Instead, a monolithic specification is given that introduces the data objects as well as the components of the production cell as objects of sort *MetalPlate* and *Component*. It defines a sort *DataflowAction* of dataflow actions and describes all possible runs of the production cell w.r.t. these actions.

In our small example, there are two components, i.e. the feed belt and the elevating rotary table, which we will simply refer to as belt and table. Data objects are metal plates, which are either blank or forged. An operation flow is introduced, that takes two components and one data object and builds a data flow action, for example flow(blank,belt,table). This action has to be read as follows: the feed belt component hands over a blank metal plate to the elevating rotary table. Obviously, the flow operation is partial, since flow of data in the production cell is defined only between

certain components. The presence of this operator is not significant, since the data
flow actions also could have been defined as constants each separately.

```
DATAFLOW_ACTION = {
    data MetalPlate = blank | forged
    data Component = crane | belt | table | robot | press | depositBelt
    data DataflowAction = flow( m:MetalPlate, A,B:Component ) strict
    axioms ∀A,B∈ Component. ∀M∈ MetalPlate .
        δ(flow(M,A,B))  ⟺(M,A,B)∈
            { (blank,crane,belt), (blank,belt,table), (blank,table,robot), (blank,robot,press),
                (forged,press,robot), (forged,robot,depositBelt), (forged,depositBelt,crane) }
}
```

Stream logical specification techniques are used to describe the behaviour of the
production cell in terms of traces over these data flow actions. Since traces are
modelled by streams of data flow actions, a stream logical predicate $PC_{Dataflow}$ is
introduced, which describes the set of possible streams. It introduces some causalities which describe the cycle of the flow of metal plates through the system. Informally spoken, this predicate claims that the system is sequential in the
following sense: a metal plate enters the system at the travelling crane, moves ar-
round in the system from one component to the following and leaves the system
again at the travelling crane. Note, that several metal plates are allowed to enter the
system and to start a new flow cycle before a metal plate has to leave the system.

```
PRODUCTIONCELL_DATAFLOW = {
    enriches NAT, STREAM, DATAFLOW_ACTION
    beltCapacity, depositBeltCapacity : Nat
    PC_Dataflow : Stream(DataflowAction) to Bool
    axioms ∀s∈ Stream(DataflowAction) .
        PC_Dataflow(s)  ⟺
            flow(blank,crane,belt) ˢcausal_beltCapacity
            flow(blank,belt,table) ˢcausal₁
            flow(blank,table,robot) ˢcausal₁
            flow(blank,robot,press) ˢcausal₁
            flow(forged,press,robot) ˢcausal₁
            flow(forged,robot,depositBelt) ˢcausal_depositBeltCapacity
            flow(forged,depositBelt,crane)
}
```

The first causality for instance claims that the feed belt may hand over a metal plate
to the elevating rotary table only after having received one from the travelling
crane. It further claims, that the feed belt has an internal buffer and so, at any time
it may have received up to a constant number beltCapacity more metal plates,
which it has handed over to the table. In general, a fixed natural number, represent-

ing the internal buffer capacity, is demanded for every component, but for most of the components this capacity is exactly one. Note, that the robot is allowed to have a maximal overall capacity of two, one for each data flow action: from the robot to the press and vice versa. This capacity allows implementations, where both arms are used simultaneously.

The data flow view is the most abstract of the presented views of the production cell. It uses data flow actions, which cannot be related to single components. The CSP description does not introduce such a kind of common actions, but splits data flow actions into separated actions, performed by each of the interacting components. For example the CSP action '$loaded_T$' has to be read as follows: the table has been loaded by an metal plate. The stream logical specification refers to this kind of actions as buffer actions, where any data flow action is refined to two buffer actions: a buffer in one component is unloaded and the corresponding buffer in the other component is loaded.

The buffer oriented view introduces explicit internal buffers of the components which are used to hand over data objects from one component to another. The buffer actions implement the data interaction between the components of the production cell, as it is described in the data flow view in the previous section. For each component, explicit buffers are specified, that are used to hold metal plates during their exchange. Each buffer is able to hold at most one metal plate at once. As the picture below illustrates, the feed belt is assumed to have two buffers, one at each end. The buffer at its front end interacts with the crane, and the buffer at its back end interacts with the elevating rotary table. The latter will be specified in our example. Although the belt is able to hold more than two plates, only those two buffers are explicitly specified, since they are necessary for the interaction with other components. The elevation rotary table has only one buffer, since its capacity is exactly one.

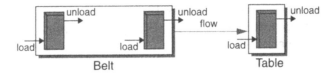

There are two operations, which can be performed on a buffer: a metal plate can be put into a buffer or it can be removed from it. We call these actions load and unload, resp. Note, that the type of buffer actions is empty here, since there is no buffer defined yet. Buffers will be defined successively in further specifications.

BUFFER_ACTION = {
    sorts *Buffer*
    data *BufferAction* =   load(b:*Buffer*)   **strict, total**
                | unload(b:*Buffer*) **strict, total**
}

A separate specification, called BUFFER, describes the behaviour of a buffer with respect to these buffer actions: a buffer cannot be unloaded unless it has been loaded and it cannot be loaded twice without being unloaded in between.

BUFFER = {
    enriches BUFFER_ACTION, STREAM
    • isLoaded •, • isUnloaded • : *Stream(BufferAction)* × *Buffer* **to Bool**
    PC^1_{Buffer} : *Stream(BufferAction)* **to Bool**
    axioms ∀s ∈ *Stream(BufferAction)* .
    PC^1_{Buffer}(s) ⇔ (∀b ∈ *Buffer* .
        load(b) ^{s}**causal**$_1$ unload(b) ∧
        s isUnLoaded b ⇔ #s < ∞ ∧ #$_{\{load(b)\}}$s = #$_{\{unload(b)\}}$s ∧
        s isLoaded b ⇔ #s < ∞ ∧ #$_{\{load(b)\}}$s > 0 ∧ #$_{\{load(b)\}}$s > #$_{\{unload(b)\}}$s)
}

Furthermore, predicates isLoaded and isUnLoaded are introduced, which indicate whether the last buffer operation in a given stream s was a load(b) or an unload(b) action, resp. Note, that the definition of these predicates relies on the alternating property of load(b) and unload(b) actions, which is stated by PC^1_{Buffer}. The buffer specification of the table introduces a buffer tableBuffer, for which the axioms of the BUFFER specification must hold.

TABLE_BUFFER = {
    enriches BUFFER
    tableBuffer : *Buffer*
}

The buffer actions load(tableBuffer) and unload(tableBuffer) correspond to the CSP actions *loaded*$_T$ and *unloaded*$_T$. The axiom load(tableBuffer) ^{s}**causal**$_1$ unload(table-Buffer) corresponds to the CSP term T = *loaded*$_T$ → *unloaded*$_T$ → T, since both descriptions restrict the set of allowed traces in the same manner.

The specification of the feed belt considers some additional capacity between its two explicit modelled buffers frontBuffer and backBuffer. The constant beltCapacity denotes the whole capacity of the feed belt and should be identical with the constant from the data flow specification PRODUCTIONCELL_DATAFLOW. This identification has to be made explicit by the semantical refinement relation between these two views.

```
BELT_BUFFER = {
    enriches NAT, BUFFER
    frontBuffer, backBuffer: Buffer
    beltCapacity : Nat
    PC²_Buffer : Stream(BufferAction) to Bool
    axioms ∀s∈ Stream(BufferAction) .
    PC²_Buffer(s) ⇔ (∀b∈ Buffer .
        frontBuffer=backBuffer ⇒ beltCapacity=1 ∧
        frontBuffer≠backBuffer ⇒ beltCapacity>1 ∧
        (beltCapacity>1 ⇒
            unload(frontBuffer) ˢcausal_(beltCapacity-1) load(backBuffer)) )
}
```

Using the buffer specifications of both, the feed belt and the elevating rotary table, the specification of the interaction of these two components can be given. It additionally states that the buffer of the elevating rotary table exactly receives those plates, which leave the buffer backBuffer of the feed belt.

```
BELT_TABLE_INTERACTION = {
    enriches TABLE_BUFFER, BELT_BUFFER
    PC³_Buffer : Stream(BufferAction) to Bool
    axioms ∀s∈ Stream(BufferAction) .
        PC³_Buffer(s) ⇔ unload(backBuffer) ˢcausal₁ load(tableBuffer)
}
```

This property of the buffer oriented view also is a property of the data flow view: the specification of the buffer oriented view is more detailed than the specification of the data flow view. It can be proven that the data flow view is an abstraction of the buffer oriented view, where two corresponding buffer actions are abstracted to a single data flow action. For example, the data flow action flow(blank,belt,table) corresponds to the join of the two buffer actions unload(backBuffer) and load(table-buffer). For a detailed discussion of this topic we refer to [1].

In a final buffer specification of the whole production cell all predicates of the previous buffer specifications of the components and their interactions are combined to one global predicate PC_{Buffer} over streams of buffer actions.

```
PRODUCTION_CELL_BUFFER = {
    enriches BELT_TABLE_INTERACTION, ...
    PC_Buffer : Stream(BufferAction) to Bool
    axioms ∀s∈ Stream(BufferAction) .
        PC_Buffer(s) ⇔ PC¹_Buffer(s) ∧ PC²_Buffer(s) ∧ PC³_Buffer(s) ∧ ...
}
```

The protocol view gives a more detailed description of the interaction between components by introducing explicit signals for the synchronization of the handing over of the data objects between interacting components. The handing over happens synchronously, i.e. both components simultaneously participate in this process. The signals indicate certain internal states of the components that can only be reached after the components have performed certain actions in a certain order. In terms of the buffers introduced in the previous section this means for example: the readyToBelt signal indicates, whether the table buffer is unloaded and is in the right physical position close to the feed belt. Once this signal is raised, the table is assumed not to change its position until the communication cycle with the belt has been finished. This will explicitly be specified in the action oriented view.

There are two operations that can be performed on a signal: it may be switched from low level to high level or vice versa. For a given signal i these actions are denoted by setHigh(i) and setLow(i). A signal must be set to high level before it may be set to low again and it never may switch to high level more than once without switching to back low level. This behaviour is defined analogously to the behaviour of buffers and, therefore, we omit the exact specification of signals here. There are two predicates over streams s of signal actions that check the internal state of a given signal i: s isHigh i and s isLow i.

Depending on the complexity of the specified protocol, a different set of signals is used to describe it. In our small example every component establishes two signals. The signal readyToBelt indicates whether the table (buffer) is ready to receive an object from the feed belt's buffer backBuffer and the signal readyToRobot indicates, whether the table (buffer) is ready to deliver an object to the robot.

```
TABLE_SIGNAL = {
    enriches SIGNAL, TABLE_BUFFER
    readyToBelt, readyToRobot: Signal
    PC¹Signal : Stream(SignalAction ∪ BufferAction) to Bool
    axioms ∀s ∈ Stream(SignalAction ∪ BufferAction) .
        PC¹Signal(s) ⇔ ∀t prefix s .
            t isHigh readyToBelt ⇒ t isUnloaded tableBuffer ∧
            t isHigh readyToRobot ⇒ t isLoaded tableBuffer
}
```

Since the table may only interact either with the feed belt or with the robot, it has to be ensured that never both signals get high simultaneously. This is ensured by using the table buffer for synchronization of these signals: the readyToBelt signal may only be high, while the table buffer is unloaded, whereas the readyToRobot signal may only be high, while the buffer is loaded. For this reason, the signal specification uses the buffer specification. The names of the signals suggest that the table does not actively participate in the interaction process: it just waits for a plate to be dropped from the belt onto it and just waits that the arm of the robot fetches it. This behaviour is formalized in the protocol specification BELT_TABLE_PRO-TOCOL given below.

The signal specification BELT_SIGNAL of the feed belt looks similar to the signal specification TABLE_SIGNAL of the table and therefore we give only an informal description here. It introduces two signals, a signal readyToTransporter concerning the interaction protocol to the transporter and a signal doneToTable which suggests that the belt is the active component in the interaction process with the table: it will perform certain actions in order to put a metal plate onto the table, while the table does not move. This behaviour of the participating components, table and feed belt, is formalized in the following specification of a protocol cycle:

BELT_TABLE_PROTOCOL = {
    enriches TABLE_SIGNAL, BELT_SIGNAL
    PC^2_{Signal} : *Stream(SignalAction $\cup$ BufferAction)* **to Bool**
    axioms $\forall s \in$ *Stream(SignalAction $\cup$ BufferAction)* .
        $PC^2_{Signal}(s)$ $\Leftrightarrow$ setHI(readyToBelt) ^{s}**causal$_1$** unload(backBuffer)
                                   ^{s}**causal$_1$** setHI(doneToTable)
                                   ^{s}**causal$_1$** setLO(readyToBelt)
                                   ^{s}**causal$_1$** load(tableBuffer)
                                   ^{s}**causal$_1$** setLO(doneToTable) $\wedge$
                setHI(readyToBelt) ^{s}**causal$_1$** setLO(doneToTable)

}

It describes a full cycle of the interaction process for handing over a metal plate from the feed belt to the elevating rotary table. The first axiom describes the dependencies within a full cycle and the second axiom claims that a new cycle must not begin before the actual one has been finished. The interaction process may be illustrated by the following signal diagram, which shows the dependencies between the changes of the signal levels. Unloading of the back end buffer happens

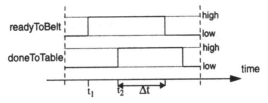

somewhere between t_1 and t_2. Note, that this specifications demands the metal plate to receive the table while the readyToBelt signal is high. Since there is no sensor on the table, and the doneToTable signal just indicates that the plate has left the belt buffer, this can only be realized by waiting a constant, but sufficient amount of time Δt after the doneToTable signal was risen, before setting the readyToBelt signal low again.

A final protocol specification of the whole system collects all the predicates of the protocol specifications and connects them to form a single PC_{Signal} predicate. Since the protocol specification is based on the buffer specifications, the predicate PC_{Buffer} also must be taken into account. Note, that the stream has to be filtered according to the appropriate set of the predicates' actions.

```
PRODUCTION_CELL_PROTOCOL = {
    enriches TRANSPORTER_BELT_PROTOCOL, BELT_TABLE_PROTOCOL, ...
    PCSignal : Stream(SignalAction ∪ BufferAction) to Bool
    axioms ∀s ∈ Stream(SignalAction ∪ BufferAction) .
        PCSignal(s) ⇔PCBuffer(BufferAction@s) ∧
                    PC¹Signal(SignalAction@s) ∧
                    PC²Signal(s),
        setLow(doneToBelt) ˢcausalbeltCapacity setHigh(doneToTable), ...
}
```

Additional axioms are introduced which state the causalities between the protocols. In our example the last action of some interaction protocol TRANSPORTER_-BELT_PROTOCOL between the transporter and the feed belt (not given here) is called setLow(doneToBelt). It is causal for the action setHigh(doneToTable) of the interaction protocol between the belt and the table component.

The signal actions, introduced in this view, do not have direct corresponding CSP actions. Signals represent states of components that can only be reached after certain actions have been performed. For example, the signal readyToBelt may only be set to high if the table buffer is unloaded and the table has reached the right position near the feed belt. These kind of dependencies are specified in the following.

The action oriented view introduces actuators and sensors, that can mechanically perform certain actions. These actuators and sensors are part of the components and are not explicitly modelled as subsystems. Their actions are synchronized within each component as well as between interacting components in order to implement the protocol described in the previous section. Sensors behave analogously to signals, but denote explicit sensors in the production cell, whereas the protocol signals denote abstract states of the components. We omit the formal specification SENSOR of sensors here and just state, that the actions and predicates of SENSOR are identical to those of SIGNAL. We distinguish different kinds of electrical powered motors, the simpliest one is used to drive the belts: it is only able to start and stop. Its formal definition does not look very interesting and, therefore, we omit it here. It introduces two actions start and stop, and two predicates isRunning and isStopped. The actions can only be performed in alternating order. A more complicated motor is allowed to start in reverse direction, too. We also omit its formal specification.

The feed belt consists of one simple motor and two sensors, one at each end of the belt, where, as mentioned above, the front end sensor neither exists in the informal specification nor in the real model.

```
BELT = {
    enriches MOTOR, SENSOR, BELT_SIGNAL
    beltMotor : Motor
    frontSensor, backSensor : Sensor
    PC_Belt : Stream(MotorAction ∪ SensorAction) to Bool
    axioms ∀s∈ Stream(MotorAction ∪ SensorAction) .
    PC_Belt(s) ⇔ ∀t prefix s .
        t^setHI(frontSensor) prefix s ⇒ t isStopped beltMotor ∧
        t^setLO(frontSensor) prefix s ⇒ t isRunning beltMotor ∧
        t isLoaded frontBuffer ⇒ t isHigh frontSensor ∧
        t isLoaded backBuffer ⇒ t isHigh backSensor ∧
        t isHigh readyToTransporter ⇒ t isStopped beltMotor ∧
        t isHigh readyToTransporter ⇒ t isLow frontSensor ∧ ...
}
```

The specifications of the action oriented view introduce a lot of implications that state the dependencies between the actions of actuators and sensors and the signals and buffers from the previous views. The first two implications define the dependencies between the frontSensor and the beltMotor: the frontSensor may only be set to low (i.e. the frontBuffer gets unloaded) while the belt is running whereas is may only be set to high (i.e. the frontBuffer gets loaded) while the belt is stopped. The next two implications claim, that a sensor must be high if its corresponding buffer is loaded. The last implications claim that the readyToTransporter signal may only be high, if the belt is stopped and the front end of the belt is unloaded. This formalizes the requirement, that a belt may only be loaded while it is stopped and its front end is empty.

The action oriented specification of the table introduces two bidirectional motors and four sensors, two for each motor.

```
TABLE = {
    enriches MOTORREV, SENSOR, TABLE_SIGNAL
    elevatorMotor, rotorMotor: MotorRev
    upperSensor, lowerSensor, leftSensor, rightSensor: Sensor
```

PC_{Table} : *Stream(MotorRevAction $\cup$ SensorAction)* **to Bool**
axioms $\forall s \in$ *Stream(MotorRevAction $\cup$ SensorAction)* .
$PC_{Table}(s) \Leftrightarrow \forall t$ **prefix** s .

    t isHigh lowerSensor $\Rightarrow$ t isStopped rotorMotor $\wedge$
    t^setHI(lowerSensor) **prefix** s $\Rightarrow$ t isRunningRev elevatorMotor $\wedge$
    t^setLO(lowerSensor) **prefix** s $\Rightarrow$ t isRunning elevatorMotor $\wedge$
    t^setHI(upperSensor) **prefix** s $\Rightarrow$ t isRunning elevatorMotor $\wedge$
    t^setLO(upperSensor) **prefix** s $\Rightarrow$ t isRunningRev elevatorMotor $\wedge$ ...

    $\forall$ init, t. (s=init^t) $\Rightarrow$ (#startRev(elevatorMotor)(init)=1) $\Rightarrow$ (
       setHI(lowerSensor) ^{t}**causal**$_1$ stop(elevatorMotor)
                               ^{t}**causal**$_1$ start(elevatorMotor)
                                 ^{t}**causal**$_1$ setLO(lowerSensor)
                                 ^{t}**causal**$_1$ setHI(upperSensor)
                                 ^{t}**causal**$_1$ stop(elevatorMotor)
                                 ^{t}**causal**$_1$ startRev(elevatorMotor)
                                 ^{t}**causal**$_1$ setLO(upperSensor)) $\wedge$
       setHI(lowerSensor) ^{t}**causal**$_1$ setLO(upperSensor) $\wedge$ ...

}

The first axiom formalizes the informal requirement "the table must not rotate when it is set at this kind of low level" (at the feed belt). The following four axioms formalize the informal requirement "electric motors are stopped immediately when the devices they power reach the boundaries of this manoeuvering space" (which is indicated by a sensor signal) using other words: a motor must not leave the critical region, indicated by a sensor signal in the wrong direction. Additionally it is assumed, that a motor changes its direction after stopping inside the critical region. The last axioms describe a full cycle of the tables elevating motion in terms of the actuators and sensors.

12.4 Evaluation

12.4.1 CSP description

Aim of this work was to build quite fast a first model of the production cell and experiment with it in order to get an insight and understanding of the problems involved by the description. We also validated somehow the informal description since we produced a model that is entirely based on the informal description. From the observations we made with the prototyping program and during the development of the description we also discovered that the efficiency (especially the capacity of the belts) of the production cell can be raised by adding additional sensors. For example the addition of sensors both at the begin and the end of the conveyor belts allows to increase their capacity.

In the development of the terms we worked in the following sequence: We first built some terms, drew their diagrams, and then tested them at first separately and then all together with the program. The diagrams and the program have helped a lot in this development. Because diagrams belong to informal development methods and the program is no verifier (at best case a prototyping program) we cannot guarantee the correctness of our description. However, it was not our goal to produce a correct product, but instead to understand the problem in order to be able to provide a formal specification in which properties can be proved, as it has been done in section 12.3.1.

12.4.2 Stream logical specification

The stream logical specification provides an abstract specification of the production cell. It formalizes the safety requirements of the informal requirement description. There are no liveness requirements specified. Assumptions according to the defined actions, i.e. actions concerning the data flow, the buffer capacity of each component, the communication protocol between interacting components and the actions performed by the actuators and sensors of each component are made explicit in the stream logical specification. Additional informal assumptions on the production cell concerning time or space dimensions are informally made. For example, sometimes it is required that between two events there exists a sufficient long time interval, or that the belts have a minimum length.

The stream logical specification is not yet complete. Therefore, it is difficult to come up with certain measurements. The development of the view oriented structure of the stream logical specification took about two weeks. Writing down the specification is straight forward and should be done within two more weeks.

We did not try to change the specification, but: changing the behaviour of one component (for example its possible actions or its buffer capacity) is strictly localized within certain specification modules and it is easy to perform such changes. To replace one component by another one seems to be a more time expensive job. The modular structure helps to localize the changes to those modules which describe the interaction to other components. But note, that this has to be done for every view.

The specification allows more efficient models than the CSP description. A maximal number for the throughput cannot be given. Most components have capacity 1, but the belts are allowed to carry up to a constant number k of plates, since a virtual sensor at the front end of each belt is introduced which may be implemented by a piece of software. Introducing virtual sensors is not strictly prohibited by other description techniques but using such an abstract description technique helps to get such ideas.

12.5 Conclusion and Remarks

This work establishes a formal specification of the production cell on different levels of abstraction. A first prototype was developed by giving a modular description in a sublanguage of CSP. A pseudo-interpreter for this sublanguage was implemented. It works interactively as well as automatically and serves to validate the CSP description against the informal description. It also can be used to search for deadlocks of the system. In order to be able to formulate and verify properties of the CSP description, a stream specification was developed, which axiomatizes traces of the production cell as streams of observable actions. This specification also is constructed in a modular way following the structure of the CSP description. Additionally, it is divided into several subspecifications each of which describes a certain aspect of the production cell. This specification formalizes safety properties but no lifeness properties, although it seems to be possible to formulate them.

A future goal is to prove the CSP description to be a model of the stream specification. To do so, it has to be shown, that its actions can be interpreted as actions of the stream specification and that any possible trace in its interpretation fulfills the stream specification. Using SPECTRUM as a common framework this can be done by translating the CSP programming constructs into predicates over streams and proving these to be derivable from the stream specification. The correctness of the translation can be proven using a calculus of Hoare (cf. [5]).

References

[1]   M. Broy et al., The Requirement and Design Specification Language SPECTRUM, An Informal Introduction, TU München, technical report TUM-I9311, 1993.

[2]   M. Broy et al., The Design of Distributed Systems - An Introduction to FOCUS, TU München, technical report TUM-I9202, 1992.

[3]   D. Dranidis, S. Gastinger, Description of a Production Cell using CSP and SPECTRUM, Ludwig-Maximilians-Universität München, technical report, in work, 1994.

[4]   R. J. van Gladbeek, Comparative Concurrency Semantics and Refinement of Actions, PhD Thesis, Centruum voor Wiskunde en Informatica, Universiteit te Amsterdam, 1990.

[5]   C.A.R. Hoare, *Communicating Sequential Processes*, Prentice-Hall, 1985.

[6]   T. Lindner, Task Description for the Case Study "Production Call", Forschungszentrum Informatik, University of Karlsruhe, 1993.

XIII. KIV

Specification and Verification of Distributed Technical Systems with Central Control

Gerhard Schellhorn, Axel Burandt

Universität Karlsruhe

Abstract

This paper presents an algebraic approach to the specification and verification of distributed technical systems, which are controlled by a central control program. The approach is demonstrated by its application to the case study "production cell". The approach uses first-order specifications to describe the possible behaviour of the system. Specifications are structured according to the physical structure of the system. A PASCAL-like program is used to enforce intended behaviour. The whole case study, including specification as well as verification of lifeness and safety conditions, is carried out using the KIV system.

13.1 Introduction

The case study "production cell" was treated by the KIV group to study how distributed systems can be modelled within first-order logic, and which requirements for the correctness of a central control program can be expressed and verified using the KIV system. We did a complete formal development, including specification, implementation of a control program and verification.

The approach uses structured algebraic specifications to model possible behaviour of the devices of the system. A separately developed control program over the specification, implemented in a PASCAL-like notation is used to enforce the intended behaviour. Verification of lifeness and some aspects of safety has been done using the KIV system ([2], [3], [4]), a system designed for the development of cor-

rect software systems. The system supports structured first-order specifications
and a tactical theorem proving approach for program verification based on Dynamic Logic ([1]).

The following section describes the intended state oriented model for the system. Section 13.3 gives an overview of the specification. Section 13.4 describes
the control program and gives the relevant correctness problems. Section 13.5 discusses the problem of verifying safety and lifeness conditions and section 13.6
concludes.

13.2 The Intended Model

The choice of the intended model was mainly motivated by the attempt to mimic
a (single) sequential control program that can be used to drive the production cell
in reality. Such a program is usually called when some sensor value has changed
(significantly), like "press has reached upper position". Activation of the control
program is done either by an interrupt to the controlling computer or by a polling
routine. The program reacts on the change by giving a number of controlling commands to the actuators, like "move press downwards". Formally speaking the program is a routine, which gets a "sensor event" as input, and reacts on it, by giving
as output some "control events" to the system. To make the program react properly,
it must keep an abstract state in a global variable, which reflects the relevant properties of the physical devices, such as "number of metal blanks on feed belt" etc.
So the control program abstracts from reality to the computer representable state
of an (not necessarily finite) automaton, where both sensor and control events correspond to state changes. Time is treated implicitly by assuming calls to the control
program, whenever sensor data signal a state change in the system. Using a control
program which abstracts reality to a state oriented representation, we have chosen
to model the devices by a specification that uses states and events too.

It should be noted that such a state oriented control program works only under
the assumption, that the reaction of the computer is fast enough to avoid significant
changes in the devices while it is running. As an example, if a robot arm signals
that it has reached an angle, where it is in front of the press, the program should
react fast enough to stop the arm at an angle, where it is still in front of the press.
Since this assumption, which can roughly be formulated as "execution time does
not matter" is widely used and did not seem critical in the concrete scenario, we
adopted it. It should nevertheless be noted, that the assumption is essential for the
abstract model described in the following. Dealing with execution times would

have required a much more complicated specification dealing with explicit time or interrupts with priorities.

13.3 Algebraic Specification of State Oriented Systems

Modelling a system by a formal algebraic specification always requires choosing a suitable abstraction from the real world and should give a specification whose structure should reflect the one of the real system.

In the case of the production cell the specification structure is naturally given by the devices of the system, so we chose to specify each component separately.

For the abstraction level, three choices are possible: The first is to model reality as close as possible, by specifying the state of a device as a tuple of all available sensor values. This would require to specify the robot as a triple of potentiometer values, and every change in those values would change the state of the device by an event "value increased" or "value decreased". The second choice is the abstract state, which the control program uses to compute its answers. Since these data types must be specified anyhow, this is the easier choice. We actually used a third, even more abstract level by simplifying the description of the moving devices robot, elevating rotary table and travelling crane. We assume that we can send the signal "move to feed belt" to the travelling crane and get the response "arrived at feed belt" when it reaches this position, stopping automatically. An approach closer to reality would have been to add a control event "stop at current position", but this would have simply increased the number of states and events, without adding anything essentially new to the problem. A more interesting point is that all three levels are linked by abstraction functions, and it would be an interesting task to study the relation between them. Several possibilities for the realization of the translation between two levels are of interest

1. Intelligent control at the devices

2. Parallel processes for every device on the computer on which the control program is executed

3. A preprocessing control program that is placed on top of the current control program

4. Direct translation of the control program to a more detailed one

The first two possibilities are beyond the concept of a sequential control program and therefore not in the scope of the approach used here, while the latter two seem to be capable by defining a suitable refinement relation between specifications and programs.

In the following we will describe the specification in detail. The specification consists of three parts, described in the following sections: On the bottom level, we have two specifications for every device, describing the events and states of the individual devices. These specifications are independent of their actual use in the production cell and are subject to reuse. At a second level we have a specification for every device that describes the restrictions imposed by the use of the device in the context of the production cell, such as where it is loaded or where it is allowed to move. This specification may vary in different contexts. On the top level the system specification of the production cell is defined by composition of the devices specification and defining the interaction between them.

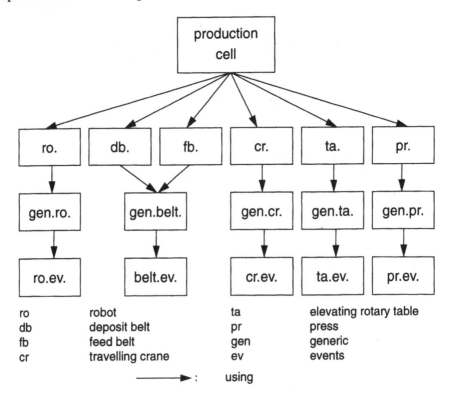

Figure 1   Global specification

13.3.1 System independent specification of devices

Since we adopted a state oriented model the specification of every device consists of two parts: a specification of the possible events and a specification of the possible states of the device. Events are classified as sensor events (events from the physical system, i.e. significant sensor value changes) and control events (actions taken by the control program, driving the actuators). For the elevating rotary table we get the specification:

TABLE_EVENTS =
    data specification
        ta_vposition =   ta_up I ta_down;
        ta_hposition =   ta_fb I ta_ro;
        ta_event   =   (* *sensor events* *)
                    ta_hat(ta_hpos : ta_hposition) I
                    ta_vat(ta_vpos : ta_vposition) I
                    ta_load I ta_unload I
                    (* control events *)
                    ta_hto(ta_hpos : ta_hposition) I
                    ta_vto(ta_vpos : ta_vposition);
    end data specification

The specification describes a free data type of ten different events represented by ground terms. Events are divided into the sensor events "table arrives at upper position" (ta_vat(ta_up)), "table arrives at lower position" (ta_vat(ta_down)), "table arrives at feed belt" (ta_hat(ta_fb)), "table arrives at robot" (ta_hat(ta_ro)), and the control events "load table" (ta_load), "unloaded table" (ta_unload), "move to upper position" (ta_vto(ta_down)), "move to lower position" (ta_vto(ta_up)), "rotate to feed belt" (ta_hto(ta_fb)) and "rotate to robot" (ta_hto(ta_ro)). It uses two boolean data types for the horizontal (i.e. the rotation angle) and the vertical position, which can be selected from the events with the selectors ta_hpos and ta_vpos. The positions "at feed belt" (ta_fb) and "at robot" (ta_fb) are independent of the concrete use of the device, but to guide intuition we have indicated their further use by appropriate identifiers. The specification of possible states of the elevating rotary table looks as follows:

GENERIC_TABLE =
    enrich TABLE_EVENTS **by**
    sorts   ta_state;
    constants ta_err : ta_state;
    functions ta_hpos : ta_state → ta_hposition;
                 ta_vpos : ta_state → ta_vposition;
                 ta_trans : ta_event, ta_state → ta_state;

predicates   ta_hmoves, ta_vmoves, ta_loaded : ta_state;
                 ta_exp : ta_event, ta_state;
variables   tas, tas1, tas2 : ta_state;
axioms

tas1 ≠ ta_err →
    (tas1 = tas2
    ↔ tas2 ≠ ta_err
        ∧ ta_hpos(tas1) = ta_hpos(tas2)
        ∧ ta_vpos(tas1) = ta_vpos(tas2)
        ∧ (ta_hmoves(tas1) ↔ ta_hmoves(tas2))
        ∧ (ta_vmoves(tas1) ↔ ta_vmoves(tas2))
        ∧ (ta_loaded(tas1) ↔ ta_loaded(tas2)))

ta_vpos(tas) ≠ ta_up ∧ ta_hmoves(tas) → tas = ta_err,

ta_exp(ta_hat(ta_fb),tas)
↔ tas ≠ ta_err ∧ ta_hmoves(tas) ∧ ta_hpos(tas) = ta_fb,
....

¬ ta_exp(ta_load),
....

ta_trans(ta_hto(ta_fb),tas) ≠ ta_err
↔ tas ≠ ta_err ∧ ta_hpos(tas) = ta_ro ∧ ta_vpos(tas) = ta_up
  ∧ ¬ ta_hmoves(tas) ∧ ¬ ta_vmoves(tas)

tas1 = ta_trans(ta_hto(ta_fb),tas) ∧ tas1 ≠ ta_err
→ ta_hpos(tas1) = ta_fb ∧ ta_vpos(tas1) = ta_up ∧ ta_hmoves(tas1)
  ∧ ¬ ta_vmoves(tas1) ∧ (ta_loaded(tas1) ↔ ta_loaded(tas)),

ta_trans(ta_load,tas) ≠ ta_err,
↔ tas ≠ ta_err ∧ ¬ ta_hmoves(tas) ∧ ¬ ta_vmoves(tas) ∧ ¬ ta_loaded(tas)

tas1 = ta_trans(ta_load,tas) ∧ tas1 ≠ ta_err
→ ta_hpos(tas1) = ta_hpos(tas) ∧ ta_vpos(tas1) = ta_vpos(tas)
  ∧ ¬ ta_hmoves(tas) ∧ ¬ ta_vmoves(tas) ∧ ta_loaded(tas),
....

end specification

The specification describes the possible states of the elevating rotary table. On events, a new state is reached by applying the transition function ta_trans to the event and the old state. According to the first axiom, states are characterized by their horizontal and vertical position and their movement. If ta_hmoves(tas) is true, the table rotates (horizontally) to the position ta_hpos(tas), if it is false, it is standing at that position. An additional state ta_err, different from all other states, is used to model crashs of the table due to incorrect control. According to the task description (part II, section 2.3) trying to rotate the table in all positions but the upper would result in such an error. This is reflected by the second axiom. The extension of the predicate ta_exp is the set of all sensor events which are expected to occur if the production cell is in state tas. They will play an important role as possible inputs for the control program. The third axiom, which represents a number of axioms for every event says that arrival at feed belt is expected if and only if the table moves towards that position. This predicate is also used to distinguish sensor and control events, since it is false in every state for control events. The rest of the axioms deal with the transition function. For every event a first axiom gives the safety conditions under which the event will not lead to the error state. A second axiom states the effects on the state, if this condition is fulfilled. For the event "rotate to feed belt" the transition will not lead to an error if and only if the table stands in upper position turned towards the robot (maybe this restriction is specific to the use of the table in the context of the production cell, so it should be moved to the following specification of context requirements, the task description does not indicate if this is the case). In the new state the table will rotate towards the angle where it is positioned towards the feed belt. Loading the table is acceptable in every position, in which the table does not move, so we have nothing assumed about the use of the device in the production cell.

The other devices are modelled in a similar way to the elevating rotary table. The specifications for the two belts coincide, they both have the same functionality. We assume a photoelectric barrier at the end of the feed belt too, which is a slight change to the task description. Some change is necessary at this point since we must be able to detect when a metal plate has moved on the rotary table. Another possibility would have been to install a timing mechanism on the computer, which would be started when starting the feed belt with a metal plate. After the time it takes the plate to arrive at the end of the belt it would signal an appropriate event. Introduction of timers or alternatively a second photo electric barrier at the front of the feed belt would have been more flexible and would have allowed to place more than one metal plate on the belts, which is not the case in the current system.

13.3.2 Specification of devices in the Production Cell context

Using the elevating rotary table in the context of the production cell yields restrictions on the positions, where the table should be loaded and unloaded. These re-

strictions are modelled by a new transition function ta_tr, which yields the error
state if the device is loaded and not in lower position turned towards the feed belt.
We get the following specification:

TABLE =
    enrich GENERIC_TABLE **by**
    functions   ta_tr : ta_event, ta_state → ta_state;
    axioms   ¬(tae = ta_load ∨ tae = ta_unload)
        →   ta_tr(tae,tas) = ta_trans(tae,tas),

            ¬(ta_hpos(tas) = ta_fb ∧ ta_vpos(tas) = ta_down)
        →   ta_tr(ta_load,tas) = ta_err,

            ta_hpos(tas) = ta_fb ∧ ta_vpos(tas) = ta_down
        →   ta_tr(ta_load,tas) = ta_trans(ta_load,tas),

            ¬(ta_hpos(tas) = ta_ro ∧ ta_vpos(tas) = ta_up)
        →   ta_tr(ta_unload,tas) = ta_err,

            ta_hpos(tas) = ta_ro ∧ ta_vpos(tas) = ta_up
        →   ta_tr(ta_unload,tas) = ta_trans(ta_unload,tas),
    end enrich

For other devices the specification contains more restrictions than that for load
and unload position, e.g. the restriction that the deposit belt may not drop a metal
blank when the blank reaches the end of the belt is formulated here (a generic belt
may do this, and in fact the feed belt does load the elevating rotary table by simply
dropping the blank in the event "fb_blankfalls"). All safety requirements that affect the use of devices in the context of the production cell are formulated in the
specification of its use in the production cell context. The only exceptions are conditions 1 and 2 in section 1.3.1 of the task description, since they are concerned
with the interaction of two devices (press and robot).

13.3.3 The system specification

The system specification composes the specifications of the devices. The states of
the whole system are the cartesian product (built by mkstate, decomposed by fbc,-
tac etc.) of the states of the individual devices, while the events of the system are
the union of the device events. In this union corresponding load and unload events
are identified. Conversion functions fbev, taev etc. are used to convert from the
events of the individual devices to the events of the system. The transition function
trans of the system calls the transition functions of the devices for their events.

Safety is formalized in the system specification by a predicate that is true for a state, if no device is in error state and interaction between robot and press is appropriate (see conditions 1 and 2 of the safety requirements given in section 1.3.1 of the task description).

PRODUCTION_CELL =
    enrich
        data specification
            using FBELT, TABLE, ROBOT, PRESS, DBELT, CRANE
            state = mkstate(fbc : fb_state, tac: ta_state, roc : ro_state,
                             prc : pr_state, dbc : db_state, crc : cr_state)
        end data specification
    by
        sorts   event;
        functions fbev : fb_event $\rightarrow$ event;
                  taev : ta_event $\rightarrow$ event;
                  roev : ro_event $\rightarrow$ event;
                  prev : pr_event $\rightarrow$ event;
                  dbev : db_event $\rightarrow$ event;
                  crev : cr_event $\rightarrow$ event;
                  trans : event, state $\rightarrow$ state;
        predicates exp : event, state;
                  safe : state;
    axioms
        (* conversion of events *)
        tae1 = tae2 $\leftrightarrow$ taev(tae1) = taev(tae2)),
        fbe $\neq$ fb_blankfalls $\rightarrow$ fbev(tae1) $\neq$ taev(tae2)),
        fbev(fb_blankfalls) = taev(ta_load),
        ....

        (* the transition function *)
        fbe $\neq$ fb_blankfalls $\rightarrow$ trans(fbev(fbe),mkstate(fbs,tas,ros,prs,dbs,crs)) =
        mkstate(fb_tr(fbe,fbs),tas,ros,prs,dbs,crs),

        trans(fbe(fb_blankfalls),mkstate(fbs,tas,ros,prs,dbs,crs)) =
        mkstate(fb_tr(fb_blankfalls,fbs),(ta_tr(ta_load,tas),ros,prs,dbs,crs)),
        ....

        (* the expected events *)
        exp(fbev(fbe),mkstate(fbs,tas,ros,prs,dbs,crs))
        $\leftrightarrow$ fb_exp(fbe,fbs),

        tae $\neq$ ta_load

$\rightarrow$ exp(taev(tae),mkstate(fbs,tas,ros,prs,dbs,crs)) $\leftrightarrow$ ta_exp(tae,tas),

safe(mkstate(fbs,tas,ros,prs,dbs,crs))
$\leftrightarrow$ fbs $\neq$ fb_err $\wedge$ tas $\neq$ ta_err $\wedge$ ros $\neq$ ro_err $\wedge$ prs $\neq$ pr_err $\wedge$
      dbs $\neq$ db_err $\wedge$ crs
$\wedge$ *(* do not crush arms with press *)*
      ((ro_pos(ros) = ro_a1atpr $\wedge$ ro_a1pos(ros) $\neq$ ro_ain) $\vee$
      (ro_pos(ros) = ro_a1atpr $\wedge$ ro_a1pos(ros) $\neq$ ro_ain)
      $\rightarrow \neg$ pr_moves(prs) $\wedge$ pr_pos(prs) $\neq$ pr_up)
$\wedge$ *(* do not rotate with extracted arms *)*
      $\neg$ (ro_moves(ros) $\wedge$ (ro_a1pos(ros) $\neq$ ro_ain $\vee$
      ro_a2pos(ros) $\neq$ ro_ain))

end enrich

The resulting system has the impressing total number of $8 \times 32 \times 1152 \times 12 \times 8 \times 12 = 339.738.624$ states (error states of the devices and identification with them by axioms not considered) and $4 + 7 + 17 + 7 + 4 + 11 = 50$ different events, 26 of them being sensor events.

The robot is clearly the most complex device with 4 relevant angles to rotate to, 3 positions per arm, 3 predicates indicating movement of the robot and its arms and 2 predicates indicating which arms are loaded. Although the specifications together have a length of 2000 lines (including comments) and contain 611 axioms, a third of which are generated automatically, they are easy to write (the only problem being copy and paste errors ...). It is easy to prove consistency of the specification, since most data types used are enumerations or record data types which are consistent by construction.

13.4 The Control Program and its Correctness

Having described the possible states, events and transitions of the production cell, we are now able to implement a procedure that controls the behaviour of the system. The procedure is a PASCAL program receiving a sensor event e as input (value parameter). It computes a list el = $(e_1, ..., e_n)$ of control events as output in its reference parameter. These are used to drive the actuators. The procedure uses a global variable v to model the current state of the physical system, that is updated according to the incoming sensor event and the computed control events. To prove correctness of the program, we have to show three characteristic properties:

1. the control program models reality adequately, i.e. the value of the variable v of the program reflects the physical state s of the system

2. the system is safe, it never reaches an erroneous state which does not satisfy the safety predicate

3. the system is life, i.e. it keeps on running forever.

Since we have chosen to model the physical system state and the abstract value of the program variable to be the same data type the "models" relation is simply expressed as equality (otherwise it would have been an abstraction function from physical states to values of the variable) and property 1 can be proved, showing that v = s is an invariant of the program. Written in Dynamic Logic the invariance assertion is

$$v = s \wedge \exp(e, s) \rightarrow \langle \text{control } (e;el) \rangle v = \text{trans}*(el, \text{trans}(e, s))$$

where control is the control procedure, trans is the transition function of the system, and exp is the predicate describing expected sensor events in state s. The assertion states formally, that if v describes the current state of the system, and e is an expected sensor event in s (i.e. a significant change in sensor values that may happen in the current state s), then the control program terminates yielding a list of control events el, and the value of variable v will reflect the induced state change by the sensor event and the control actions computed. This state change is computed formally by applying the transition function to first the event e and then to the list of events el (for el = $(e_1, ..., e_n)$, $trans*(el,s)$ is inductively defined to be $trans(e_n, trans(... trans(e_1,s) ...)))$. Proving the correctness assertion is trivial in this case, where abstract system state and the model of the physical system coincide, since the program simply updates the variable v at the beginning with $trans(e,v)$ and at the end with $trans*(el,v)$ using the same transition function as the physical system (if the models would differ the proof would become nontrivial).

To show how safety of the system can be guaranteed, we should first have a look at the program (160 loc). Apart from these assignments described above, it is a case distinction on the 26 possible sensor events. In every case an answer is computed, sometimes with, sometimes without looking at the value of v. The part of the control procedure, as presented in Figure 2, gives two cases.

Depicted is the reaction on the event "robot arrives at upper position", which is simply "rotate to robot" and the reaction on the event "robot arm2 has been retracted". This is the most complicated case, since depending on where the robot is standing (in front of elevating rotary table or in front of deposit belt) control events must be computed, that decide where the robot has to be rotated to next and whether the press has to be moved. The program makes assumptions about the state it is in. E.g. it assumes that if arm2 has been retracted from the deposit belt, the arm has

```
procedure control(e; var el)
begin
v := trans(e,v); el := nil;
case e of
    ....
    ta_vat(ta_up):                              (* table arrives in upper position *)
        begin el := cons(taev(ta_hto(ta_ro)),el) end      (* rotate to robot *)
    roev(ro_a2at(ro_ain)):                      (* arm2 has been retracted *)
        begin
            if ro_pos(roc(v)) = ro_a2pr          (* arm2 in front of press *)
            then begin
                el := cons(prev(pr_to(pr_mid)),el);   (* move press to loading position *)
                if ro_a1loaded(roc(v))            (* robot arm1 is loaded *)
                then el := cons(prev(pr_to(ro_a1pr)),el)
                                                  (* move with arm1 to press *)
                else    if db_loaded(dbc(v))      (* dep. belt is loaded *)
                        then el := cons(roev(ro_to(ro_a1ta)),el)
                                                  (* move with arm1 to e. r. table *)
                        else el := cons(roev(ro_to(ro_a2db)),el)
                                                  (* move with arm2 to deposit belt *)
            end
            else                                 (* robot is with arm2 at dep. belt *)
                if pr_loaded(prc(v))             (* press is loaded *)
                then el := cons(roev(ro_to(ro_a2pr)),el)
                                                  (* move with arm2 to press *)
                else    if ro_a1loaded(roc(v))    (* arm1 is loaded *)
                        then el := cons(roev(ro_to(ro_a1pr)),el)
                                                  (* move with arm1 to press *)
                        else el := cons(roev(ro_to(ro_a1ta)),el)
                                                  (* move with arm1 to e. r. table *)
        end
    ...
end case;
v := trans*(el,v)
end
```

Figure 2   Procedure control

just been unloaded. These assumptions must hold, if the resulting state after the program should be safe. In the example, if arm2 were still loaded, and press were loaded too, arm2 would move to the press, and would try to pick up a second metal plate, resulting in an error. The assumptions can be intuitively formulated as "loaded devices move forward until they reach the position to unload, and unloaded devices move backward until they reach the position to load". Formalizing this safety invariant, we get a predicate safe-inv(s), and we have to prove that

$$v = s \land \text{safe}(s) \land \text{safe-inv}(s) \land \exp(e;s) \rightarrow$$

$$\langle \text{control}(e;s) \rangle \, (\text{safe}(s) \land \text{safe-inv}(s))$$

holds. Then, started in an initial state, which satisfies *safe(s)* ∧ safe-inv(s), our program will work in the sense that it will never reach an unsafe state.

To guarantee lifeness of the system, we must assure that in any state reached by the system we still have an expected event e, for which *exp(e,s)* holds, which means that not all devices are stopped. This is not the case for all states satisfying *safe(s)* and safe-inv(s), so we must impose further restrictions. Two obvious ones are that the production cell must contain metal plates, and that there must be a possibility to move at least one metal plate, which requires that one position is not loaded. Since the current description has seven positions where a metal plate may be (one for every device except the robot, who has two arms, that may be loaded), the number of plates the cell may handle, is restricted to 6. Another less obvious restriction is "loaded devices are that have reached their 'unloading position' are unloaded immediately if the device to load is ready", for they would wait forever otherwise.

A very intricate problem here are the two robot arms, since their movement is not independent of each other. The two critical events are just the ones, when one robot arm has been retracted. One is shown in the program above. If we would follow the task description for the robot (part II, section 2.1.3) literally, we would get a deadlock, if there were just one metal in the production cell, after picking up a piece from the elevating rotary table (step 1), since there is no blank in the press to get. Even if we assume a second piece in the press, we would get a deadlock after step 3, if all positions in the production cell except the arms of the robot are filled (then we now should first pick up a blank from the elevating rotary table). Formulating all these restrictions by a predicate life-inv(s), we have to show that they are left invariant by the control program too, i.e.

$$v = s \land \text{safe}(s) \land \text{safe-inv}(s) \land \text{life-inv}(s) \land \exp(e, s)$$

$$\rightarrow \langle \text{control}(e;s) \rangle \, (\text{safe}(s) \land \text{safe-inv}(s) \land \text{life-inv}(s))$$

and that these restrictions guarantee indeed lifeness, i.e.

$$\text{safe}(s) \land \text{safe-inf}(s) \land \text{life-inv}(s) \rightarrow (\exists e) . \exp(e, s)$$

13.5 Verification

Verification of the safety and lifeness condition of the production cell is trivial from the theoretical point of view, since both conditions can be viewed as formulas of the propositional calculus (for the lifeness condition the existential quantifier can be replaced by an explicit disjunction on the 50 possible events), which is decidable. So all that is required to prove the goals is an efficient prover for propositional logic. Our first impression of the case study was, that we could simply use the propositional rules of the sequent calculus built into the KIV-System together with appropriate rewrite rules. Then proofs could be done automatically.

Unfortunately, this attempt is not feasible, since proofs grow exponentially, when case distinctions (via the usual rules of sequent calculus for conjunctions in the succedent and disjunction in the antecedent) are applied without restrictions. The same proof trees simply get duplicated by doing case distinctions that are irrelevant to the current subgoal. As an example, if we want to prove a subgoal that states: "After some events initiated by the control program, press is loaded", a case distinction on the initial position of the travelling crane will simply duplicate the subgoal. Since we wanted to do proofs with the KIV-System, we decided to give up the attempt to prove correctness automatically. To avoid doing all case distinctions by hand, we used the "module specific" heuristic to model the typical situations where a case distinction is appropriate (the name stems from the fact, that the heuristic was originally designed to handle typical situations in proving module correctness). The "typical situations" used here are quite simple, they specify situations where at least one of the premises can immediately seen to be an axiom. About 60 – 70% of the case distinctions can be done automatically using the heuristic, the rest must be given by hand.

To increase the readability of the sequents of the proof we used the "module specific" heuristic to weaken unnecessary preconditions from the proof too.

Apart from case distinctions and weakening steps the proof consists mostly of simplification steps, that reduce applications of the global transition function to applications of their local counterpart, eliminate applications of the local transition function by giving appropriate preconditions and deal with the inequality of constants (such as $ta\_fb \neq ta\_ro$). A typical lemma used for simplification is shown in Figure 3.

The Lemma is used as a rule to rewrite every instance of the premise of the implication to the conclusion. Altogether we used over 1000 such rules, which demands a very efficient simplification strategy. All the lemmas used for simpli-

$$ta\_loaded(ta\_tr(ta\_vto(ta\_up),tas)$$
$$\rightarrow \quad ta\_tr(ta\_vto(ta\_up),tas) = ta\_err$$
$$\vee \quad (ta\_tr(ta\_vto(ta\_up),tas)) \neq ta\_err$$
$$\wedge \quad tas \neq ta\_err \wedge ta\_vpos(tas) = ta\_down$$
$$\wedge \quad ta\_hpos(tas) = ta\_fb \wedge ta\_loaded(tas))$$

Figure 3   An example of a simplification lemma

fication are axioms or propositional reformulations of axioms, so they can be proved easily. 200 of them are axioms from data specifications, which are used automatically.

A first version of the specification, the implementation of the control program, and an initial definition of the simplifier rules and the "typical situations" for the "module specific" heuristic can be derived in about two weeks of work.

But now we encounter the typical problems of verification: Trying to prove the lifeness goal we directly run into an unprovable subgoal. Since the subgoal explicitly shows a state that is not life, i.e. one where no event is expected, the decision which part of our system is incorrect (specification, simplifier rules, program or one of the invariants safe-inv and life-inv) is easy, but as it turns out, the error we detected is not the only one.

Altogether we discovered about 30 errors during the verification process we have done so far. Most of the errors were discovered during the proof of the lifeness goal. Some of the errors were purely syntactical, some concerned the strategy of the program but most of them resulted from missing properties in the invariants safe-inv and life-inv. One of the most intricate lifeness properties is that the robot does not rotate with arm1 to the elevating rotary table, if arm1 is not loaded, arm2 is loaded and *all* the other devices except press are not loaded.

If we would have to start to prove lifeness from scratch every time we discovered an error, we would have never reached any success. But fortunately the KIV-system can reuse the proofs of corrected goals, which saves a vast amount of time. Nevertheless proving the lifeness is still quite a lot of work to do, and unfortunately we have not found a way to make the proof modular, i.e. we did not find a set of lemmata sufficient to prove the lifeness goal, such that changes in the invariants would affect only *some* of the lemmata.

Starting with an initial version of the goal we arrived at a complete proof for the lifeness assertion after about two weeks of work. The statistic for the lifeness proof is depicted below.

Safety still requires some more work. The initial goal first splits into 26 goals, one for every sensor event. Further conditionals in these cases (with a maximum of 6 for the event "robot arm2 has been retracted") give 51 cases to prove. The proof for every goal goal requires about a day of work, so we did only 11 exemplary cases including the 6 cases of the event "robot arm2 has been retracted". Proofs are so simple and tedious (and they all look very similar), that it is possible to deal with several proofs simultaneously (which was never possible in other case studies). Three typical statistics for the lifeness proof and the first two cases of the "robot arm2 has been retracted" look as follows:

	lifeness	roa2atin-case1	roa2atin-case2
proof steps	1316	731	951
simplification	603	432	525
weakening	295	100	174
case distinction	395	199	252
interactions	81	72	74

The diagram shows the number of proof steps required to proof the goal, which splits in simplification, weakening and case distinction steps (as described above). The interactive steps are all case distinctions.

13.6 Conclusions and Further Work

We have done a case study in specification and verification of a distributed technical system with a central control program. The algebraic approach was suitable to derive a structured specification of the system as a composition of reusable device specifications. We implemented a control program, that although it is not suitable to drive the production cell model of the FZI due to the chosen abstraction level, seems not too far away from a realistic application. Here the connection between different abstraction levels seems to be an interesting topic for further research. The tactical theorem proving approach used in the KIV-System was sufficient to prove the goals, although it seems that work has to be done in the modularization of correctness proofs to make them more feasible. For the finite state space used here, it should also be possible to use techniques of symbolic model checking, which would do proofs automatically and therefore seem to be more adequate.

Maybe the invariants safe-inv life-inv could be derived automatically too from the control program.

Tactical theorem proving may again become relevant, if we change towards a more realistic scenario adding suitable sensors or timers, to get rid of the "only one plate on each belt" restriction. Allowing any (finite) number of plates on a belt would turn the problem from a propositional logic one to a problem of predicate logic, where techniques operating on finite state spaces would be no longer applicable. Other steps towards a more realistic scenario include the introduction of a startup routine, or the possibility to add and subtract metal plates from the cell. A final point completely missing here is the comparison with other approaches to the specification of distributed systems. Connections to functional specifications using streams as data type should be clarified as well as the connections to specifications using temporal logic. Finally we wish to thank our students Markus Friedel and Farzad Safa for their work on this case study.

References

[1]   D. Harel: *First Order Dynamic Logic.* Springer LNCS 1979.

[2]   M. Heisel, W. Reif, W. Stephan: *A Dynamic Logic for Program Verification.* "Logic at Botik" 89, Meyer, Taitslin (eds.), Springer LNCS 1989.

[3]   M. Heisel, W. Reif, W. Stephan: *Tactical Theorem Proving in Program Verification.* 10th International Conference on Automated Deduction, Kaiserslautern, FRG, Springer LNCS 1990.

[4]   W. Reif: *Verification of Large Software Systems.* Conference on Foundations of Software Technology and Theoretical Computer Science, New Dehli, India, Shyamasundar (ed.), Springer LNCS 1992.

[5]   G. Schellhorn: *Specification and Verification of Distributed Technical Systems with Central Control,* Technical Report 3/94, Fakultät für Informatik, Universität Karlsruhe.

XIV. Tatzelwurm

Verification of Safety Requirements
with a Program Verification System

Stefan Klingenbeck, Thomas Käufl

Universität Karlsruhe

Abstract

In this paper the results of the use of the verification system *Tatzelwurm* are presented for the verification of the safety requirements of a program. The program considered controls the machinery of a production cell. Two versions of the program have been implemented. The safety properties of the simpler version could be established fully automized. This was not so for the more elaborate version, but after some work on the specification of the program and the safety requirements we are convinced that it is possible to find the proofs of the verification conditions in a reasonable amount of time.

14.1 Introduction

The foundations of the verification for sequential and imperative programs are well known [7] and a variety of tools for the verification are available. (See for example [6].) But only a few programs having more than 100 lines of code have been verified until now. The main reason for this is the enormous effort needed to verify programs. Therefore it is necessary to develop program verification systems which work with a high grade of automation.

After a sketch of the verification system *Tatzelwurm* and the software development method used, the main ideas behind the specification of the problem and the program code are presented. The following sections summarize the results and some conclusions obtained during the project.

14.2 The Program Verifier *Tatzelwurm*

The verification system *Tatzelwurm* is designed to develop and verify sequential programs written in an imperative language. A Hoare like calculus is used to generate proof obligations, called *verification conditions* in the following, which are sufficient for the (partial) correctness of the program. Termination can be verified with the well founded set method. For the proof of validity of the verification conditions, a theorem prover specialized on program verification is available.

14.2.1 The programming language

The system accepts a subset of Pascal [1] as programming language. It comprises

- procedure calls, assignments, conditionals, while and repeat statements
- function calls
- constant and type definitions, variable declarations, procedure and function declarations
- integers, reals, arrays and records, subrange types and scalar (enumeration) types
- modules as in UCSD Pascal

In [3] or [4] the subset of Pascal accepted by the verification system is defined. The use of Pascal as programming language is not essential. The Pascal parser may be replaced by a parser for any other language fulfilling the requirements sketched above.

14.2.2 The specification language

An order sorted first-order logic with equality is used as specification language. The data types of Pascal and the functions and predicates operating on them have a predefined meaning in the specification language. The subtype relationship corresponds to the subsort relationship. (Integer is considered as subsort of real e.g.)

14.2.3 The combination of specifications and programs

Tatzelwurm works on inputs consisting of Pascal statements and logical formulae termed annotations.

A symbol c occurring in an annotation and as identifier in a program is assumed to denote the same object. If for example x occurs in a statement *sta* followed by $P(x)$ as annotation, then $P(x)$ is assumed to hold after the execution of *sta*. To be more precise, *Hoare triples* of the format X | *sta* | Y where X and Y are annotations and *sta* is a sequence of statements are treated. X is said to be the *precondition* and Y the *postcondition* of *sta*. X | *sta* | Y expresses that the truth of Y is implied by X, provided the execution of *sta* terminates. $y > 0 \mid y := x + y \mid y > x$ for example states that $y > x$ is true, if $y > 0$ holds before the execution of the assignment.

Annotations may occur in the program following the keywords **entry, exit, invariant, assert** and **assume**. The entry and exit annotations specify the behaviour of programs, procedures and functions. They act as pre- and postcondition. The behaviour of each loop must be specified by means of an invariant condition. Properties necessary for the proof of the correctness of a program can be added following the keyword **assume**. The proof of their validity is not forced by the verifier, since they may express properties assumed to hold for specific symbols appearing in the specification. A logical formula X following **assert** is termed an assertion. X in *sta*1; **assert** X; *sta* 2 acts as postcondition for *sta*1 and as precondition for *sta* 2. Thus by use of assertions the proof of the correctness of programs can be split into independent parts. This allows to avoid too complicated correctness proofs by a suitable insertion of intermediate assertions.

The verification system decomposes the program into paths beginning or ending with an assertion, invariant, entry of exit condition. These paths (being Hoare triples) are used for the generation of verification conditions that are sufficient for the partial correctness of the program part. For example, the verification condition for the triple $y > 0 \mid y := x + y \mid y > x$ is the formula $y > 0 \rightarrow x + y > x$.

14.2.4 The Automated Theorem Prover

The prover [5] uses the analytic tableaus developed by Smullyan [8]. The set of rules is enlarged by rules for equivalences and for a generalization of the modus ponens. Additional features allow to use function symbols and the equality predicate. A decision procedure for the theory of quantifier free formulae of not interpreted function symbols with equality is part of the prover, too. It allows to prove theorems containing equations.

The prover can be used fully automaticaly without interaction with the user. But it is also possible to do proofs interactively. Here the user can suggest the subgoals to be proved and the instantiations to be made. The user can also define new inference rules and combine them into tactics in order to guide the prover problem dependently.

The prover uses reduction procedures for theories. At present there is a reduction procedure for the linear inequalities and equations over the rationals, for the Presburger-Arithmetic and for the enumeration, array and record types of Pascal. In order to support reasoning with arbitrary equations demodulation is available.

As the prover is tableau based, it combines the advantages of automated provers with proof checkers. The user can decide whether a formula is to be proved fully automatedly or interactively indicate appropriate subgoals, lemmata or rules. The user can also enter instantiations for bound variables. The reduction procedures for theories are set up so that the user need not bother with details.

14.2.5 Software development

The verification system supports an evolutionary life cycle concept. The user starts by designing or modifying a module. The program need not be completely present when the verification starts. It is possible to verify the correctness of the logical structure implied by the annotations first. In subsequent steps, code implementing functions or procedures can be entered. Now the proof that the code fulfills the annotations must be done.

In this method not implemented code is represented by its specification. Therefore, the method can be used to develop software in a top-down approach. But the user is free to verify first the most critical program part completely. This organization of the work is the fastest way to decide whether a costly redesign is necessary when verification has failed.

As the refinement of the implementation is speculative, the use of old modules that are not yet verified harmonizes with this approach. (The behaviour of the old software specified the user can decide whether he first verifies the use of the software or its correctness.)

A major problem in software development with formal methods is the validation of the formal specification (answering the question whether informally stated requirements are fulfilled. This query is usually answered by inspection or rapid prototyping.) In our approach the implementation is not the product of a formal transformation process. So shortcomings of the specification are not propagated systematically. Also the character of the specification is descriptive whereas the implementation is algorithmic. These two kinds of formalizations of the problem introduce in some sense a diversity allowing to find out flaws at an early stage.

The proof of a verification condition or a lemma may fail due to one of the following reasons. The specification or the implementation contains an error or in the case of a valid condition the user did not load necessary definitions and lemmas. Proofs for a valid verification condition or lemma may also be too complex to be found without any human guidance. Then the prover must be used interactively to develop complex proofs or to analyze failed proof attempts to localize the error in the program or its specification. (This is an advantage of tableau based theorem provers and will be a topic of our future work.) Developing complicated proofs or analyzing failed proof attempts needs much time. Much time can be saved using well structured programs and specifications, and organizing the proofs leading to a high level of automation.

14.3 Formal Description of the Production Cell

We implemented a program controlling the machinery of a production cell and proved that it fulfils some safety requirements. These requirements, the effect of the commands sent by the program, and their adequacy must be specified formally. First the temporal aspects of the behaviour of the production cell must be formalized.

The protocol for the transmission of sensor values and commands states that sensor values are always transmitted as a record of the values of all sensors of the production cell. Hence the control program has to read these sensor values and to calculate and send commands while the production cell is working (See figure 1.).

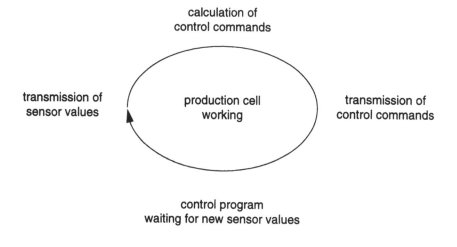

Figure 1 The control process

For the sake of clarity, time is modelled explicitly in our approach. (The alternative had been the abstract description of the production cell as finite automaton.) A discrete time is sufficient, if one considers completing one cycle of figure 1 as a single time step. (See figure 2.)

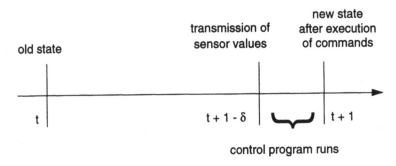

Figure 2   The time model

According to the time model and our goal, namely that the proof of the control program satisfies safety properties, the subsequent assumptions are made. The sensor values are transmitted to the control program correctly at time $t - \delta$.

1. The computation and transmission of the commands take place in the interval between $t - \delta$ and t. The production cell will obtain its new state at time $t + 1$.

2. The input and output routines of the control program work correctly.

3. The commands are transmitted to the machinery correctly and each device executes the commands received according to the formal description of the production cell.

The second assumption requires the period δ to be long enough for the computation and transmission of the new commands to the machinery. It is obvious that these actions must be executed as fast as possible, but there is a least value for δ. This value has influence on the design of the apparatus of the production cell: For example the angular motion of the robot must be such that it cannot pass the interval 2° to -2° during the period δ. Otherwise it could collide with the feed belt before receiving the stop command. Similar restrictions hold for the period $1 - \delta$. Both kinds of restrictions are expressed explicitly in the formal description of the production cell.

The production cell is specified in sorted first order logic with equality and function symbols (the specification language of our system). Its state is modelled by a set of functions, predicates, and state variables, corresponding to the position and motion of each device. Additionally predicates expressing wether a device is loaded with a blank have been defined. For example the crane is specified using two predicates and four functions describing its motion and position.

(**type** *Crane-is-Loaded(Time)*))

(**type** *Crane-Mag-on(Time)*))

(**type** *crane-has-vert-pos(Time): Real)*

(**type** *crane-has-hor-pos(Time): Position)*

(**type** *crane-has-vert-mo(Time): V-Motion)*

(**type** *crane-has-hor-mo(Time): H-Motion)*

The set of axiom is split into two subsets: The axioms describing the behaviour of the production cell in the period t to $t + 1 - \delta$ and those for the period $t + 1 - \delta$ to $t + 1$.

The following example of an axiom states that the crane is not loaded at the time t, if the control program transmitted the command *crane-magnet-off* (= *hhger-mag-aus*) during the period $t + 1 - \delta$ to $t + 1$ (disregarding the position of the crane and the charge of the crane).

axiom: *Crane-is-Loaded-c*

$\forall t$: *Time hhger-mag-aus* $\in$ *command(t)* $\rightarrow$ ¬*Crane-Is-Loaded(t)*

The deposit belt is divided into two regions:

1. *end* (A blank is in the photoelectric barrier)

2. *begin* (A blank is at the part of the belt before the end region)

The position of a blank at the deposit belt may be none (no blank at the belt), begin or end. The axiom below states, that a blank at the beginning of the deposit belt at the time t may be transported to its end or stay in the begin region of the belt at time $t + 1 - \delta$, if the belt motor is working. (But the blank cannot drop or disappear from the belt.)

axiom: *Dbelt-Blank-Has-Pos*

$\forall t$: *Time*

   Dbelt-in-Motion(t) $\land$ *dbelt-pos(t) = begin*
   $\rightarrow$ *(dbelt-pos(t + 1 - δ) = begin* $\lor$ *dbelt-pos(t + 1 - δ) = end)*

The specification reflects in a way the fact that the production cell can be considered a finite automaton. It is, however, not necessary to provide a complete state transition table. It suffices to consider those transitions that occur in a specification or a proof. For example it is not necessary to formalize the situation when the magnet of the crane puts down a blank but is not exactly in the position where the safe placement of a blank is guaranteed. This formalization is not needed, if it is verified that the crane places a blank on the belt in a safe position only. If during a proof a formalization of a situation not covered up to that point becomes necessary, it can easily added. On the other hand axioms never needed in a proof can be discarded. Once the work on the proofs is completed, a version of the formal description can be obtained which contains exactly those axioms that are necessary for the safety of the machinery.

14.4 The Implementation

We have written two control programs for the production cell: a simple version CP1 dealing with only one blank at a time in the whole production cell and an enhanced version CP2 dealing with at most one blank per device.

The program CP1 has 400 lines of code, CP2 has 250 lines of code. The implementation of CP1 took three days. The implementation of CP2 took 14 days. The programs consist of a while-loop that reads all sensor values, calculates commands using the sonsor values and additionally stored state information, and sends the commands.

```
if ([crane_at_dbelt_s] <= sensors)
then     if ([crane_bottom_ ] <= sensors)
         then    if ([fbelt_blank_at_end_s] <= sensors)
                 then command := hhger_stop_v
                 else command := hhger_heben
         else    if ([crane_top_s] <= sensors)
                 then    if ([fbelt_blank_at_end_s] <= sensors)
                         then command := hhger_senken
                         else command := hhger_stop_v
                 else command := hhger_stop_v
```

Figure 3   The control of the vertical crane movement in CP1

CP1 uses nested conditional statements to calculate the commands. It may send superfluous commands. Since only one blank may travel through the production cell, it is not necessary to verify that a blank cannot be put on an already loaded device. This is a very important simplification, because in this case most of the internal stored information about the state of the production cell does not concern safety aspects. Additionally fewer state variables than in CP2 are required. It turned out that even under these assumptions the nested conditional statements became inscrutable.

The enhanced implementation CP2 has two objectives:

1. To deal with several blanks in the production cell and to guarantee that at most one blank is on each device.

2. To avoid nested conditional statements to get a more lucid program.

To guarantee that at most one blank is on each device, we use a more sophisticated idea of representing the state of the production cell. For each device a set of CP2 states is defined reflecting the steps of processing a blank.

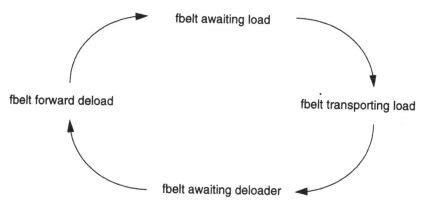

Figure 4   The states of the feed belt in CP2

For example the transportation of a blank on the feed belt consists of the following steps: Waiting for the blank (*fbelt-awaiting-load*), transporting the blank to the end of the belt (*fbelt-transporting-load*), waiting at the end of the belt until the rotating table is in the correct position to be loaded with a blank (*fbelt-awaiting-deloader*), transporting the blank from the belt to the rotating table (*fbelt-forward-deload*).

The main advantage of the CP2 states is that each state has exactly one successor and one predecessor state. Each CP2 state represents a set of states in the specification of the production cell. This representation is expressed e.g. for the state *dbelt-awaiting-deloader* by:

definition: *Corr-Fbelt-Awaiting-Deloader*
∀t: *Time*
    (*Corr-Repr(fbelt-awaiting-deloader t*)
    ↔ *Fbelt-Blank-Has-Pos(t) = end* ∧ ¬*Fbelt-Is-Moving(t)*)

The CP2 states at time t together with the sensor values of time t + 1 - δ are used to calculate the commands and the new CP2 states at time t + 1. The correctness of this representation for every point t of time must be verified.

Instead of conditional statements with many sub-categories the control information is stored in a decision table that is interpreted by the program at run time. The table consists of two rows, a condition part and an action part. The action part of the row are performed, if its conditions are fulfilled. The conditions of different rows are allowed to be true simultaneously. The condition part consists of conditions on the sensor values and the CP2 state of the production cell. The action part consists of commands to be sent and the new CP2 state. A decision table of 60 rows is included in the program CP2. Pascal sets are used for implementation of the table.

Condition part:
st_con [15] := [*fbelt_awaiting_deloader, rotable_awaiting_load*];
se_con [15] := []
Action part:
st_act [15] := [*fbelt_forward_deload*];
co_act [15] := [*band1_los*]

Figure 5   A row of the decision table

In addition to the decision table and the read/write procedures, the program CP2 contains only the interpreter for the table as outlined in the subsequent figure.

```
while i <= maxrule do
begin    if    (decision.state_conditions[i] <= state_messages) and
               (decision.sensor_conditions[i] <= sensor_message)
         then    begin    new_st_i := new_st_i + decision.state_actions[i];
                          comm_i := comm_i + decision.command_actions[i];
                 end;
         i:= i + 1;
end
```

Figure 6   The interpreter for the decision table

The main reason for using a decision table was the greater flexibility of the program. Using assertions, it is possible to give the program a structure allowing to change entries of the table without repetition of the entire correctness proof of the program.

14.5 The Safety Requirements

Instead of formalizing error situations, we focussed on conditions that necessarily lead to the occurrence of an error. We did not specify that the robot arms do not crash against the press. We have specified the safety requirement that the robot rotates only with retracted arms, which must be fulfilled by the control program.

The safety requirements can be divided into three groups. The first group contains eight restrictions on the motions of the rotating table (vertical and horizontal), the robot (arm1 extraction, arm2 extraction, rotation), press, crane (vertical and horizontal). For example

definition: *Rob-Rot-Error*

$\forall t$: *Time*

    (*Rob-Rot-Error*(t)

    $\leftrightarrow \neg robot\text{-}has\text{-}mo(t) = none \wedge \neg Arm1\text{-}Retracted(t) \wedge \neg Arm2\text{-}Retracted(t)$)

The second group consists of four restrictions on the capacity of the feed belt, the rotating table, the press and the deposit belt. For example

definition: *Rotable-2blanks-Error*

$\forall t$: *Time*

    (*Rotable-2blanks-Error*(t)

    $\leftrightarrow$ *Fbelt-Is-Moving*(t) $\wedge$ *fbelt-blank-has-pos*(t) = *end* $\wedge$ *Rotable-Is-Loaded*(t)

      $\wedge$ *Rotable-Is-Ready*(t))

The operations executed when a blank is transported from a device to the next one are subject to five restrictions on the safe deposition of blanks. For example

definition: *Arm1-Drop-Error*

$\forall t$: *Time*

    (*Arm1-Drop-Error*(t)

    $\leftrightarrow \neg Arm1\text{-}Mag\text{-}on(t) \wedge \neg(Arm1\text{-}Ex\text{-}Press(t) \wedge Arm1\text{-}Ang\text{-}Press(t))$)

Besides the 18 safety restrictions mentioned above there is a special one. As the program can transmit several commands simultaneously, it must be verified that such a set of commands does not contain contradictory ones.

14.6 Results

It has been verified that the control program CP1 fulfils the safety restrictions imposed on the motions and placements of blanks. (The restrictions of the second group were of no interest, since CP1 handles only one part.) Some errors in the program and the specification removed all proofs have been found automatically within three weeks. This because the program CP1 has almost no state variables affecting the safety requirements.

The approach taken for CP1 cannot be used for CP2, since this program has to keep a record of the positions of the blanks. Additionally, the correct calculation of the CP2 states must be verified. About 110 proofs are necessary to show that CP2 fulfills the safety requirements. After some work it turned out that it is necessary to express properties of the decision table in lemmas to obtain a simpler structure of the proofs. (The lemmas must be established, too, but using a good structure these proofs can be found automatically.) Having inspected the specification more closely we found out that the sort hierarchy was not fine-grained enough. After correcting this, we proved a verification condition of average size automatically, without using the more advanced features of the prover. So we are convinced that it is possible to find the proofs of the verification conditions in a reasonable time.

14.7 Conclusions

In an experiment we have demonstrated how to specify and verify properties of programs containing 400 (CP1) lines of code. (CP2 has 250 lines of code.) The specification including the axioms for the production cell and the definition of the safety requirements contains 600 lines. We have shown that it is possible to do the verification with reasonable effort straightforwardly, provided the properties to be established are not too complex. If this does not hold, the specification and proof process must be organized carefully.

The structure introduced by definitions and lemmas must have the consequence that each proof can be done nearly automatically. The major part of the proofs must be found without any human assistance, since it usually happens that shortcomings of definitions or lemmas must be corrected, with the consequence that other proofs must be repeated. The specification, if modularized appropriately, the number of these repetitions will be small.

The verification in mind led to definitions and lemmas applicable to only one device. The specifications for the decision table can be divided into two groups.

The first expresses some global properties of the table and the second consists of definitions and lemmas for single rows. Only one row changed, therefore, its properties, and additionally global properties of the table, must be proved again.

The number of simple errors in the specification (typing errors e.g.), the detection of which is too costly using a formal proof, can be reduced considerably using walkthroughs and inspections. (This work is necessary for the validation anyway.)

For proofs carried out interactively with our system, in the average 20% of the inferences are due to human interaction. This ratio must be decreased considerably. Additionally, it is necessary to increase the number of automatically found proofs in order to decrease the time necessary for the proof management.

Liveness properties have not been proved although this does not seem too difficult for the control program CP2. Economical properties of the machinery, a high output of blanks e.g., are out of the scope of the verification of functional correctness.

References

[1] K. Jensen, N. Wirth: Pascal: User Manual and Report. Berlin, Heidelberg, New York, Tokyo: 1983; Springer

[2] Th. Käufl: The Program Verification System Tatzelwurm: User Manual. Unpublished working paper

[3] Th. Käufl: Program Verifier Tatzelwurm: The Correctness and Completeness of the Generation of the Verification Conditions. Interner Bericht 9/89. Institut für Logik, Komplexität und Deduktionssysteme, Universität Karlsruhe: 1989

[4] Th. Käufl: The Program Verifier Tatzelwurm. in [6].

[5] Th. Käufl, N. Zabel: Cooperation of Decision Procedures in a Tableau-Based Theorem Prover. Revue d'Intelligence Artificielle, Vol. 4, no. 3: 1990, pp. 99 - 126

[6] Sichere Software: Formale Spezifikation und Verifikation vertrauenswürdiger Systeme. H. Kersten (Hrsg.). Heidelberg: 1990

[7] J. Loeckx, K. Sieber: The Foundations of Program Verification. Stuttgart, Chicester 1984.

[8] R.M. Smullyan: First Order Logic. Berlin, Heidelberg, New York: 1968

XV. HTTDs and HOL

*On the use of a graphical specification language
and an interactive theorem prover for the formal
development of a real-time production cell*

Rachel Cardell-Oliver

University of Essex

Abstract

This paper illustrates the use of hierarchical timed transition diagrams (HTTDs) and the HOL theorem prover for the formal specification and verification of a production cell. The specification generalizes the geometries and component speeds of the production cell, real-time behaviour is modelled, and verification is by partially automated deductive proof using the HOL system.

15.1 Introduction

This paper is a contribution to the case study project "production cell" which seeks to evaluate different approaches to the problems of formal and semi-formal software construction. The paper illustrates the use of hierarchical timed transition diagrams (HTTDs) and the HOL theorem prover [4][9] for the formal specification and verification of the production cell. The costs and benefits of this approach and some lessons learnt from the example are discussed.

This version of the case study is distinctive in three ways. The *real-time* behaviour of the controller and its environment is modelled. The specification *generalizes* the task description and its simulation by allowing, for example, many different geometries and component speeds. Verification is by *partially automated deductive proof* using the HOL theorem prover. This method has the potential to verify larger systems than those, such as the production cell, at the limits of current model checking and symbolic model checking technology.

The mathematical theory upon which this case study is based is that of timed transition systems (TTSs): a language for the specification and verification of real-time, reactive computer systems [5][6]. The theory of TTSs is not sufficient on its own for the formal development of the production cell since the non-trivial size of this case study requires further support. In particular, there is need for new *notation* for expressing large specifications, *mechanized support* for making specifications and performing verification proofs, and a *method* for developing specifications that are both verifiable and from which a faithful implementation can be generated.

For the formal specification of the production cell we have used the graphical notation of hierarchical timed transition diagrams (HTTDs), a minor extension of TTDs [5] which shares the TTS semantics of TTDs. Mechanical support is provided by the HOL theorem prover for higher order logic (Church's simple type theory with the addition of polymorphic types, and secure principles of definition) [2][4]. HOL is used both to express graphical HTTD specifications in a form suitable for formal verification and also to perform verification proofs interactively.

The remainder of the paper is organized as follows. Section 15.2 outlines the "formal software construction method" that has been used and describes the current status of the case study and tools. Specification and verification examples from the case study are described in Sections 15.3 and 15.4. Section 15.5 presents results of this study, and Section 15.6 conclusions on the use of HTTDs and HOL for the formal development of systems such as the production cell.

15.2 Method

The method used here to produce a formally verified design for the production cell can be described in seven stages. Although presented as a simple sequence, in practice the repetition of earlier stages is often required to clarify assumptions, correct mistakes and simplify specifications and proofs.

Specification is carried out in four stages:

S1: decide how the controller will be organized as a set of communicating processes;

S2: for each of these processes, outline the actions it must perform and their ordering;

S3: define an interface for each process; what variables will be used to communicate with other processes and with the controller's environment? what system constants should be specified?

S4: refine the processes from stage **S2** into graphs of timed transitions (see Section 15.3) using the variables and constants identified in **S3**.

Verification of the system follows:

V1: express the requirements given in the task description in terms of the variables and constants of the specification; Requirements are stated as formulae of real-time temporal logic (see Section 15.4);

V2: outline verification proofs for each requirement, identifying the proof rules and proof strategies to be used;

V3: construct the proofs of **V2** in the HOL system; details of the proofs will be performed automatically.

The implementation of controller software from its formal specification may proceed in two ways. HTTD specifications have an alternative representation as imperative programs in the Statext language [7]. Each HTTD-Statext process could be executed on a dedicated microprocessor with access to shared sensors and actuators. For example, the implementation environment proposed in the Esterel production cell case study [8] would be suitable. If the timing constraints of the specification are chosen to reflect those of the implementation hardware then such an implemenation would represent faithfully the behaviour of the specification except in one detail. The implemenation would display true concurrency while the HTTD specification assumes interleaving concurrency and a synchronised initial state for all processes. This difference would not, however, affect the validity of the great majority of verified requirements.

On the other hand, if the timing constraints of the specification only represent the relative ordering of events and not "real" times such as seconds or processor cycles, then the specification can be animated according to the operational semantics of HTTDs, providing a single processor implementation of the controller. State machine implementations such as those generated for Lustre processes are of this type [8].

Current Status

The production cell has been formally specified and verified using the method above. The specification is completely defined as a HOL-HTTD theory (**S1-S4**), all of the task description requirements have been verified by hand (**V1-V2**), and a representative sample (some of each class of requirements) have been verified in HOL using proof programs developed within a HOL-HTTD tool (**V3**).

The HOL-HTTD tool is a research vehicle still under development. We have not implemented links from the tool to graphical software for inputting specifications, or to simulators, implementation compilers or additional proof tools such as symbolic model checkers. In particular, we have not generated an implementation in hardware or software for the production cell controller, or animated the specification with the simulator. Successful links between the HOL system and such tools have been demonstrated in other work [1].

15.3 Specification

S1: Identify Communicating Processes

The production cell is a reactive system comprising a controller and its environment. For this case study, we have chosen to model the controller by eight concurrent, communicating processes. The environment is also modelled as a communicating process. We write [Init](C ‖ E) to denote a system whose initial state is characterised by the predicate Init and whose behaviour is given by the concurrent execution of processes C and E. The production cell is specified by the system ProdCell where,

$$ProdCell = [Init](Controller ‖ Env)$$
$$Controller = Rotate ‖ Arm1 ‖ Arm2 ‖ Press ‖$$
$$FBelt ‖ ERT ‖ DBelt ‖ Crane$$

S2: HTTD Processes

Each of the processes of the controller has been described by a hierarchical timed transition diagram (HTTD). An HTTD is a graphical representation for behaviours such as sequences of actions, the non-deterministic choice between actions, repeated actions and so on. We shall illustrate the HTTD notation with an example: the behaviour of the robot's right arm.

Our specification of the robot's arms is derived from the task description and the simulator provided. Once the right arm is pointing towards the elevating rotary table, it unloads the table by turning its hand magnet on, extending to the table, picking up a part when it becomes available, and finally retracting the arm. The arm process then waits until the loaded arm is pointing in the direction of the press and the press is ready for loading. It extends its arm to the press, deposits its part and retracts. When the robot and table are again ready for unloading, this sequence is repeated.

The sequence of actions we have just described is represented by the HTTD of Figure 1. Labelled circles in the diagram are states of the process, each identified by the process name (Arm1) and location number. The starred location (1*) is the initial state of the process. Each labelled double arrow between states (a hierarchical transition - HT) represents one or more timed transitions (TTs). The label on each arrow is a name identifying its set of TTs. Timed transitions will be defined shortly, but first we describe the states of the HTTD for Arm1.

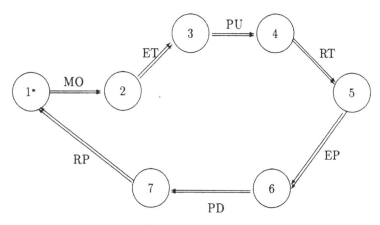

MO	=	once pointing to table, turn hand magnet on
ET	=	extend to table
PU	=	when table ready, pickup part
RT	=	retract arm
EP	=	when pointing to press, extend arm
PD	=	put down part
RP	=	retract arm

Figure 1   HTTD Specification for Arm1

S3: Process Interfaces

The states of Arm1 form part of the global states for ProdCell. State variables may be classified as sensors, actuators, exported or imported variables. Sensors and actuators read and write values between the controller processes and the environment. Exported and imported shared variables, each of which may be read-only or read-write, are used for communication between controller processes.

Figure 2 summarizes the interface for the process Arm1. Arm1 has two actuators: its own motor a1mot, which may be on for extension (push) or retraction (pull) or off (stop), and the hand magnet h1, which may be turned on (magon) or off (magoff). The environment reacts to these actuators by moving the arm in the appropriate direction or stopping it, and by activating or deactivating an electromagnet at the end of the arm which picks up or drops parts.

Actuators	a1mot, h1, rmot
Sensors	a1pos, rpos, tpos, trot, ppos
Export R-O	h1load
Import R-W	tload, pload
Constants	table_ext, press1_ext, extension_tol, table_rot, table_tol, press1_rot, press1_tol, at_arm, to_arm

Figure 2   Interface Variables and Constants for Arm1

The sensors of Arm1 return the current positions of five objects of the environment: arm1's extension distance, the robot's angle of rotation, the height and rotation of the elevating rotary table, and the position of the press. Arm1 also reads the actuator, rmot, which belongs to the rotation process. Like the arm motor, rmot may be pushing, pulling or stopped. Each of the state variables h1load, tload and pload describe whether the hand, table and press are carrying a part (full) or not (empty).

In order to decide when to turn its motor on and off, Arm1 makes use of a number of numerical constants. The position of the arm is compared with the distance it must be extended to reach the table (table_ext) or press (press1_ext), within a tolerance (extension_tol) which defines the depth radius of the loading platform. Using tolerances also makes an implementation more robust to different sensing frequencies. The current rotation of the robot is compared with degrees of rotation needed for arm1 to be pointing at the table (table_rot) or the press (press1_rot), within a tolerance for each (table_tol, press1_tol) which describes the width radius of the loading platform. For unloading, the elevating rotary table must be at arm height (at_arm) and rotated towards the arm (to_arm).

S4: Refining HTTD Processes

Given an HTTD description of the behaviour of Arm1 and having identified the arm's interface, the behaviour of this process can now be described in more detail. Each HTTD is simply a graphical shorthand for an underlying graph of timed transitions (TTs) [5][6]. For example, the seven HTs of Arm1 (see Figure 1) abbreviate eleven TTs (see Appendix). We use the term HTTD to refer to both the shorthand and fully expanded versions of a specification diagram, or a diagram mixing both HTs and TTs.

The HT from location 2 to 3, whose expansion is shown in Figure 3, specifies the extension of the robot's right arm to the table. The extension is effected by turning the motor on and then, once the arm is within a tolerance of the table, off again.

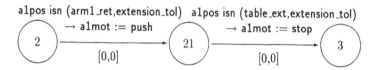

Figure 3   TTs to extend arm1 to the table

Each labelled single arrow in Figure 3 with its input and output locations is called a timed transition (TT). A timed transition has five parts: an input and output location, an enabling condition, an action and a timing constraint. A transition is written c → a. When the enabling condition, c, is true the transition is said to be enabled. For example, the enabling condition of the first TT of Figure 3 is that arm1's position is near arm1_ret. The predicate x isn (m,n) means m-n ≤ x ≤ m+n. The action, a, specifies the value of variables in the state reached by performing the transition. For example, the action a1mot := push on the first TT of Figure 3 means that when state 21 is first reached the value of the variable a1mot is push. It is assumed that variables remain unchanged across a transition unless explicitly changed by an action. With each transition is associated a timing constraint, [l,u], comprising a lower bound l and an upper bound u. The transition can only be taken after it has been continuously enabled for l time units and must be taken once it has been enabled for u time units. The timing constraint [0,0] on each of the transitions above means that the action is taken as soon as its enabling condition becomes true. In this specification none of the other production cell HTTDs can falsify these enabling conditions before the transitions can be taken.

Not all HTs represent a sequence of TTs. The pickup HT is shown in Figure 4. Its timing constraint [1,1] means that the enabling condition must hold for one time unit before the transition can be taken. That is it takes one time unit after extension to pick up a part. A constraint such as [1,3] could be used where the time taken is at least one time unit, but no more than three time units.

The fully expanded HTTD for the Arm1 process is given in the appendix.

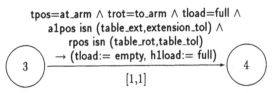

Figure 4   TT to pick up a part from the table

Behaviour of the Controller

The production cell is specified as a set of communicating processes, each of which is an HTTD, together with initial conditions that must be satisfied by the first state of the system. We call the specification an HTTD system. In the HTTD system ProdCell the initial condition is a large predicate which includes the conjunct, (Arm1=1) ∧ (a1mot=stop) ∧ (h1=magoff) ∧ (h1load=empty). Arm1=1 means that the process Arm1 starts execution in location 1. Thereafter, the system progresses from one state to the next by taking any transition that is enabled and has been waiting for at least its lower bound. That is, concurrency is modelled by the interleaving of transitions from all the processes in a system, in this case the controller processes and the environment. If no transition can be taken, then time advances by one step. There may, of course, be more than one transition that can be taken, in which case one is chosen non-deterministically.

15.4 Verification

A requirement is satisfied by a particular HTTD system if all allowable behaviours of that system satisfy the requirement. There are certain constraints on behaviours besides those dictated by the initial condition and the transitions themselves [5]. For example, to avoid ambiguity about the times of state changes, time and state may not change together. Other constraints ensure fairness and progress.

We write S ⊨ R when an HTTD system S satisfies a requirement R, and ⊢ S ⊨ R when we have proved satisfaction. There are logically sound proof rules for verifying a class of requirements stated in real-time temporal logic (RTTL) [5][6]. This class of requirements includes bounded invariance and bounded response properties, action conditions, and system invariance. Examples of requirements from these classes and their representation in RTTL are given below. All of the requirements given in the task description can be expressed within this class, as well as some extra requirements about the liveness of the production cell.

Having a set of suitable proof rules is not sufficient for the verification of the production cell because it is generally infeasible to check all the preconditions of a rule by hand. For example, a typical proof of the 50 or so required to verify the production cell, generates 70 preconditions. So proof programs have been written in the HOL system to perform checks automatically and sufficiently fast to support interactive proof. Although significant parts of proofs have been automated in the HOL-HTTD system, a human expert must still guide the proof by choosing invariants for some rules, identifying sequences of transitions which establish certain properties and so on.

The remainder of this section illustrates the use of our proof theory and HOL-HTTD tool to verify properties of the production cell.

Robot Never Drops Blanks on the Floor

The robot's right arm never drops blanks on the floor because whenever its right hand releases a load, by turning off the magnet, the hand is over the press' loading platform, and the platform is empty and in the middle position. Formally, we have proved,

$\vdash$ ProdCell $\models$

$\square$ (h1=magon $\wedge$ $\bigcirc$ h1=magoff) $\Rightarrow$

((rpos isn (press1_rot,press_tol)) $\wedge$
(a1pos isn (press1_ext,extension_tol)) $\wedge$
(pload = empty) $\wedge$ (ppos = middle))))

The RTTL requirement reads "it is always true that if (h1=magon) now and (h1=magoff) in the next state ($\bigcirc$ is the next state operator), then rpos is near the press rotation for arm1, a1pos is near the press extension, and the press is unloaded and in its middle position".

This property can be proved by the ActionOnlyIf rule which states,

- For any HTTD system S, and temporal predicates A and B, IF there is a set of timed transitions ActS from S such that,

  1. Taking any transition of ActS implies A,
  2. Taking any transition except those of ActS implies not A,
  3. For every transition of ActS, if the transition is enabled then B is true,

- THEN S satisfies A $\Rightarrow$ B at all times (S $\models$ $\square$ A $\Rightarrow$ B).

To perform this proof with the HOL-HTTD tool, the user calls a function which returns a proved theorem given - the goal to be proved, the action A, the condition B, and the definition of the production cell HTTD system. The set ActS is found automatically. To prove that Arm1 never drops blanks, we have used A = (h1=magon $\land$ $\bigcirc$ h1=magoff), and B the conclusion shown above. ActS is found to be the set containing only the Arm1 transition from location 6 to 7 which releases a part. The proof function must effectively check that for every other transition of ProdCell (there are approximately 70), that taking that transition implies $\neg$ A. However, the transitions of a process need not be checked individually when that process never changes the variables of A. For example, the Press process never changes h1 and so taking any transition of the press implies $\neg$A.

The ActionOnlyIf rule, although simple, is sufficient to prove nearly three quarters of the requirements stated in the task description from "the robot must not be rotated clockwise if arm1 points towards the minimum rotation needed to reach each part" to "do not put blanks on the table or press if they are already loaded".

Real-Time Liveness

Although liveness requirements are not treated in any detail in the task description, there are many requirements of interest for the production cell, in particular those which identify upper bounds on the time taken for certain sequences of actions.

We have proved, for example, that if arm1 is about to unload the table, then eventually within a given time the press will be loaded. In RTTL $\Diamond_n$ means eventually within n time steps. The sequence of actions which establish this property are (time taken given in brackets): unload the table (1), retract arm (TR = (table_ext-arm1_ret)*Sarm), rotate to press (TPR = (press1_rot-table_rot)*Srot), extend arm to press (PE = (press1_ext-arm1_ret)*Sarm), and finally load the press (1). Formally, we have proved,

$$\vdash \text{ ProdCell} \models \Box \text{ arm1\_unloads\_table} \Rightarrow$$
$$\Diamond_{1+TR+TPR+PE+1} \text{ arm1\_has\_loaded\_press}$$

The predicates arm1_unloads_table and so on abbreviate a conjunction of statements on the values of state variables and locations of processes. The full predicates for this theorem and those below are listed in the appendix.

To simplify the proof, we have assumed the press is ready to be loaded and thus each transition in the sequence above is enabled as soon as its input location is reached. More complex properties can be proved by the method shown here combined with case analysis.

Three assumptions have been made about the behaviour of the environment. Each motor move is assumed to take a time bounded above by the distance travelled multiplied by the speed of the motor (Srot, Sarm). For example, we assume the theorem,

⊢ ProdCell ⊨ □ arm1_starts_extending_to_press ⇒
$$\Diamond_{PE} \text{ arm1\_reaches\_press}$$

The proof of real-time liveness uses transitivity of the $\Diamond_n$ operator, three environment motor assumptions, and eleven bounded response lemmas for particular transitions. The bounded response proof rule states,

- For any HTTD system S, which contains a transition τ with upper time bound u, IF for the temporal predicates p and q,

  1. p implies that τ is enabled

  2. Taking any transition of S *except* τ maintains p. That is, if p is true and the transition is taken then in the next state p is true.

  3. If p is true and τ is taken then in the next state q is true.

- THEN at all times S satisfies that p implies eventually q within u time steps $(S \vDash \Box p \Rightarrow \Diamond_u q)$.

To perform this proof in the HOL-HTTD system the user assumes the environment theorems, and identifies the 11 bounded response lemmas. Each of these is proved automatically by a function similar to the ActionOnlyIf rule function. Another function automatically applies transitivity laws to prove the main liveness theorem. The user interaction required to identify the individual lemmas is tedious and work is in progress to automate this task too.

Other Requirements

There is not space in this paper to describe more verification examples in detail. Instead the two remaining classes of proof rule are summarized, and the reader is referred to a technical report [3].

The bounded invariance rule is used to verify that after arm1 has loaded the press then it is either retracted, or pointing towards the table until such time as it extends to load the press again. A similar proof shows that after the press begins to close it is either moving or in its upper or lower position until it is ready to be loaded by arm1. Together with corresponding properties for arm2, these lemmas can be used to show that the the press never collides with the robot's arms. Bounded invariance proofs are constructed from bounded invariance lemmas for particular transitions, assumptions about the environment, and temporal transitivity, in the same way as for the bounded response proof described above.

The system invariance rule is used to prove properties which always hold during the execution of an HTTD system. Such properties are often required as lemmas which identify predicates with certain locations of a process. For example, it is always true that either, Arm1 is in locations 1, 2 or 5 and its arm is retracted, or Arm1 is in locations 21, 3, 4 or 41 and its arm motor is on or arm extended, and the robot's rotation is with arm1 towards the table, or Arm1 is in locations 51, 6, 7 or 71 and its arm motor is on or arm extended, and the robot's rotation is with arm1 towards the press.

15.5 Results

The HOL-HTTD specification of the production cell controller consists of 650 lines of commented HOL-readable code. The HTTD for the controller contains about 70 timed transitions and uses 34 state variables and 40 system constants. The state space of this specification for a particular instantiation of the geometry and component speeds has in the order of 10^{24} states. Our specification is, thus, at or beyond the limits of current model checking and symbolic model checking technology (see StateCharts case study [8]).

Applying a typical HTTD proof rule in HOL takes about 35 seconds of run time, 50 seconds garbage collection, and generates around 6,000 intermediate theorems. The elapsed time at the terminal is 1 to 2 minutes. A typical proof of this type requires the proof of 350 (70 times, say, 5 conjuncts in the main predicate) subgoals of the form $ProdCell \models x=X \land Taken\ \tau \Rightarrow \bigcirc x=Y$. For theorems involving transitivity over a sequence of transitions these figures should be multiplied by the number of transition lemmas in the sequence. For example, the liveness theorem discussed in Section 15.4 required 11 bounded response lemmas, and used 3 environment assumptions. It takes around 25 minutes elapsed time to perform this proof in HOL, given a list of lemmas to be proved. HOL maintains a theory file of proved theorems which can be reused, and so no lemma or theorem need be proved more than once.

We found that writing a generalized specification, and performing verification proofs in HOL, uncovered errors or ambiguities which were unlikely to have been discovered by a specification and simulator animation alone. For example, in the course of verification it was necessary to make explicit assumptions about the environment (e.g. when a motor is turned on it will eventually reach its destination within a bounded time). Also, in order to verify that blanks are never dropped by the belts it was necessary to assume that the feed and deposit belts are able to accomodate at least two blanks.

Generalizing the task description identified some problems. The strategy of extending arms during rotation, and stopping motors if there is danger of a collision, is not safe in a generalized specification, although it works for the simulator settings. In general, an arm must be able not only to stop, but also withdraw again if it has extended too far before passing an obstacle on the way to its destination. This could occur if, for example, the press could close faster than arm2 deposits, perhaps because of a slow crane process. The problem is avoided in our specification by requiring the arms to retract to a safe distance before rotation, and extend afterwards, as stated in the task description. This solution works for all geometries and component speeds. The ProdCell specification also removes unnecessary rotation restrictions suggested in the task description: both the rotation sequences (table,press1,press2,deposit) and (press1,table,press2,deposit) are used.

From the simulation[1], we found that we had made a number of incorrect assumptions in our specification: the robot's arms need not extend from its centre of rotation; detecting the exact position of the robot's arms or rotation when it is moving may not be possible; each loading platform has depth and width so you don't know exactly where a part is located; and to pick up an object, the hand magnet should be turned on before the arm reaches the loading platform.

Approximately two person months has been spent on this case study. Most of this time was spent, not on the production cell itself, but on developing proof support in HOL, and identifying a specification style which we believe will lead to faithful implementations. The verification of all the requirements in the task description has not been completed in HOL although proofs for each class of requirements given have been performed formally. With hindsight, and our current tool, we now estimate that for a similar problem it would take up to four person weeks to write, perhaps rewrite, and document a specification and to verify it completely using our HOL-HTTD theory.

15.6 Conclusions

The idea of using HTTDs and real-time was to make our specification as realistic as possible, thus increasing the probability that properties verified of the specification also hold for an implementation. HTTD specifications are easy to read be-

---

1. Roger Hale from SRI International, identified these incorrect assumptions in this specification, based on his experience using a Tempura program version of an interval temporal logic specification of the production cell to drive the simulator.

cause they are well structured and look similar to imperative programs. This should encourage the evaluation of a specification by different groups of people - implementors, designers, customers and so on - increasing the chances of detecting differences between the view of the formal development team and those who will build or use an implementation.

Writing a generalized specification, rather than one for a specific implementation environment, makes the specification and its eventual implementation more robust because the specifier must examine assumptions which might otherwise be overlooked. The time spent by the specifier to develop a general specification is offset by the specification being re-usable.

As verification by deductive proof in HOL is expensive in the time taken to discover and perform proofs, developers should make good use of animation and simulation to give fast feedback on specification errors. Our HOL-HTTD tool demonstrates that the cost of formal verification can be reduced substantially by proof automation, and we plan to build on this in future work. It is an open question whether we can automate sufficient of the proof effort to make feasible the verification of specifications much larger than the production cell of this case study.

References

[1]   Andersen, F., Petersen, K.D., Pettersson, J.S., A Graphical Tool for Proving Progress To appear in *Proceedings of the 7th International Conference on Higher Order Logic Theorem Proving and Its Applications*, Springer Verlag.

[2]   Church, A. A Formulation of the Simple Theory of Types. *Journal of Symbolic Logic*, 5, 1940.

[3]   Cardell-Oliver, R.M. *A Case Study Using Timed Transition Diagrams and the HOL theorem prover for the Formal Development of a Production Cell*, University of Essex Technical Report, In preparation.

[4]   Gordon, M.J.C., Melham, T.F. *Introduction to the HOL System*, Cambridge University Press, March 1994

[5]   Hale, R. W. S., Cardell-Oliver, R. M., Herbert, J. M. J. An Embedding of Timed Transition Systems in HOL. *Formal Methods in System Design*, 3(1&2),pages 151-174, Kluwer, September 1993

[6]   Henzinger, T. A., Manna, Z. and Pnueli, A. Temporal proof methodologies for real-time systems. In *Proceedings of the 18th Symposium on Principles of Programming Languages*. ACM Press, 1991.

[7] Kesten, Y. and Pnueli, A. Timed and Hybrid Statecharts and their Textual Representation. In *Lecture Notes in Computer Science, number 571*. Springer-Verlag, 1992.

[8] Lewerentz, C. Lindner, T. (eds), *Case Study "Production Cell" A Comparative Study in Formal Software Development*, FZI-Publication 1/94, 1994.

[9] SRI International and DSTO Australia. *The HOL System*. Cambridge Computer Science Research Center, SRI International, Revised version, 1991. Four volumes.

Appendix

The Expanded HTTD for Arm1

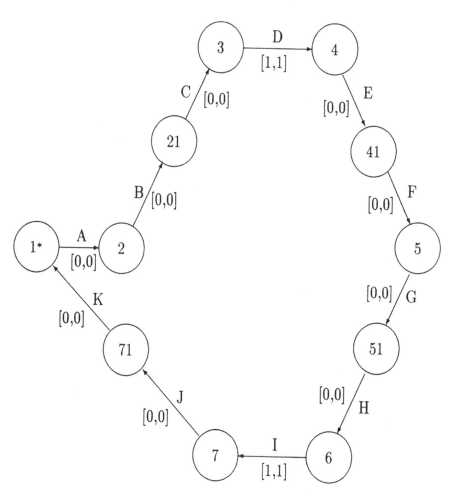

A = rpos isn (table_rot,table_tol) ∧ rmot=stop ∧

      tload=full ∧ h1load=empty ∧ h1=magoff

      → h1:=magon

B = a1pos isn (arm1_ret,extension_tol) → a1mot:=push

C = a1pos isn (table_ext,extension_tol) → a1mot:=stop

D = a1pos isn (table_ext,extension_tol) ∧ rpos isn (table_rot,table_tol) ∧

      tpos=at_arm ∧ trot=to_arm ∧ tload=full

      → h1load:=full, tload:=empty

E = a1pos isn (table_ext,extension_tol) → a1mot:=pull

F = a1pos isn (arm1_ret,extension_tol) → a1mot:=stop

G = rpos isn (press1_rot,press_tol) ∧ rmot=stop ∧

      ppos=middle ∧ pload=empty ∧ h1load=full

      → a1mot:=push

H = a1pos isn (press1_ext,extension_tol) → a1mot:=stop

I = a1pos isn (press1_ext,extension_tol) ∧

      rpos isn (press1_rot,press_tol) ∧

      pload=empty ∧ h1=magon

      → h1:=magoff, h1load:=empty, pload:=full

J = a1pos isn (press1_ext,extension_tol) → a1mot:=pull

K = a1pos isn (arm1_ret,extension_tol) → a1mot:=stop

Predicate Abbreviations used to Verify Real-Time Liveness

arm1_unloads_table =

      Arm1=3 ∧ Arm2=1 ∧ Press=1 ∧ Rotate=1 ∧

      a1pos isn (table_ext,extension_tol) ∧ pload=empty ∧

      rpos isn (table_rot,table_tol) ∧ tload=full ∧

      tpos=at_arm ∧ trot=to_arm

arm1_has_loaded_press =

      Arm1=7 ∧ Arm2=1 ∧ Press=1 ∧ Rotate=2 ∧

      a1pos isn (press1_ext,extension_tol) ∧

      rpos isn (press1_rot,press_tol) ∧

      h1load=empty ∧ pload=full ∧ ppos=middle

arm1_starts_extending_to_press =

      Arm1=51 ∧ Arm2=1 ∧ Press=1 ∧ Rotate=2 ∧

      h1load=full ∧ pload=empty ∧

      rpos isn (press1_rot,press_tol) ∧

      a1pos isn (arm1_ret,extension_tol)

arm1_reaches_press =

      Arm1=51 ∧ Arm2=1 ∧ Press=1 ∧ Rotate=2 ∧

      h1load=full ∧ pload=empty ∧

      rpos isn (press1_rot,press_tol) ∧

      a1pos isn (press1_ext,extension_tol)

XVI. RAISE

A Rigorous Approach Using Stepwise Refinement

François Erasmy, Emil Sekerinski

Abstract

We present the process-oriented RAISE contribution, beginning with a short presentation of the RAISE method, tools, and specification language. Then we show how the production cell software is developed using a rigorous approach with successive refinements of the specification. We present extracts from the production cell specifications. Finally, we give an evaluation of the RAISE contribution: the properties that could be specified and proved, the assumptions made, the flexibility of the specifications and proofs and some measures concerning them.

16.1 Goal of the RAISE Contribution

The main goal of this contribution is to show a simple algebraic and process-oriented specification and verification approach. Specifications are developed in a top down fashion using a refinement strategy. This approach leads to compact specifications and simple proofs. Our basic, intuitive idea is that each machine is controlled by its own process, and all these processes run in parallel. Each process communicates with the processes controlling the neighbouring machines and with its own sensors and actuators. In this way, the structure of the production cell is reflected by the structure of the controlling processes: we get a modular, extensible and reusable set of control components.

Here, we focus on the methodological aspects of this approach. We do not aim at identifying and verifying all safety and liveness conditions. Particularly, we have the following goals:

- to specify the production cell system in a process-oriented style,

- to apply the development paradigm of stepwise refinement to the production cell example,

- to prove safety properties of the system at a high level of abstraction (implying compact specifications and, therefore, smaller proofs),

- to have a minimum of effort to prove the safety properties of the implementation (by proving at each refinement step only the development relation between the two successive refinement levels),

- to evaluate the use of the RAISE method, language and tools in this example.

16.2 Introduction to RAISE

RAISE stands for *Rigorous Approach to Industrial Software Engineering*. RAISE is a software development method that comprises a specification language (RSL) and tools supporting the development method.

Initially, RAISE was the name of an industrial research and development project inside the ESPRIT programme. Its goal was to develop notations, techniques and tools to enable the usage of formal methods in industrial software development. This first step was carried out during the years 1985 through 1990. Now, in a second step, the ESPRIT project LaCoS (Large scale Correct Software using formal methods) helps to refine the RAISE technology. This phase is planned to last from 1990 to 1995.

The development method

The development method[1] is characterized by *stepwise refinement* of the specification (until an implementation is reached) and by *rigorous proof* of the development relation (e. g. an implementation relation) between two consecutive refinement steps. Rigorous means: it should be a formal proof, but it might also be some thorough informal argument. Therefore, RAISE uses the term of *justification*

---

1. See [1] for a full description of the RAISE development method.

instead of 'proof': a justification is a proof that may contain informal parts. The RAISE toolbox provides a tool to support rigorous proving: the justification editor.

The specification language

RSL (RAISE Specification Language) is a VDM-like (see [5]) specification language with many enhancements, the main additions being modularity, concurrency, and algebraic specifications (see [7] for a RSL tutorial). RSL is a wide-spectrum language, so it is usable at all development stages. It allows algebraic and model-oriented, applicative and imperative, sequential and concurrent specifications. RSL has mathematical semantics and comes along with proof rules.

A RSL specification consists of module definitions. A module contains definitions of types (sets of values), channels, values (constants and functions), variables, axioms and modules. Modules are either *schemes* or *objects*. A scheme denotes a class of models, whereas an object denotes a single model.

For specification of concurrency, RSL provides the following CSP-like (see [4]) constructs:

- typed channel definitions: **channel** ch: Type
- communication primitives for
  - input from channel ch: ch ?
  - output on channel ch: ch ! expr
- concurrent functions (accessing the enumerated channels):
  value f: $T_1 \rightarrow$ **in** ch_1 **out** ch_2 T_2
- the sequential combinator: $expr_1$; $expr_2$
- the parallel combinator: $expr_1$ ‖ $expr_2$
- the interlock combinator: $expr_1$ ⫲ $expr_2$
  This is a a special concurrent combinator. To explain its meaning, let exp_1 and exp_2 be expressions. Then exp_1 ⫲ exp_2 is evaluated by evaluating exp_1 and exp_2 concurrently and allowing them only to communicate with one another (i. e. no other process can interfere)[1].

---

1. It can easily be seen that the interlock combinator is not associative: compare
$$(ch1 ! () ⫲ ch1?) ⫲ ch2 ! ()$$
which can communicate through channel ch1 and
$$ch1 ! () ⫲ (ch1? ⫲ ch2 ! ())$$
which produces a deadlock. Therefore, the evaluation of an expression of the form **exp1** ⫲ **exp2** ⫲ **exp3** is done from the left to the right.

The proof rules

RAISE provides proof rules for the RSL constructs. They may be used in justifications (see [2]). There are two types of rules: equivalence rules and inference rules[1].

A simple example of an equivalence rule[2] is the rule [is_annihilation] for the equivalence operator '≡':

[is_annihilation]
  expr ≡ expr ≅ **true**

It says that an expression is equivalent to itself i. e. the '≡' operator applied to the same expressions on both sides may be reduced to **true**.

The following two equivalence rules will be use later:

[interlock_communication4]
  ch ! () ; expr ╫ ch? ≅ expr

[interlock_communication11]
  ch? ; expr ╫ ch ! () ≅ expr

They both deal with the ability of the interlock combinator to communicate. They provide rules for the special case when only signals are exchanged and the right hand side of the interlock combinator is a single communication expression. The rule [interlock_communication4] says: if the expressions ch ! () ; expr and ch? are interlocked, then they will communicate and only expr will be left.

The toolbox

As shown in fig. 1, the RAISE toolbox contains the following sets of tools:

- library tools for listing, deleting from, showing dependencies within and propagating changes to the library containing the RAISE entities,

- editors for the different RAISE documents or entities (modules, theories, development relations and justifications),

- the justification editor supporting proofs of theories and development relations,

- translators to generate Ada and C++ code (with certain restrictions)

- the pretty printer producing the LaTeX representation corresponding to a given RAISE document.

---

1. As the later will not be used in this presentation, we will not explain them in more detail.
2. Note that the symbol '≅' separates the left side of an equivalence rule from the right side.

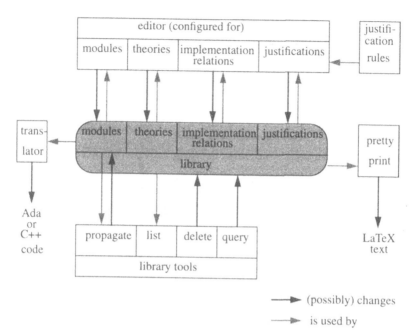

Figure 1   The RAISE tool box

16.3 Development with RAISE

In our project the stepwise refinement process yielded the following (intermediate) results (cf. fig. 2):

1. description of the machine states (ProductionCell_1),

2. specification of the safety conditions with CSP-like RSL constructs by extension of ProductionCell_1 (ProductionCell_2),

3. axiomatic specification of the control processes by refinement of ProductionCell_2 (ProductionCell_3),

4. proof that ProductionCell_3 verifies the safety conditions defined in ProductionCell_2,

5. explicit recursive definition of the control processes (refinement of ProductionCell_3) (ProductionCell_4),

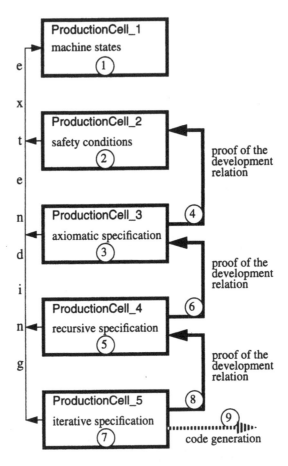

Figure 2 The development process with RAISE

6. formal proof that ProductionCell_4 is an implementation of ProductionCell_3 (proof of the *implementation relation*),

7. iterative specification of the machine control processes (ProductionCell_5),

8. rigorous proof that ProductionCell_5 is an implementation of ProductionCell_4,

9. generation of executable code by the RAISE translator tool (as soon as the RAISE tools allow to).

This section presents these nine steps for a simplified version of the production cell. It consists of the robot, the press and two simple machines feeding the robot with items resp. allowing the robot to placethe pressed items.

16.3.1 Specification of the machine states

Our first development step is to identify the states of the different machines in the production cell and to define their state transition functions. We use state transition functions to define explicitly which sequences of state changes are allowed and that the state transitions are deterministic.

```
scheme ProductionCell_1 =
  class
    type
      /* State definitions for the robot. We abstract from the movement. */
      StateRobot ==
        arm1_over_table | arm1_loaded_over_table |
        arm1_loaded_near_table | arm2_near_press |
        arm2_in_press | arm2_loaded_in_press |
        arm2_loaded_near_press | arm1_loaded_near_press |
        arm1_loaded_in_press | arm1_in_press |
        arm1_near_press | arm2_loaded_near_dbelt |
        arm2_loaded_over_dbelt | arm2_over_dbelt |
        arm2_near_dbelt | arm1_near_table

    type
      /* State definition for the press. */
      StatePress == down_loaded | down | middle | middle_loaded | up

    value
      /* State transition function for the robot states. */
      ...

    value
      /* State transition function for the press states. */
      δ_press : StatePress →ₘ StatePress =
        [
          down_loaded |→ down,
          down |→ middle,
          middle |→ middle_loaded,
          middle_loaded |→ up,
          up |→ down_loaded
        ]

  end
```

Figure 3   ProductionCell_1 *(extracts)*

We specify the states and state transition functions in a module (*scheme* in RAISE terminology) called ProductionCell_1. It defines the production cell machine states in an algebraic manner. Abstract types such as StatePress (cf. fig. 4) are given as enumeration and state transition functions (such as δ_press) are given by maps.

Remarks:

— The notation T == a | b | c | ... is an abbreviation for: 'the type T is generated by the constants a, b, c, ...'.

— δ_press is declared as a map ($\rightarrow_m$) from StatePress to StatePress. Its value is defined by pairs like (down_loaded |→ down) .

16.3.2 Specification of the safety conditions

In ProductionCell_2 (cf. fig. 4) we define

- one control process for each production cell machine,
- channels through which they communicate and
- axioms describing the safety conditions that the machines must meet.

Nearly all of these safety conditions are used to ensure that the press does not damage the robot (by colliding with one of its arms) and vice versa.

ProductionCell_2 refines ProductionCell_1 by extending it. The class corresponding to the scheme ProductionCell_2 consists of all declarations of ProductionCell_1 and the additional declarations of the class expression following the **with**-clause (cf. fig. 4).

In our specification of the production cell, channels are only used to send signals from one control process to another. No other information is needed[1].

We are using the following naming convention for signals: signal names are composed from the name of the sending process concatenated with the signal name and the receiver name.

As already mentioned in section 16.1, we define one communicating process for each machine that has to be controlled. This models the intuitive concept of machines controlling themselves and interacting through defined interfaces (here through channels) with their environment. This modularization also facilitates extension or reuse of the specification.

In this process-oriented approach the safety conditions are expressed in the local context of the particular processes, i.e. machines.

---

1. Therefore, we define the channels to be of the pre-defined RSL type **Unit**. The type **Unit** has one single element called (). The type **Unit** is also used to give a type to void function domains or co-domains in a value declaration (as for the function Robot).

scheme ProductionCell_2 =
  extend ProductionCell_1 **with**
    class
        ...
      channel
        RobotTakenTable, RobotTakenPress, RobotLoadedPress,
        RobotPlacedDBelt
        :
      Unit
      channel PressAvailRobot, PressFreeRobot : **Unit**
      value
        /* Process controling the production cell. */
        ProductionCell : **Unit** → **in** ... **out** ... **Unit**,
        /* Process controling the robot. */
        Robot : StateRobot → **in** ... **out** ... **Unit**,
        /* Process controling the press. */
        Press : StatePress → **in** ... **out** ... **Unit**,
        /* Interface to the production cell hardware. */
        PCInterface : **Unit** → **in** ... **out** ... **Unit**,
        ...
      axiom
        /* Safety conditions for the robot and the press. */
        [safety_robot_press1]
          Press(down_loaded) ‖ PressAvailRobot?
          ‖
          RobotTakenPress ! () ≡
            Press(down)
        [safety_robot_press2]
          ∀sp : StatePress •
            sp ∈ {down, middle, middle_loaded, up} ⇒
            (Press(sp) ‖ PressAvailRobot? ≡ **stop**),
        ...
    end

Figure 4 ProductionCell_2 *(extracts)*

As an example we specify that the press only moves up to the middle after the robot arm number 2 is out of danger. To formalize this we split the condition in one condition for the press and one for the robot, using their respective interfaces. Here, we use the channels PressAvailRobot and RobotTakenPress provided at the interfaces. Two conditions have to hold:

- The press only sends the signal PressAvailRobot in the state down_loaded and, after sending PressAvailRobot, it accepts the signal RobotTakenPress.

- The robot only accepts the signal PressAvailRobot in the state arm2_near_press and only sends a signal RobotTakenPress in the state arm2_loaded_near_press (that means that arm 2 is safe outside the press).

To express the first condition, we define two axioms for the press:

- Axiom [safety_robot_press1] states: when the press is in the state down_loaded, it sends the signal PressAvailRobot and then accepts the signal RobotTakenPress[1].

- Axiom [safety_robot_press2] states: if the press is not in the state down_loaded, it does not send the signal PressAvailRobot.

To express the second condition, we define two more axioms for the robot to ensure that

- the robot accepts the signal PressAvailRobot *iff* it is in the state arm2_near_press and

- the robot sends a signal RobotTakenPress *iff* it is in the state arm2_loaded_near_press.

Together these axioms specify that the press only moves up to the middle position after the robot arm number 2 is out of danger!

16.3.3 The axiomatic specification

In ProductionCell_3 we give an axiomatic specification of the control processes; we declare an axiom for each possible state of each machine.

As an example we show two of the five axioms for the press in fig. 5.

```
axiom
    /* Axiomatic definition of the Press process. */
    [Press1]
        Press(down_loaded) ≡
            PressAvailRobot ! () ; RobotTakenPress? ; Press(down),
    [Press2]
        Press(down) ≡ PressMovePCl ! middle ; PClReadyPress? ; Press(middle),
    ...
```

Figure 5 ProductionCell_3: *axiomatic definition of the press (extracts)*

---

1. The axiom additionally expresses: if the press is in the state down_loaded and the above mentioned communications take place, then the press changes its state to down

16.3.4 The first implementation relation

As we want ProductionCell_3 to be a refinement of ProductionCell_2, we have to define the implementation relation ProductionCell_3 ≤ ProductionCell_2 (i. e. ProductionCell_3 implements ProductionCell_2) and prove this relation[1].

For this purpose we use the RAISE justification editor. It helps us by splitting the proof of the implementation relation into several easier (sub-)proofs. Most of these new proofs serve to verify that the axiomatic specifications of the control processes meet the safety conditions.

As an example for a RAISE justification, we prove (see fig. 6) that the axiomatic definition of the press satisfies the axiom [safety_robot_press1] (cf. fig. 5).

 Press(down_loaded) ‖ PressAvailRobot? ‖ RobotTakenPress ! () ≡ Press(down)
Press1
 PressAvailRobot ! () ; RobotTakenPress?; Press(down) ‖ PressAvailRobot?
  ‖
  RobotTakenPress ! () ≡
   Press(down)
interlock_communication4
 RobotTakenPress ?; Press(down) ‖ RobotTakenPress ! () ≡ Press(down)
interlock_communication11
 Press(down) ≡ Press(down)
is_annihilation
 true
qed

Figure 6 A Simple RAISE Justification

We expand Press(down_loaded) in the first line of the proof with the axiom [Press1] from ProductionCell_3 (see fig. 5). Then we apply the communication rules interlock_communication4 and interlock_communication11. Finally we eliminate the equivalence (≡) with the rule is_annihilation.

16.3.5 The recursive specification

We derive a recursive specification of the control processes as shown in ProductionCell_4 in a rigorous manner from ProductionCell_3 by applying the axioms of ProductionCell_3 and by eliminating the machine state.

Take for example the press. We define the new control process Press() by

---

1. the implementation relation is a special case of the refinement relation

Press() ≡ Press(down_loaded).

With axiom [Press1] from ProductionCell_3 (see fig. 5) we get

Press() ≡ PressAvailRobot ! () ; RobotTakenPress? ; Press(down).

Then we apply the axioms [Press2], [Press3], [Press4] and the above definition of Press() and we finally get the following recursive definition of the press.

```
Press() ≡
    PressAvailRobot ! () ; RobotTakenPress? ;
    PressMovePCl ! middle ; PCIReadyPress? ;
    PressFreeRobot ! () ; RobotLoadedPress? ;
    PressMovePCl ! up ; PCIReadyPress? ;
    PressMovePCl ! down ; PCIReadyPress? ; Press()
```

Figure 7 ProductionCell_4: *recursive definition of the press*

This way, we have formally derived the recursive definition of the press control process from its axiomatic specification.

16.3.6 The second implementation relation

The new specification is correct by construction. However, since we invented the recursive definition with paper an pencil and not using the RAISE tools, this is not known to the RAISE system and we still must prove the implementation relation.

Basically, we have to prove that, if we abstract from the press and robot control processes, ProductionCell_4 implements ProductionCell_4, denoted in RSL by

hide Press, Robot **in** ProductionCell_4 ≤ **hide** Press, Robot **in** ProductionCell_3

The important fact is that the old and new control processes for the press and the robot behave the same way in their environment, i. e. inside the process ProductionCell().

16.3.7 The iterative specification

It is now very easy to transform the recursive specification ProductionCell_4 of the control processes into an iterative specification (ProductionCell_5, see fig. 5) with an infinite while loop. We simply apply a standard technique for recursion removal[1].

---

1. as done in [3], p.381

```
Press() ≡
    while true do
        PressAvailRobot ! () ; RobotTakenPress? ;
        PressMovePCl ! middle ; PCIReadyPress? ;
        PressFreeRobot ! () ; RobotLoadedPress? ;
        PressMovePCl ! up ; PCIReadyPress? ;
        PressMovePCl ! down ; PCIReadyPress?
    end
```

Figure 8 ProductionCell_5: *iterative definition of the press*

16.3.8 The third implementation relation

As the justification editor does not provide rules for recursion removal, we cannot prove formally the implementation relation between ProductionCell_4 and ProductionCell_5. However, we can provide a rigorous proof by 'manually' applying the recursion removal technique mentioned above. This means we write down the corresponding theorem in a comment of the justification and provide the result of its application to the justification editor. In the RAISE terminology, this is called a 'replacement argument' and is accommodated RAISE method of rigorous development.

16.3.9 Code generation

At present time, we cannot generate executable code, because the current RAISE toolbox is unable to translate the CSP-constructs (like the parallel and interlock combinators or the input and output expressions) into executable code. With future versions of the RAISE tools it will be possible to generate of C++ or Ada code. We are thinking about translating the iterative specification manually into OCCAM code.

16.4 Evaluation of the Refinement Approach with RAISE

After having described the development process of the production cell software, we want to evaluate our approach by discussing the following aspects:

- the properties of the production cell that could be formally described and verified,

- the assumptions made about the environment,

- the flexibility of the specifications and proofs,
- the size of our specifications and the effort for proofs.

Properties

The first property that should be guaranteed is *safety*. As shown in 16.3.2 we have defined safety conditions to make sure that the press does not damage the robot and vice versa. We have also proved that all following refinements of the specification satisfy the safety conditions (by proving that the implementation relation holds for each pair of successive specifications). Furthermore we have taken into account the worker's safety by forcing the robot to retract both of its arms before turning. So there is only a small dangerous area, that can be locked off.

The second property is *liveness*. It is possible to express and prove liveness of each machine separately, but we have been unable to express or prove the liveness of the production cell. This is due to the annoying fact that the justification editor lacks rules for the parallel combinator and its interaction with other combinators, rules for hiding schemes[1], and so on.

The last property is the *efficiency*[2] of the code produced by the development. This issue was not addressed. So we have no results concerning the possibility to express or prove such efficiency conditions in RAISE.

Assumptions

When writing a specification, you have to make (implicit or explicit) assumptions about the environment in which the software will work. In our case we suppose that the hardware interface procedures terminate. This implies that the hardware has no break-down. We can make this explicit by stating for each of these procedures the post-condition 'true'. The second assumption is that these procedures (which are not fully implemented) do what their names suggest. This assumption is implicit. In ProductionCell_2 we have also defined axioms that describe the liveness and the proper behaviour of the hardware interface process. These axioms are explicit assumptions about the behaviour of the hardware.

Flexibility

It is easy to extend our specifications to cover the whole production cell. We need simply to revise the specifications ProductionCell_1 to ProductionCell_5. We add the corresponding definitions for the crane, the feed belt, the deposit belt. These are

---

1. These are schemes that hide some of the declarations inside.
2. E. g. the number of items that can be simultaneously treated in the production cell.

similar to those of the other machines. For example in ProductionCell_1, we add the state definitions and the state transition functions for the new machines to the existing scheme. If it becomes necessary, it is possible to modularize the specifications, e. g. by describing each machine in its own module.

Unfortunately, the RAISE tools do not allow to propagate these changes to the justifications, so that we have to do the whole justification again, even those parts that are not affected by the changes. An escape from this dilemma is the possibility to make a rigorous justification. In an informal argument we explain that we have already proved parts of the new justifications for a previous version and, since the changes do not affect these parts, they still hold.

Measures

Table 1 shows the *length* of the specifications in lines of code[1]. Note that ProductionCell_2 to ProductionCell_5 are extensions of ProductionCell_1, so strictly calculating one must add 82 lines of code to their lengths.

Module	length [lines of code]
ProductionCell_1	82
ProductionCell_2	128
ProductionCell_3	203
ProductionCell_4	134
ProductionCell_5	129

Table 1   Specification Lengths

As to the measure of *effort*, the main difficulties were to specify the safety conditions in a process-oriented manner and to find a form for our specifications that facilitates the justifications.

The process-oriented specification of the safety conditions took us two days of intense work and resulted in the schemes ProductionCell_1 and ProductionCell_2. The reason for this difficulty was that we had no experience in process-oriented specification.

The second problem was to express the axioms the 'right way'. As we already mentioned, the justification editor does not provide rules for all RSL constructs or

---

1. We indicate the length of the listings formatted by the RAISE pretty printer.

combinations of RSL constructs. So if you use them, you are unable to make a fully formal justification. But this was exactly our intention for the proof of the first implementation relation. This led us to revise our specifications several times. However, with the experience gained from this project, this problem should only occur rarely in future projects.

To be complete, we still need to mention the measures concerning the justifications. The proof that the axiomatic specification verifies the safety conditions (i. e. the proof of the first implementation relation) is very straightforward, although it is very long. This is due to the fact that we have to do many similar sub-proofs. The time spent for a fully formal proof of the first implementation relation is about 4 hours work for a somehow experienced person.

16.5 Conclusion

The main outcomes of this work up to now are:

- specifications of the press, the robot and the production cell hardware interface processes at different abstraction levels,

- an implementation (ProductionCell_5) of the press, the robot and the production cell interface processes,

- rigorous proof that the implementation satisfies the safety conditions.

The specifications and proofs for the crane, the feed belt and the deposit belt still need to be done. As we have already mentioned in Section 16.4, the existing specifications are easily extended to comprise also the specifications of the new machines, but the existing justifications cannot simply be re-used or extended. We must prove all implementation relations again, even those parts already proven in previous justifications.

This point is also our main critique of RAISE. We find it very annoying that it is impossible to propagate changes in RAISE modules to justifications the way it is done for modules. As a matter of fact RAISE provides a tool to propagate changes that have occurred inside a module to all modules depending on that module. We are aware that extending this tool to handle justifications is all but a trivial matter.

A second point we have to criticize is the incomplete set of rules for the justification editor. There are no rules for the parallel combinator, especially rules for the interaction of channel hiding and the parallel combinator (like the expansion law in CCS, see [6] chapter 3.3 p. 69).

The overall impression is however very positive. The RAISE method of rigorous stepwise refinement has proved to be very useful in deriving a correct implementation from a very abstract specification.

RSL is very expressive and allows flexible, modular specifications. RSL supports stepwise refinement, as it provides constructs of different abstraction levels. There are low-level (procedural) constructs so that it is possible to write an implementation in RSL and then translate it automatically in Ada and C++ code. However at the time being, not all RSL features are translatable (e. g. the channels).

There are only positive points to mention about the toolbox: its user interface is uniform, it is easy to use and allows fast working (for example the justification editor) and, last but not least, its set of tools is complete. We particularly appreciated the excellent syntax editor and, apart from the above critique, the fast justification editor.

The Korso group at FZI provided many stimulating discussions. We are in particular indebted to C. Lewerentz and Th. Lindner for their encouragement and help in producing the final version.

References

[1]   S. Brock and C.W. George: *The RAISE Method Manual.* LACOS/CRI/DOC/3, 1990, Computer Resources International A/S.

[2]   C. George and S. Prehn: *The RAISE Justification Handbook.* LACOS/CRI/DOC/7, 1992, Computer Resources International A/S.

[3]   Anne Haxthausen and Chris George: A Concurrency Case Study Using RAISE. In *Proceedings of FME'93, Lecture Notes in Computer Science*, pages 367 - 387. Springer-Verlag, 1993.

[4]   C. A. R. Hoare: *Communicating Sequential Processes.* Prentice-Hall International, 1985.

[5]   C. B. Jones: *Systematic Software development — Using VDM, 2nd Edition.* Prentice-Hall International, 1989.

[6]   R. Milner: *Communication and Concurrency.* Prentice Hall, 1989.

[7]   The RAISE Language Group: *The RAISE Specification Language.* BCS Practitioner Series, Prentice-Hall International, 1992.

[8]   *RAISE Tools Reference Manual.* LACOS/CRI/DOC/13, 1992, Computer Resources International A/S.

XVII. Deductive Synthesis

Applied to the Case Study Production Cell

Jochen Burghardt

GMD Berlin

Abstract

Using the synthesis approach of Manna and Waldinger, a TTL-like control circuitery has been derived from a predicate logic specification with explicit time and space. Emphasis has been put on a modelling close to the informal requirements description, including mechanical and geometrical issues, and stating the liveness condition of the cell as the topmost goal. The experience has been made that using a non-discrete time model neccessarily implies modelling continous time and continous motion functions. In a second attempt, two domain-specific logical operators have been defined as higher-order predicates that schematise frequent patterns in specification and proof and hence allow a concise and expressive presentation.

17.1 Aim

Our aim is to write a formal specification of the production cell that is as close as possible to the informal requirements description, and to show how a verified TTL-like circuitery can be constructed from this, using deductive program synthesis. The main emphasis lies on the formal requirements specification which also covers mechanical aspects and thus allows to reason not only about software issues but also about issues of mechanical engineering.

Besides an approach confined to first order predicate logic with explicit, continuous time, an attempt is presented to employ application specific user defined logical operators to get a more concise specification as a well as proof.

17.2 Deductive Program Synthesis

The deductive program synthesis approach due to Manna and Waldinger [3] is a method for program development in the small. No methodological support e.g. for decomposition into modules is provided, instead, it concentrates on deriving one algorithm from a given specification and some given axioms of background knowledge.

Axioms and specifications are given as first order predicate logic formulas. One tries to prove the specification formula, thereby simultaneously constructing a correct functional program from the answer substitutions arising from unification. The proof rules include resolution for formulas in non-clausal form and some generalisations of e-resolution and paramodulation described in [4].

A program is purely functional and represented as a term built up from function symbols including a ternary *if · then · else · fi*, recursive programs arise from induction in the proof.

Figure 1 shows as a very easy example the synthesis of a program x satisfying the specification $p(x)$. Formulas 1 and 2 provide the assumed background knowledge, formula 3 states the proof goal. The resolution proof in step 4. - 6. actually proves the formula $\exists x\ p(x)$. The "output" column of formula 6 contains the synthesized program.

	Assertions	Goals	Output	
1.	$q(x) \rightarrow p(f(x))$			Axiom
2.	$q(c) \vee q(d)$			Axiom
3.		$p(x)$	x	Specification
4.		$q(x_1)$	$f(x_1)$	3 res 1
5.		$\neg q(d)$	$f(c)$	4 res 2
6.		$true$	if $q(d)$ then $f(d)$ else $f(c)$ fi	5 res 4

Figure 1   Example synthesis proof

17.3 Modelling the Production Cell

It is well known that the transition from an informal requirements description to a formal specification is the most critical step wrt. correctness within the formal scenario, since the formal specification can of course not be mathematically verified against the informal description. We tried to adopt an approach to defuse this problem as far as possible: to create a formal language level in which the informal description from part II can be expressed almost "1:1" and thus be easily validated. The specification obtained this way is a requirements specification, not a design specification; due to its high degree of implicitness it does not admit rapid prototyping, nor an immediate stepwise refinement into executable code.

First, a suitable terminology has been fixed, consisting of predicate and function symbols together with their informal explanations. Figure 2 shows some example explanations. Time has been modelled by explicit parameters in order to cope with the restriction to first order predicate logic, and to be able to talk about deadlines explicitly. Space is modelled by three-dimensional cartesian coordinate vectors, transformation into, resp. from, polar coordinates are axiomatized as far as needed. The desired "program" will consist in an asynchronous circuitery built up from TTL-like components, modelled as time-dependent functions. Switching times are ignored within this setting. No explicit feedbacks are allowed in the circuitery, since this would amount to deal with infinite terms which is not supported by the proof tool. Instead, circuitery feedbacks are hidden in circuits like flip flops.

Then, a collection of obvious facts about the behaviour of the machines could be formalised. See figure 3 for some examples; the full specification and the synthesis proof are contained in [1].

The formal specification consists of four parts:

- the description of behaviour required from each machine,
- the description of behaviour required from each control circuit,
- background facts from geometry, arithmetics and physics, and
- the actual specification of the production cell's goal.

The specification has the property of locality in the sense that in order to validate a certain axiom it is only necessary to check this single axiom against its informal description, using the terminology description.

The specification has been modularized in the obvious way, having for each machine type one module that formally describes its required behaviour, and three

Prediactes:

$robot(r, x)$ $\leftrightarrow$ r is a two armed robot placed at coordinates x

$extends_i(r, t)$ $\leftrightarrow$ at time t, the robot r is extending its ith arm

Functions:

$pos_i(r, t)$ = coordinates of the electromagnet of robot r's i^{th} arm at time t

$dist\_xy(x, x_1)$ = distance of the xy projections of coordinates x and x_1

$r(\vec{c}, \vec{s})$ = a two armed robot with control inputs $\vec{c}$ and sensor outputs $\vec{s}$

$val(c, t)$ = value of the time-dependent function c at time t

$trigger(c, v)$ = output of a Schmitt trigger circuit with input c and threshold v

Constants:

d_3 = coordinates of the elevating rotary table (turning center)

d_4 = coordinates of the robot (turning center)

$maxlg_i$ = maximum length the ith arm of a robot can extend to

Figure 2   Informal meanings of some predicate, function, and constant symbols

additional modules describing the control circuits' behaviour, the overall design of the production cell, and some necessary mathematical and physical background knowledge. One should note that none of these specification modules is related to a part of the implementation in the sense that the latter is obtained by a series of refinements of the former. Instead, each specification module describes a different aspect of the modelled reality, not of the implementation.

The adopted approach also discovers the senselessness of a "production" cell whose purpose solely consists in circulating metal blanks, since it is not possible to provide a goal formula that would not be also satisfied by an empty cell. Therefore, we had to assign the travelling crane an ability to "consume" metal blanks, that is, to retransform them into unforged ones, and to pose two separate specification goals: one for the consumer, the travelling crane, and one for the producer, the rest of the production cell, the latter saying in a formal notion "If an unforged metal blank lies on the feed belt, it will eventually appear forged on the deposit belt".

The approach of predicate logic with explicit time as specification language allows for inclusion of given technical/physical frame requirements and thus for the

Module "Robot":

11: If the first arm is extending long enough, it will eventually reach each length between its current and its maximal one.

$$\forall r, x, t, d \; \exists t_2 : \quad robot(r, x)$$
$$\wedge \; dist\_xy(x, pos_1(r, t)) \leqslant d \leqslant maxlg_1$$
$$\rightarrow (\quad (\forall t_1 : \; t \leqslant t_1 < t_2 \; \rightarrow \; extends_1(r, t_1))$$
$$\rightarrow \; dist\_xy(x, pos_1(r, t_2)) = d \quad)$$
$$\wedge \; (\forall t_3 : \; t \leqslant t_3 < t_2 \rightarrow dist\_xy(x, pos_1(r, t_3)) < d)$$

12: Only if the first arm extends, its length can grow.

$$\forall r, x, t, t_2 : \quad robot(r, x)$$
$$\wedge \; t \leqslant t_2$$
$$\wedge \; dist\_xy(x, pos_1(r, t)) < dist\_xy(x, pos_1(r, t_2))$$
$$\rightarrow \; \exists t_1 : \; t < t_1 < t_2 \wedge extends_1(r, t_1)$$

13: Motor control and sensors (c_1: extend first arm, s_1: length of first arm)

$$\forall c_1, c_2, c_3, c_4, c_5, c_6, c_7, c_8, s_1, s_2, s_3, x, t :$$
$$robot(r(c_1, c_2, c_3, c_4, c_5, c_6, c_7, c_8, s_1, s_2, s_3), x)$$
$$\rightarrow \; (extends_1(r(c_1, c_2, c_3, \ldots, c_8, s_1, s_2, s_3), t) \; \leftrightarrow \; val(c_1, t) = 1)$$
$$\wedge \; dist\_xy(x, pos_1(r(c_1, c_2, c_3, \ldots, c_8, s_1, s_2, s_3), t)) = val(s_1, t)$$

Module "Factory":

21: A two armed robot is placed at d_4.

$$robot(r(c_1, c_2, c_3, c_4, c_5, c_6, c_7, c_8, s_1, s_2, s_3), d_4)$$

22: The elevating rotary table is reachable by the first arm of the robot.

$$dist\_xy(d_4, d_3) \leqslant maxlg_1$$

Module "Circuits":

31: Trigger circuit

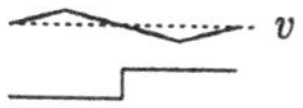

$$\forall c, v, t : \; val(trigger(c, v), t) = 1 \; \leftrightarrow \; val(c, t) < v$$

Figure 3 Some axioms from the specification

treatment of systems with control loops partly outside the hardware/software area. For example, the robot control in some situation starts its motor to extend an arm until it reaches a certain length, cf. figure 4. A verification of the subgoal that the robot arm will in fact reach the desired length and then stop is impossible without considering the mechanical properties of the arm involved. The same holds for the whole production cell: to verify the ultimate specification goal that it will produce forged metal blanks from unforged ones requires the formal consideration of its mechanical behaviour in the proof; it does not suffice to restrict the proof to the software, resp. hardware, aspects.

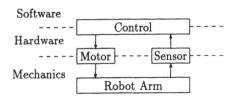

Figure 4   Closed control loop including mechanical feedback

It is also possible to derive necessary requirements concerning issues outside the hardware/software area. For example, it has been derived that the angle between the deposit belt's starting point, the robot's turning center, and the press has to be 90 degrees in order to deposit the forged blanks in the right alignment angle. Thus, the deductive approach can be extended to serve as a framework for the engineering of the whole production cell including mechanical aspects. In a future scenario, a mechanical engineer could be provided from his customer with a formal requirements description of a production cell, and from the manufacturers of the cell's machines with their formal behaviour description. He could then develop a verified overall configuration of the cell including its control, using the deductive approach to integrate classical mechanical engineering tasks and software engineering.

Finally, there was a rather surprising experience concerning the time modelling, showing how much care is needed in formalizing the background knowledge for the requirements specification. Consider again the control loop of figure 4. It is necessary at some point of proof to show that at some time t_2 the robot arm will be extended to a given length provided its length at time t_1 has been smaller. Assume that from the behavioural requirements description of the robot we know that the arm will eventually extend to any given length (within its limits) if its extension motor is running long enough.

What is needed for the correctness proof of the feedback arrangement above is, however, that there is a *minimal* time in which the desired length is reached, in order to stop the extension motor just at that point. Thus, it is not sufficient to have rational numbers as time domain since they are not closed wrt. infima. In fact, if the desired length the arm is to be extended to happens to be such that it is reached if $(t_2 - t_1)^2 = 2$, then at each $t_2 > t_1 + \sqrt{2}$ the length has been reached, but there is no minimal (rational) t_2. The problem has been circumvented by including the existence of minimal times into the requirements specification, cf. axiom 11 in figure 3.

17.4 Synthesis

Two approaches have been made to synthesize a control circuitery for the production cell. The first approach used only the level of first order predicates, starting from the specification as described above, and proving its satisfiability. Figure 5 shows an example proof of a very simple control circuitery, figure 6 shows the circuitery.

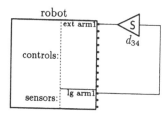

Figure 6   Control circuitery from figure 5

One main difficulty in finding a proof was to make explicit the necessary assumptions about continuity of certain functions involved in simple feedback loops. They were "forgotten" in first versions of the specification and were not recognized before the analysis of failed proof attempts. Consider, for example, the safety requirement that the first robot arm may enter the press only if the latter is in its middle position. Assume the control circuiteries will stop the robot arm if approaching the press to a certain distance d_s when it is not in middle position, and prevent the press from moving off the middle position as long as the arm remains within the distance d_s. The proof that this control meets the safety requirement, however, has to be based on the intermediate value theorem from calculus. Figure 7 provides a counter example if the arm movement was not continuous, assuming the press in upper position. We had to add one instance of the intermediate value theorem for each function required to be continuous.

Find a control circuitery to extend the robot's first arm to a given length d_{34}.

Conjecture:
$$\exists r_0 : \forall t_0 : \exists t \quad dist\_xy(d_4, pos_1(r_0, t_0)) \leqslant d_{34}$$
$$\rightarrow dist\_xy(d_4, pos_1(r_0, t)) = d_{34}$$
where $d_{34} = dist\_xy(d_4, d_3)$

Proof (skolem functions indicated by "$"):

assumption: $dist\_xy(d_4, pos_1(r_0, t_0^\$)) \leqslant d_{34}$

goal: $dist\_xy(d_4, pos_1(r_0, t)) = d_{34}$

51 = 11 res assumption,21,22:
$$(\quad (t_0^\$ \leqslant t_1 < t_2^\$ \;\rightarrow\; extends_1(r_0, t_1))$$
$$\rightarrow dist\_xy(d_4, pos_1(r_0, t_2^\$)) = d_{34})$$
$$\wedge \quad t_0^\$ \leqslant t_3 < t_2^\$ \;\rightarrow\; dist\_xy(d_4, pos_1(r_0, t_3)) < d_{34}$$

52 = split 51:
$$(t_0^\$ \leqslant t_1 < t_2^\$ \;\rightarrow\; extends_1(r_0, t_1))$$
$$\rightarrow dist\_xy(d_4, pos_1(r_0, t_2^\$)) = d_{34}$$

53 = split 51:
$$t_0^\$ \leqslant t_3 < t_2^\$ \;\rightarrow\; dist\_xy(d_4, pos_1(r_0, t_3)) < d_{34}$$

54 = 52 res 13:
$$(t_0^\$ \leqslant t_1 < t_2^\$ \;\rightarrow\; val(c_1, t_1) = 1)$$
$$\rightarrow dist\_xy(d_4, pos_1(r_0, t_2^\$)) = d_{34}$$

55 = 54 res 31:
$$(t_0^\$ \leqslant t_1 < t_2^\$ \;\rightarrow\; val(c, t_1) < d_{34})$$
$$\rightarrow dist\_xy(d_4, pos_1(r_0, t_2^\$)) = d_{34}$$
$$\text{where } r_0 = r(trigger(c, d_{34}), c_2, c_3, \ldots, c_8, s_1, s_2, s_3)$$

56 = 55 rep 13:
$$(t_0^\$ \leqslant t_1 < t_2^\$ \;\rightarrow\; dist\_xy(d_4, pos_1(r_0, t_1)) < d_{34})$$
$$\rightarrow dist\_xy(d_4, pos_1(r_0, t_2^\$)) = d_{34}$$
$$\text{where } r_0 = r(trigger(s_1, d_{34}), c_2, c_3, \ldots, c_8, s_1, s_2, s_3)$$

57 = 56 res 53:
$$dist\_xy(d_4, pos_1(r_0, t_2^\$)) = d_{34}$$
$$\text{where } r_0 = r(trigger(s_1, d_{34}), c_2, c_3, \ldots, c_8, s_1, s_2, s_3)$$

Figure 5   Example proof of a simple control circuitery

length of robot arm 1:
(press in upper position)

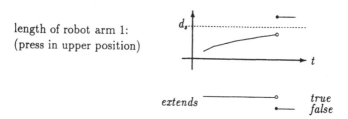

extends ———o   true
        •—   false

Figure 7   Neclecting a safety requirement by incontinuous motion

The control circuitry was not really "synthesized" in the sense that an actual intermediate proof goal would provide many hints which program resp. circuitry constructs to insert. Instead, a previously constructed circuitry was in fact verified. Moreover, to reuse earlier parts of the proof is much easier if proofs are conducted bottom-up, while true synthesis would require top-down (backward) proofs. For this reason, large parts of the proof have been conducted in a bottom-up manner, like e.g. in figure 5.

The second approach used the experience gained during the first one to identify higher level concepts which turned out to be valuable in lifting specification and proof to a higher level of expressiveness. Two new ternary logical operators were defined in terms of a restricted second order predicate logic, see figure 8.

$$unt(t_0, P, Q) :\leftrightarrow \forall t_1 : \; t_1 < t_0 \; \lor \; (\exists t : \; t_0 \leqslant t \leqslant t_1 \land Q(t)) \; \lor \; P(t_1)$$
$$ldt(t_0, P, Q) :\leftrightarrow \exists t_1 : \; t_0 \leqslant t_1 \; \land \; (\forall t : \; (t_0 \leqslant t \leqslant t_1 \rightarrow P(t)) \; \rightarrow \; Q(t_1))$$
$$\land \; (\forall t : \; t_0 \leqslant t \leqslant t_1 \rightarrow \neg Q(t))$$

Figure 8   Application specific logical operators

The concepts are borrowed from Mishra/Chandy's language *Unity* [2]. The formula $unt(t_0, P, Q)$ means that from time t_0, the unary predicate Q holds until the unary predicate P becomes true, or for ever ("P *until* Q"). Formula $ldt(t_0, P, Q)$ means that from time t_0, if P holds long enough, then Q will eventually become true at a minimal time t_1 ("P *leads to* Q").

A background theory of useful axioms about *unt* and *ldt* has been proved, including the monotonicity of *unt* in the second and third and of *ldt* in the third argument, and the anti-monotonicity of *ldt* in the second argument, which enabled us to include both operators into the polarity-based non-clausal resolution rule.

Since *unt* and *ldt* reflect frequent patterns of the specification and the proof, both can be made shorter and easier to understand by using these operators. Figure

9 shows the analogon to the proof of figure 5 using *ldt*. One fact (61) from the background theory about *unt* and *ldt* is used.

11':
$$\forall r, x, t, d : robot(r, x)$$
$$\land \ dist\_xy(x, pos_1(r, t)) \leqslant d \leqslant maxlg_1$$
$$\rightarrow \ ldt(t, \lambda t_1 : extends_1(r, t_1),$$
$$\lambda t_2 : dist\_xy(x, pos_1(r, t_2)) \geqslant d)$$

61:
$$ldt(t_0, \neg P, P) \ \rightarrow \ \exists t \ \ t_0 \leqslant t \land P(t)$$

71 = 11' res **assumption,21,22**:
$$ldt(t_0, \lambda t_1 : extends_1(r, t_1),$$
$$\lambda t_2 : dist\_xy(x, pos_1(r, t_2)) \geqslant d_{34})$$

72 = 13 res **31**:
$$dist\_xy(x, pos_1(r_0, t)) < d_{34} \ \rightarrow \ extends_1(r_0, t_1)$$
where $r_0 = r(trigger(s_1, d_{34}), c_2, c_3, ..., s_3)$ as before

73 = 71 res **72**:
$$ldt(t_0, \lambda t_1 : dist\_xy(x, pos_1(r_0, t_1)) < d_{34},$$
$$\lambda t_2 : dist\_xy(x, pos_1(r, t_2)) \geqslant d_{34})$$

74 = 73 res **61**:
$$\exists t_2 : \ dist\_xy(x, pos_1(r, t_2)) \geqslant d_{34}$$

Figure 9   Proof analogon of figure 5, using application specific logical operators

Note that chains of *ldt*s can simulate state transitions like in the finite automaton paradigm; figure 10 shows an example.

$$pos(s, t_0) = d_5$$
$$\rightarrow ldt(t_0, press\_up, \lambda t_1 : ldt(t_1, press\_down, \lambda t_2 : state(s, t_2) = forged))$$

Figure 10   Modelling state transitions by chains of ldt

In part VI, the production cell has been modelled as a kind of finite automaton written in *Lustre*, allowing fully automated verification of the main requirements by state exploration using binary decision diagrams. However, in that approach verification cannot deal with issues that are not formalizable as automaton properties. It would be interesting to investigate whether the *unt/ldt* approach can achieve a vertical decomposition of the model in the sense that on the higher level only automaton properties need to be dealt with while on the lower level the remaining properties are covered.

17.5 Evaluation

17.5.1 Provable properties

The adopted approach makes it easy to formulate and prove all desired liveness and safety properties. The liveness property says that each unforged metal blank entered into the production cell will eventually leave it forged, it has been discussed in section 17.3. The safety requirements are comprised in the additional goal "Never any damage occurs", where a necessary condition for any damage is provided by enumerating all critical combinations of machines (e.g. robot/press).

A disadvantage of this approach consists in the risk of overlooking certain possible conflict situations when writing the specification. For example, in the first version of the informal safety requirements given in part II it was not required that the feed belt may transport metal blanks only if the elevating rotary table is empty.

Each informal safety requirement in part II is a consequence of one of the following principles:

- the avoidance of machine collisions (1, 2, 5, 6),
- the limitations of machine mobility (3, 4, 5),
- the demand to keep metal blanks from falling from great height (7, 9), or
- the necessity to keep the metal blanks sufficiently separate (8).

It is principally possible to base the productions cell's safety requirements on these four principles. However, formalising the first principle needs a complete description of the machine shapes and motion tracks, and, moreover, a proof for each of $n \cdot (n-1)$ pairs of machines that they will not collide, no matter how far they in fact are separated. Since both is very expensive, we have chosen to state the possible collision situations explicitly.

17.5.2 Explicit assumptions

Assumptions about the behaviour of a single machine as well as about the overall configuration of the production cell are explicitly stated in the corresponding specification module. Moreover, it is possible to derive additional requirements on behaviour resp. configuration during the proof, cf. section 17.3.

17.5.3 Statistics

The specification comprises eight modules with total eighty axioms, see figure 11, however, not all axioms are actually used.

Module	No. of Axioms
Press	9
Robot	24
Elevating rotary table	12
Belt	5
Overall configuration incl. travelling crane	9
Mathematics	14
Circuits	8
Total	80

Figure 11   Length of specification modules

It is difficult to estimate the effort for the proof, since in parallel to its conduction the support tool "Sysyfos" had to be improved in order to be able to cope with the proof at all. As a result of the engagement in the case study, a semi-graphic user interface and a proof replay mechanism have been built into the support tool; in the later phase, the restricted higher order unification for *unt* and *ldt* required some implementation work. With this relativization, the effort e.g. for finding resp. verifying the sub-circuitery to move a metal blank from the elevating rotary table into the press can be stated as about 1-2 man weeks. The proof includes 210 steps without any use of *unt* and *ldt* and was the first subproof of the case study. A later proof of a comparable task is shortened to the order of magnitude of about 1-2 man days, due to the experience gained, especially concerning the continuity issues discussed above.

17.5.4 Maintenance

The main effort when developing a control circuitery for a different, but similar production cell is the conduction of a new proof. It should be easy to obtain the new formal specification, building up on the formal terminology provided. Using the pure first order predicate logic approach, only few parts of the original proof may be reused, depending on the degree of similarity between both tasks. However, using the *unt*/*ldt* approach, a large amount of proof effort is dedicated to the schematisation of controlling principles as background theorems which need not be proved again, cf. e.g. theorem 61 in figure 9 which is the heart of the proof there. It is expected that the remaining proof effort to "instantiate" the background theorems taylored to the new cell configuration is rather small. In any case, the necessary effort to obtain a new verified control circuitry is still much greater than to reconfigure an object oriented controller program, say.

17.5.5 Efficiency

The paradigm of deductive program synthesis does not make any statements about efficiency of the constructed programs. In the setting of the production cell, moreover, efficiency does not mean short software reaction times, but a high overall throughput rate. Following the approach of extending software engineering methods to include also mechanical engineering, one could estimate the "algorithmic complexity" of the whole production cell. This would need the generalisation of a complexity calculus for reactive systems. Since no recursion is involved, the maximal work time of a metal blank could be calculated exactly. However, a proof that the specific configuration and control of the cell guarantee maximal thoughput seems to be as difficult as complexity lower bound proofs for algorithmic problems.

17.5.6 Mechanical requirements

As mentioned in section 17.3, during the synthesis proof a couple of additional requirements to the configuration of the production cell were deduced. They mostly state that the limitations of machine mobility allow to reach the necessary points, e.g. that the elevating rotary table can be reached by the robot's first arm, cf. axiom 22 in figure 3. Another group of requirements concerns the fitting of dimensions and angles, e.g. that the upper position of the elevating rotary table, the robot's first arm, and the middle position of the press must be all at the same height.

Some conditions need not really be required but their validity would lead to a simpler control circuitry, e.g. if it is known that the distance from the robot's turning center to the elevating rotary table's is greater than to the press, it is sufficient to contract the first arm during its way to the press, otherwise the circuitery had to be prepared for both retracting *and* extending.

When operating the production cell in an "open" mode, i.e. without the travelling crane, additional requirements on the loading resp. unloading behaviour arise, e.g. the feed belt may be loaded only if there is suffient free space available at its start. The latter condition makes the existence of an additional feed belt sensor necessary, either at its start or (leading to an easier and more robust control) at its end.

Our modelling is based on the idealizing assumption that there are no impreci-
sions in geometrical sizes. In practice, however, this won't be the case, e.g. the feed
belt will not deliver each metal blank exactly to the elevating rotary table's turning
center, d3. A model of the production cell that takes this fact into account would
have to deal with admissable tolerance intervals, stating e.g. that the robot's first
arm will safely grab the metal blank if it lies within the area d_3+x with $\|x\| < \varepsilon_3$.
Each machine may add its own inaccuracy to the tolerance interval, but may also
decrease the interval in some respect due to some alignment effect, e.g. at photoelectric cells. Then, one has to require additionally that the tolerance intervals are
small enough to allow proper operation. E.g. the tolerance interval of a metal
blank's position in the press contains the sum of tolerances of the robot's first arm,
the elevating rotary table, the feed belt, and the (external) feed belt loading device;
it must be ensured that this deviation is small enough to allow safe pressing of the
blank.

17.6 Conclusion

Our experiences with the production cell case study seem to confirm the following
theses:

- *A good requirements specification should consist of a collection of almost
  obvious facts in formal notation.*
  The absense of need for executability provides the freedom to state formal requirements as an almost direct translation of natural language formulation. The former can be validated against the latter in a local manner.

- *Requirement specification modules describe different aspects of the modelled reality, not of the implementation.*
  In contrast to design specification modules, the former do not refine into
  implementation modules, they are rather orthogonal to them.

- *Predicate logic can be seen as "assembler language" for specifications.*
  It is desirable to build higher language constructs upon it in order to come
  to more concise specifications as well as proofs.

- *The level of formal description can be lifted as high as purely technical
  issues are involved.*
  There seems to be no reason to stop within the level of software engineering, rather, the logic-based methods can serve as a framework for a verified overall engineering. This has been demonstrated by our treatment of

the production cell which lies entirely in the technical area and whose specification included the topmost goal (production of forged metal blanks). On the other hand, if the topmost goal is non-technical, like e.g. in a medical information system, our approach is not fully applicable.

- *There are only a couple of adequate levels of description.*
  Our experience has shown that the decision to choose a non-discrete time modelling necessarily implies a description based on continuous time and continuous functions; there seems to exist no intermediate level (e.g. of rational time and arbitrary functions). A more realistic approach could use differentiable functions. While in the former approach, for example, a motor is assumed to run with full speed immediately after it has been started, the latter approach allows to reason about accelerations and starting velocities. While not urgently required for the production cell case study, this level of description becomes unavoidable when dealing with time critical applications eg. from the area of vehicle control systems where it is vital to talk about acceleration and brake times.

References

[1]   Jochen Burghardt, Deduktive Synthese der Steuerung einer Fertigungszelle, GMD working papers, to appear 1994

[2]   Misra, J., Chandy, J., Parallel Program Design, A Foundation, Addison-Wesley, 1988

[3]   Zohar Manna, Richard Waldinger, A Deductive Approach to Program Synthesis, in: ACM Transactions on Programming Languages and Systems, Vol. 2 No. 1, p. 90-121, Jan 1980

[4]   Zohar Manna, Richard Waldinger, Special Relations in Automated Deduction, in: Journal of the ACM, p. 1-59, Jan 1986

XVIII. Symbolic Timing Diagrams

*Synthesis of a Production Cell Controller using
Symbolic Timing Diagrams*

Franz Korf, Rainer Schlör

Universität Oldenburg and OFFIS Oldenburg

Abstract

We present a novel method for controller synthesis and verification from high-level interface specifications. Specifications are expressed in the form of requirements using *Symbolic Timing Diagrams*, which is a visual formalism based on the notion of timing diagrams commonly used by hardware designers.
We suggest a top-down design methodology where specifications are created incrementally and discuss two different specification styles found while we applied the method to synthesize a distributed controller for the production cell.
We then introduce the ICOS[2] system, which has been used to carry out the practical work. ICOS[2] allows to analyze specifications for consistency and completeness and to synthesize C- and VHDL-code from the given specifications automatically.

18.1 Introduction

This paper presents a novel method for controller synthesis from requirement specifications. While the state-of-the-art high level synthesis path starts from operational descriptions (which have to be constructed from typically informal requirement specifications by hand) we have developed a method and according tools to capture and support synthesis from requirement specifications directly (Figure 1).

In order to capture requirements for components of a system-level controller design, we have designed a graphical specification language called *Symbolic Tim-*

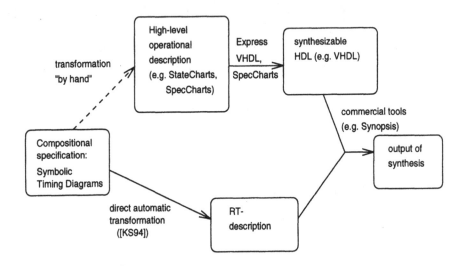

Figure 1 Standard high-level synthesis path compared to synthesis from requirement specifications

ing Diagrams (STDs in short). This language, which was first introduced in [22], allows to capture individual safety- and liveness-requirements by separate diagrams (e.g. collision freedom, guaranteed access to a critical processing step) , which are stored and managed in a design library similar to the concepts which VHDL provides for design entities.

A Symbolic Timing Diagram specification is a collection of diagrams, which is associated with an interface description (entity declaration, in VHDL terminology)[1]. Together with a variant of the StateChart language Symbolic Timing Diagrams are part of VHDL/S (a language developed in the ESPRIT-project 6128 FORMAT [9]) as a specification language on top of VHDL, which allows formal specification and verification of designs described in VHDL. In this context Symbolic Timing Diagrams have already been applied successfully to describe medium-complexity hardware systems and verify selected properties ([3]).

The task to perform interface controller synthesis from requirement specifications is substantially harder than the standard high-level synthesis methods which start from operational descriptions and are already well supported by commercially available tools. This is due to two basic observations:

---

1. At this point, we assume familarity with basic concepts of VHDL [10].

1. *Different requirements stated about the behavior of the same component in a Symbolic Timing Diagram specification may interfere.* This is not a flaw of the approach, but typical of early design steps, where each requirement will focus on one aspect of the behavior of a component at a time. The price to be paid for this flexibility is that requirement specifications expressed using Symbolic Timing Diagrams must be analyzed for *consistency*, which can fortunately be done automatically (provided that all data types are finite).

2. *There is no a priori control structure in a requirement specification.* Again, this is not a flaw but a desired feature of the specification language, since at early steps of the design of a complex system the control structure will either be unknown or change during the design process. It turns out that a control structure can be automatically obtained as a side effect of the consistency analysis procedure (again, provided that all data types are finite).

The main goal of the work described in this paper was to achieve a complete design of a distributed controller for the production cell starting from a requirement specification expressed by Symbolic Timing Diagrams and to obtain an implementation of a distributed controller by automatic synthesis (Figure 2).

The methodology supported by ICOS² entails two basic phases for the synthesis process from requirement specifications:

1. **Decomposition phase**: Partition global system requirements as stated in [15] into local requirements to be guaranteed by the system components (compositional specification). In order to establish the *correctness* of this decomposition, ICOS² supports two independent methods: (1) *Testing* is supported by waveform monitoring similar to standard VHDL-simulation environments. (2) *Automatic formal verification* can be performed using a so-called tautology-checker. A report on the verification tasks arising during this design phase is under preparation.

   In addition, for this case study a special graphical simulation facility was available [4]. ICOS² provides a mechanism to link such a simulation model to the synthesis output, which gives an excellent graphical animation of the global system behaviour. Both the testing approach supported by the simulation model and the automatic verification supported by ICOS² led to the identification of subtle errors in the specification (e.g. glitches, global deadlocks).

2. **Synthesis phase**: Check the decomposed specification for consistency, which includes the checks, that local requirements are satisfiable and that the specification is compositional, e.g. asynchronous inputs are never restricted. If this check fails, then the specification must be revised. The detailed description of the synthesis phase is given in this report.

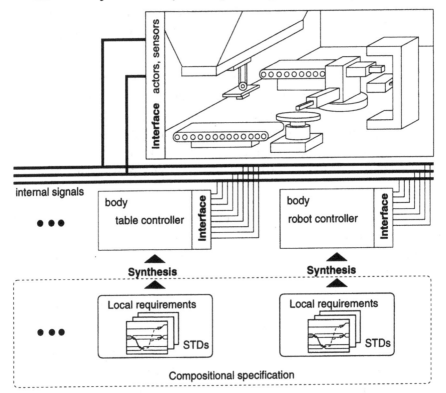

Figure 2   Production cell and controller interface

Related work. Most of the known approaches to use timing diagrams in a formal sense ([5], [11], [16]) differ from our approach in that they have built-in means to specify control structures such as iteration and concatenation (sequencing). This implies that timing diagrams are interpreted in a way similar to the statements of an imperative language such as *C* or even VHDL. In contrast, the semantics of the STD language associates with each timing diagram a *constraint* on the set of admissible behaviors of a component, which is analogous to the statement of facts in PROLOG.

To improve the quality of the synthesis results it is important to build up the synthesis tools on formal methods. An approach similar to our is reported in ([12],

[23], [8]). In this context, timing diagrams are also used to specify the controller components of a design. In contrast to our synthesis process, their approach is based on the formal description technique T-LOTOS as intermediate representation for the semantics of timing diagrams. Since the semantics of STDs with finite data types can be expressed using the linear time temporal logic PTL ([22], [6]), our synthesis approach is based on the theory of temporal logic. Several approaches to synthesize components from PTL specifications have been investigated ([7], [18], [19], [20], [2]). While these techniques in principal can also be used for the STD specification language, their practical applicability is limited by the fact that the decision procedure for PTL formulae is exponential in the size of the formula. As shown in [14] our approach reduces this complexity drastically by exploiting the special structure of the characteristic formulae describing the semantics of a STD specification.

The rest of the paper is organized as follows: Section 18.2 introduces the specification language Symbolic Timing Diagrams and describes two specification styles applied for specifying a distributed controller of the production cell. Section 18.3 explains the design methodology supported by the ICOS^2-system. Section 18.4 describes the components of ICOS^2, which have been used in the case study. Section 18.5 explains the main steps of the synthesis part. The complexity of our synthesis approach is discussed in Section 18.6. Section 18.7 summarizes the results of the case study. Finally, Section 18.8 reports future extensions of this work.

18.2 Symbolic Timing Diagrams

In order to specify requirements we use a graphical formalism called Symbolic Timing Diagrams, which is derived from the notion of classical timing diagrams as used by hardware designers. In contrast to these diagrams, which are often used informally or with an ad-hoc semantics in mind, Symbolic Timing Diagrams have a precise semantics, which is defined by translation into a characterizing temporal-logic formula ([22], [6]).

Symbolic Timing Diagrams are intended to be used for a declarative specification of the behaviour of individual system components as observable at their interface. While in the realm of hardware design specifications often take the form of protocols with concrete timing constraints, Symbolic Timing Diagrams are directed towards the early steps of a complex hardware design, where requirements are abstract system properties wherein only the *order* of events and their (under given premises) guaranteed occurrence matters.

In the following we give a short introduction to Symbolic Timing Diagrams; a detailed exposition can be found in [9] and [6]. A Symbolic Timing Diagram consists of a number of *symbolic waveforms*. Each waveform has several *regions* which are separated by *edges*. The regions are labelled by predicates (denoted by boolean expressions over the variables observable at the components interface). Between any two edges on different waveforms there may be *constraints* (denoted by respective arrow shapes) used to express a required partial ordering of these edges. Figure 3 shows the different forms of constraints. A *causality#* constraint (denoted by an emphasized arrow) expresses that (1) an event matched by edge *e1* must occur before an event matched by edge *e2* and (2) if the event matched by edge *e1* occurs, then an event matched by edge *e2* will occur eventually. A *precedence#* constraint is weaker than a causality constraint and expresses only condition (1).[1] A precedence# constraint also has a weak form (denoted by a dashed arrow shape), which denotes a premise (a constraint expected to be satisfied by the environment) rather than a constraint. The semantics behind premise constraints is explained together with the concept of activation and deactivation of diagrams below.

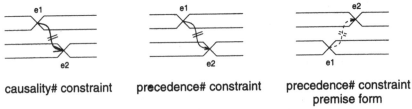

causality# constraint   precedence# constraint   precedence# constraint
                                                                              premise form

Figure 3 Basic types of constraints and their graphical denotation

At a given time instance, a Symbolic Timing Diagram can be either activated (in effect) or deactivated. A Symbolic Timing Diagram can be regarded as a dynamic pattern called into effect either initially (initial form) or dynamically (invariant form), i.e. whenever the values of the interface variables conform to the conjunction of the initial predicates on each waveform of the diagram (called the *activation condition* of the diagram). In the following part of the run of the system the constraints have to be obeyed. If a premise is violated or the diagram is matched completely, the diagram becomes deactivated, i.e. it does not restrict the run of the system any longer. Note that overlapping instances of activated diagrams may be created (so-called concurrent activation, Figure 4).

---

1. The #-symbol indicates that *e1* and *e2* must not occur simultaneously; this is graphically depicted by two short bars crossing the arrow.

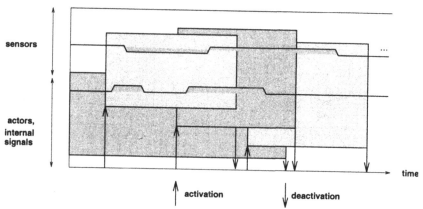

Figure 4 Activation and deactivation of Symbolic Timing Diagrams

Recall that our goal was to synthesize a distributed controller for the production cell. Given the natural composition of the system of separate entities (feed belt, table, robot, press, deposit belt and crane), we partitioned the specification into according sub-specifications expressing local requirements (Figure 2). This is possible thanks to the capability of STDs to distinguish between constraints to be guaranteed by a system component and assumptions about the behaviour of other components or the physical environment by means of premise constraints. The synthesis procedure described in Section 18.5 automatically decomposes the sub-specifications into maximal independent sets; e.g. the specification for the table can be partitioned into three sets (for vertical movement, rotation and load state).

We identified two styles of specification suitable for synthesis. The first one, which we call "axiomatic", employs a single type of diagrams, denoted [*P* **causes** *sig*] (Figure 5,(a)), where *sig* is an assertion about a boolean-valued output signal *x* (either *x = true* or *x = false*) and *P* is an assertion about input signals and other internal signals of a component. The diagram states that

1. activation occurs whenever the component is in a state which satisfies $(\neg P \wedge \neg sig)$

2. From then on, a change from $\neg sig$ to *sig* (e2) may happen only after the state of the component has changed to satisfy *P* (e1)

3. From then on, the change from $\neg sig$ to *sig* is guaranteed to occur, if the assertion *P persists to hold* (e3) until the change has occurred; otherwise, the diagram is deactivated.

[P causes sig]   **T_R_StopLeft**

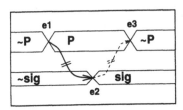   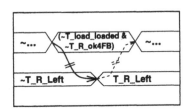

(a)   (b)

*Figure 5   STD for the idiom [P **causes** sig] (a) and example
diagram taken from the specification of the table (b)*

Note that this specification style relies on mutual stability guarantees (assumptions) in order to achieve a correct system function. Given that these stability guarantees have been established, a synthesized self-timed hardware design will operate free of hazards. If the assertion *P* depends on signals from the physical environment, then stability intervals are real-time conditions; in that case we have to rely on some meta-level reasoning in order to establish that the controller reacts "fast-enough" e.g. to stop the table on its way down when reaching its bottom position. In our current verification methodology, we have to take such real-time dependent knowledge as assumption. The axiomatic specification style yields a maximal concurrent controller. In this version, our specification is able to process 5 blanks simultaneously.

In contrast to this style, the second style is more operational in nature and may introduce sequential behaviour where concurrent operation could be possible. In this style, we use an additional internal signal "robotPhase" to indicate the local execution phase of a component. This signal is used to control activation of the diagrams in the component specification. A second waveform is used to specify which signals are kept stable during the phase described by the diagram.

The example diagram R_init shown in Figure 6[1] shows an initialization sequence for the robot using this style. Note that at the end of the diagram signals

---

1. The double-line left border of the diagram denotes the activation form 'initial'. The dotted arrow leading from the cross on the left border to the edge labelled (*robotPhase = 1*) indicates that this event is guaranteed to happen. An alternating waveform associated with a boolean-valued signal named *sig* denotes the predicate ¬*sig* for level 'low' and *sig* for level 'high'.

R_Init

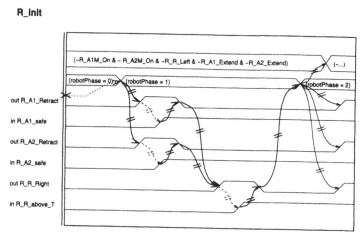

Figure 6   Specification of an initialization phase of the robot

R_A1_Retract, R_A2_Retract and R_R_Right are kept stable until robotPhase 2 is reached, where their further behaviour is controlled by another diagram (which is responsible for robotPhase 2).

18.3 Synthesis Path and Design Methodology

As pointed out in the introduction, Symbolic Timing Diagrams are collections of individual diagrams describing different aspects of the required behavior of a component. The synthesis path performs an analysis of the Symbolic Timing Diagram specification for consistency. Since each timing diagram in the specification constrains the set of admissible behaviors of a component and a collection of timing diagrams is interpreted as conjunction over all these constraints, the resulting semantics for a Symbolic Timing Diagram specification may be not satisfiable. If this is the case, the specification must be weakened by either removing one (or more) diagrams from the collection or by weakening some of the diagrams in the collection. If the specification is consistent, analysis proceeds to check if the specification is *input deterministic*, i.e. whether the entirety of all diagrams ensures that the output values to be observed at the interface of the component are uniquely determined by the input values. If this is not the case, the specification will usually be strengthened by adding one (or more) requirements (diagrams) to the specification or by strengthening some of the diagrams in the collection (Figure 7). Both checks are explained in detail in [14].

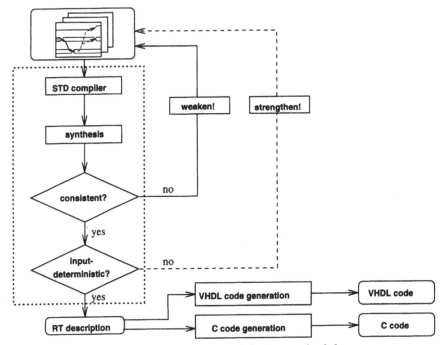

Figure 7   Synthesis path and design methodology

18.4 Using ICOS² for the Production Cell Case Study

ICOS² - an acronym for Interface Controller Synthesis and Verification System - has been used to synthesize a distributed controller of the production cell. This Section explains the components of ICOS², which have been used within the case study.

Design Capture As shown in Figure 8 ICOS² is built around three data bases:

- a data base for Symbolic Timing Diagrams
- a data base for PTL formulae
- a data base for Rabin automata

The designer primarily works on the Symbolic Timing Diagram data base. Using an integrated editor for Symbolic Timing Diagrams the designer can create and modify a diagram. The design browser connects a diagram to an entity. Each entity has an interface specification given as VHDL entity declaration [10].

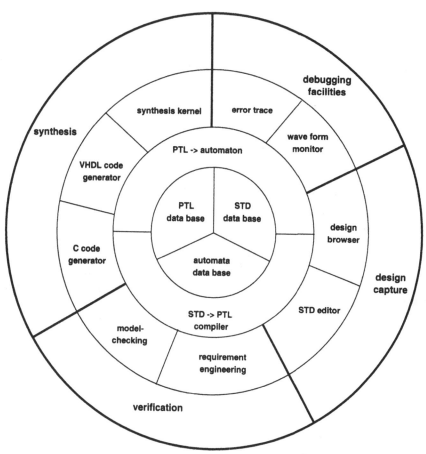

Figure 8 The structure of ICOS^2

As depicted in Figure 2 we synthesize for each physical component of the production cell a controller module. Each controller module is an entity listed in the design browser. The interface of a controller module collects the actors and sensors of the production cell, which will be used by this controller module. Moreover internal signals, which will be used for the communication between controller modules, are listed in the interface.

Synthesis From the Symbolic Timing Diagrams specifying a controller module ICOS^2 synthesizes a set of submodules, which satisfy the requirements given by these diagrams. Each submodule is a FSM (finite state machine). The synthesis works as follows: First each Symbolic Timing Diagram is translated into a PTL formula which characterizes the semantics of the diagram. These PTL formulae serve as input for the synthesis kernel which generates the set of FSMs. Section 18.5 explains the synthesis path in more detail.

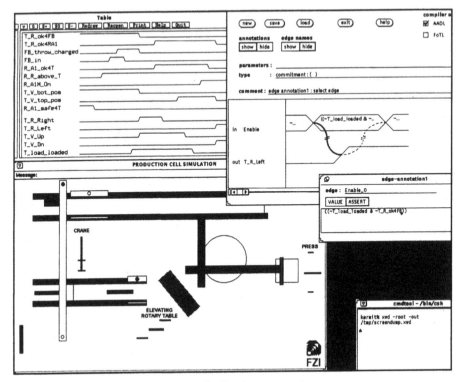

Figure 9   Design scenario

C-Code Generation For each controller module ICOS^2 generates a C module, which encodes the behaviour of the FSMs of this module. The execution model ensures that all FSMs will be executed in parallel.

Execution Model The FSMs of all controller modules will be executed in parallel according to the S/R model of COSPAN [13]. This execution model splits each step of each FSM into a selection - and a resolution phase. Within the selection phase each FSM selects the values which should be visible at its output ports within the next step. The decision, which values should be assigned to the output ports, is based on the actual state only. After all FSMs have finished the selection phase, the selected output port values are visible for all FSMs of all controller modules. Now all FSMs starts the resolution phase synchronously. Each FSM steps one transition. This transition uses the values generated in the selection phase.

The screen dump of Figure 9 displays a typical situation during the controller design of the production cell. The synthesized controller drives the production cell. Using the wave form monitor of ICOS^2 the interface of the table will be observed

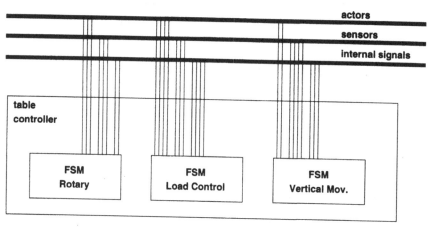

Figure 10   The Controller module for the table

during execution time. The background of the screen displays one Symbolic Timing Diagram for the table.

18.5 Synthesis Process and Synthesis Techniques

Figure 7 shows the integration of the synthesis kernel into the design flow. Using the synthesis path depicted in Figure 11 as guideline the current Section explains the principles of the synthesis procedure. A more technical and detailed description of a similar synthesis path is given in [14].

Input/output of the synthesis process A local requirement specification, which may be used as input for the synthesis process, is a set of individual requirements. In general requirements may be represented e.g. as PTL formulae [17], ω-automata (Rabin, Büchi automata [24]), or Symbolic Timing Diagrams. If Symbolic Timing Diagrams are used as input format the synthesis process can be efficient due to the inherent structure of Symbolic Timing Diagrams. The output of the synthesis is a set of parallel executing FSMs coded in C or VHDL. The controller executes according to the S/R model of COSPAN [13].

Abstract intermediate model We use Rabin automata [24] as abstract intermediate model within the synthesis process. Rabin automata recognize the class of ω-regular languages. A Rabin automaton $\mathcal{A}$ is a finite automaton on infinite words. $L(\mathcal{A})$ is the set of infinite words accepted by $\mathcal{A}$.

Synthesis step 1 This synthesis step generates out of each Symbolic Timing Diagram a Rabin automaton, which accepts the semantics of the diagram. This will be done in two steps:

1. The Symbolic Timing Diagram will be translated into a PTL formula, which defines the semantics of this diagram [22].

2. The PTL formula will be translated into a Rabin automaton, which accepts the semantics of the formula.

In the average case the size of the Rabin automaton is linear in the number of edges of the Symbolic Timing Diagram [14]. For all Symbolic Timing Diagrams of the production cell case study this complexity result holds.

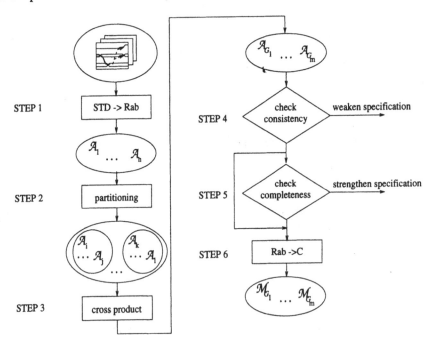

Figure 11   Synthesis Path

Synthesis step 2 The controller module generated by the synthesis process must satisfy all requirements specified by the Symbolic Timing Diagrams. Thus an ad hoc realization of the synthesis would do the "cross-product" of $A_1...A_n$ and use the result automaton $A_{\sqcap}$ where $L(A_{\sqcap}) = \bigcap_{i=1}^{n} L(A_i)$, as input for the consistency check. The number of states of $A_{\sqcap}$ grows exponentially with n.

To overcome this state explosion problem we generate a partition of $A_1...A_n$ such that the Symbolic Timing Diagrams belonging to automata of different sets of the partition are independent. Let O be the set of output ports defined in the interface of the controller module.

- Based on a partition $O_1...O_k$ of O a partition $P_1...P_k$ of $A_1...A_n$ will be generated, such that the automata of P_i restrict only the output ports of O_i. The Symbolic Timing Diagrams belonging to automata of different sets of $P_1...P_k$ do not interfere. Since a Symbolic Timing Diagram defines the set of output ports, whose values are restricted by this diagram, the generation of $O_1...O_k$ is simple.

The axiomatic specification technique (cf. Section 18.2) supports this partition technique. As shown in Figure 10 ICOS[2] generates for the table controller module a partition of three sets:

- **Set 1** All automata/diagrams which specify the rotary movement of the table.

- **Set 2** All automata/diagrams which specify the vertical movement of the table.

- **Set 3** All automata/diagrams which control the load status of the table.

Synthesis step 3 Using the standard operation for the "cross-product of Rabin automata" this step generates for each set G_i of the partition a group automaton A_{G_i} such that $L(A_{G_i}) = \bigcap_{A \in G_i} L(A)$.

Synthesis step 4 This step checks for the consistency of the specification. A *consistent* specification must satisfy three conditions:

1. The specification must be satisfiable.

2. The specification must guarantee that values assigned to output ports do not depend upon the values read from input ports at the same moment.

3. The specification must not make any assumptions about future values of input ports at any time.

	exploit structural properties of STDs	
	yes	no
states	4	17
edges	9	61

Table 1   Space reduction gained by synthesis step 1

In [14], two theorems concerning the consistency check are given. The first one gives some structural properties for checking the consistency of the language of a Rabin automaton. The second one yields that the consistency of all group automata guarantee the consistency of the specification. Thus the consistency check can be done on the automata generated by the partition.

Synthesis step 5 This synthesis step checks whether the specification is input deterministic (complete). This test is optional because for some applications it may be out of interest, whether the controller module is complete/ input deterministic.

Synthesis step 6 This step generates out of each group automaton $\mathcal{A}_{G_i}$ a FSM $\mathcal{M}_{G_i}$, which is extended by the accepting condition of $\mathcal{A}_{G_i}$. This extension guarantees that each word generated by $\mathcal{M}_{G_i}$ is a prefix of an infinite word accepted by $\mathcal{A}_{G_i}$. These FSMs, which form the controller module, are coded in C. Figure 10 depicts the controller module of the table of the production cell.

18.6 Discussion of the Complexity

The synthesis process is based on the decision procedure for PTL which is exponential in the size of the formula. In general this would limit the practical application of our approach. However the techniques developed for synthesis step 1 and 2 reduce this complexity drastically.

Ad synthesis step 1: This synthesis step exploits the property that two instantiations of the same Symbolic Timing Diagram cannot be active in parallel. Without exploiting this property an operator similar to the complementation construction of Büchi automata [21] must be used twice. This operator generates an automaton with $2^{O(n \log n)}$ states, where n is the number of states of the input automaton [21].

For a typical Symbolic Timing Diagram of the production cell case study table 1 compares the size of an automaton generated with the technique of synthesis step 1 and an automaton generated without this technique. Hence, without the construction of synthesis step 1 a state explosion problem would arise.

Ad synthesis step 2: Without using the partitioning techniques of this step ICOS^2 has to build the 'cross-product' of $\mathcal{A}_1 \dots \mathcal{A}_n$ and use the result automaton $\mathcal{A}_\Pi$ as input for the consistency check. The number of states of $\mathcal{A}_\Pi$ grows exponentially in n. Using a partition of $\mathcal{A}_1 \dots \mathcal{A}_n$, which is based on independent Symbolic Timing Di-

agrams, we overcome this state explosion problem, because it is sufficient to build
the 'cross-product' of the automata of each set of the partition.

controller module	#independent groups	automata			
		group	#STDs	states	trans.
feed belt	3	c1:	2	8	22
		c2:	4	50	303
		c3:	2	8	22
table	3	c1:	4	45	224
		c2:	4	45	224
		c3:	4	16	44
robot	5	c1:	4	45	224
		c2:	2	8	22
		c3:	4	45	224
		c4:	2	8	22
		c5:	4	45	224
press	3	c1:	4	50	303
		c2:	2	8	22
		c3:	2	8	22
deposit belt	3	c1:	2	8	22
		c2:	4	50	303
		c3:	2	8	22
crane	3	c1:	4	46	229
		c2:	2	8	22
		c3:	4	45	224

Table 2   Space reduction gained by synthesis step 2

Important for this attack against the state explosion problem is, that for realistic
examples a partition, which distributes $A_1 \ldots A_n$ over several sets of the partition,
exists. Table 2 shows how the partition technique works for the production cell.
The first column gives the controller module. The second column of table 2 lists

Entity	# of TDs	TDs → automata time (sec.)	size of automata		automaton → time (sec.)	C object code (kByte)
			nodes	edges		
feed	8	2.72	66	347	5.19	176
table	12	2.74	106	492	2.25	107
robot	16	6.19	151	716	30.12	1269
press	8	2.13	66	347	2.51	84
deposit belt	8	2.42	66	347	2.53	69
crane	10	2.50	99	775	2.18	94

Table 3   Over all statistics of the case study

the number of group automata, and the last one the number of STDs, states, and transitions for each group automaton.

18.7 Experimental Results of the Production Cell

The experimental results given in this paper show the feasibility of the approach. Table 3 shows the synthesis results for the production cell. The first column gives the controller module. The second one lists the number of Symbolic Timing Diagrams (axiomatic style) which have been used to specify the controller. The third column of table 3 lists the time which has been used by the synthesis kernel, the fourth one the size of the synthesized automata, the fifth one the time which has been used to generate the C code, and the last one the size of the object code. ICOS[2] has been implemented in the functional programming language **ML** [1] on a SUN SPARC 10 workstation.

The next version of ICOS[2] will contain additional optimizations techniques which will reduce the object code size. Using ICOS[2] a student designed the production cell controller within 30 hours. This proves that ICOS[2] is suitable to reduce the distributed controller design time significantly. The verification techniques embedded in ICOS[2] allow to prove global properties of the synthesis result.

18.8 Conclusion

Within this paper we showed the application of a novel design methodology integrated in the ICOS^2 system to synthesize a distributed controller for the production cell. The case study demonstrated:

- Symbolic Timing Diagrams are a natural method to specify a controller.

- Due to the compositional specification style the synthesizing procedure could cope with the complexity of the case study thanks to automatic decomposition technique. Without the decomposition the complexity of this case study could not have been mastered.

- Specifications where easy to write, test and debug. In an industrial context we expect that this method can reduce the time to market significantly.

Current implementation activities provide a link to feed the FSM output as described in the paper to a silicon-compiler. Future conceptual research will extend the synthesis to data path integration.

Acknowledgements Swen Masuhr provided the specification of the distributed controller of the production cell using ICOS^2. We would like to thank W. Damm, H. Hungar, B. Josko, P. Kelb, and T. Lindner for fruitful discussions.

The work of the second author has been supported by the European Community ESPRIT project No. 6128 (FORMAT).

References

[1] A. Appel and D. MacQueen. Standard ML of New Jersey. Technical report, AT&T BELL Laboratories, 1993.

[2] A. Anuchitanucul and Z. Manna. Realizability and synthesis of reactive modules. In CAV 94, LNCS 818, pages 156-168, 1994.

[3] M. Bombana, P. Cavalloro, and G. Zaza. Specification of a medium complexity device. Technical report, ITALTEL, 1993.

[4] A. Brauer and Th. Lindner. Implementation and usage of a simulation with graphical visualization of the production cell. In C. Lewerenz and T. Lindner, editors, *Case Study "Production Cell": A Comparative Study in Formal Specification and Verification*, pages 273-284. FZI Publication 1/94, Forschungszentrum Informatik, Haid-und-Neu-Straße 10-14, D-76131 Karlsruhe, 1994.

[5]   G. Boriello. Formalized timing diagrams. In Proceedings, *The European Conference on Design Automation*, pages 372-377, Brussels, Belgium, March 1992.

[6]   W. Damm, B. Josko, and R. Schlör. Specification and verification of VHDL-based hardware design. In E. Börger, editor, *Specification and validation methods for programming languages and systems*. Oxford University Press, 1994. To appear.

[7]   E.A. Emerson and E.M. Clarke. Using branching time temporal logic to synthesize synchronization skeletons. In Sci. Comp. Prog., volume 2, pages 241-266, 1982.

[8]   W. Grass, M. Mutz, and W.D. Tiedemann. High level synthesis based on formal methods. In Euro Micro 94, Liverpool, September 5-8, 1994.

[9]   J. Helbig, R. Schlör, W. Damm, G. Doehmen, and P. Kelb. VHDL/S - integrating statecharts, timing diagrams, and VHDL. *Microprocessing and Microprogramming 38*, pages 571-580, 1993.

[10] IEEE. *IEEE Standard 1076-1987: VHDL Language Reference Manual*, 1987.

[11] Khordoc, Dufresne, and Czerny. A Stimulus/Response System based on Hierarchical Timing Diagrams, Publication #770. Technical report, Universite de Montreal, 1991.

[12] C. Delgado Kloos, T. de Mignel, T. Robles, and G. Rabay. VHDL generation from a timed extension of the formal description technique LOTOS within the FORMAT proect. In *Proc. 19th EUROMICRO Conf., Microprocessing and Microprogramming*, pages 589-596, 1993.

[13] B. Kurshan and J. Katzenelson. S/R: A language for specifying protocols and other coordinating processes. In *Proc. 5th Ann. Int. Phoenix Conf. Comput. Commun., IEEE*, 1986.

[14] F. Korf and R. Schlör. Interface controller synthesis from requirement specifications. In *Proceedings, The European Conference on Design Automation*, pages 385-394, Paris, France, 1994. IEEE Computer Society Press.

[15] T. Lindner. Task description. In C. Lewerenz and T. Lindner, editors, *Case Study "Production Cell": A Comparative Study in Formal Specification and Verification*, pages 9-21. FZI Publication 1/94, Forschungszentrum Informatik, Haid-und-Neu-Straße 10-14, D-76131 Karlsruhe, 1994.

[16] Ph. Moeschler, H.P. Amann, and F. Pellandini. High-level modelling using extended timing diagrams. In *proceedings of EURO-DAC'93*, September 1993.

[17] Zohar Manna and Amir Pnueli. *The Temporal Logic of Reactive and Concurrent Systems. Specification*. Springer-Verlag, 1992.

[18] Z. Manna and P. Wolper. Synthesis of communicating processes from temporal logic specifications. In *ACM Trans. Prog. Lang. Sys.*, volume 6, pages 68-93, 1984.

[19] A. Pnueli and R. Rosner. On the synthesis of a reacitve module. In *Proc. 16th ACM Symp. Princ. of Prog. Lang.*, pages 179-190, 1989.

[20]   A. Pnueli and R. Rosner. On the synthesis of an asynchronous reactive module. In *16th International Colloquium on Automata, Languages, and Programming, LNCS 372*, pages 652-671, 1989.

[21]   S. Safra. On the complexity of ω-automata. In *Proc. 29th Symp. on Found. of Comput. Sci*, pages 319-327, 1988.

[22]   R. Schlör and W. Damm. Specification and verification of system level hardware design using timing diagrams. In *Proceedings, The European Conference on Design Automation*, pages 518-524, France, February 1993.

[23]   W.D. Tiedemann, S. Lenk, C. Grobe, and W. Grass. Introducing structure into behavioral descriptions obtained from timing diagram specification. In *Proc. 19th EUROMICRO Conf., Microprocessing and Microprogramming*, pages 581-588, 1993.

[24]   J. van Leeuwen, editor. *Handbook of Theoretical Computer Science*. MIT Press, 1990.

XIX. LCM and MCM

Specification of a Control System using Dynamic Logic and Process Algebra

Roel Wieringa

Vrije Universiteit Amsterdam

Abstract

LCM 3.0 is a specification language based on dynamic logic and process algebra, and can be used to specify systems of dynamic objects that communicate synchronously. LCM 3.0 was developed for the specification of object-oriented information systems, but contains sufficient facilities for the specification of control to apply it to the specification of control-intensive systems as well. In this paper, the results of such an application are reported. The paper concludes with a discussion of the need for theorem-proving support and of the extensions that would be needed to be able to specify real-time properties.

19.1 Introduction

LCM 3.0 (Language for Conceptual Modeling version 3.0) is a formal language developed for the specification of the external behavior of data-intensive systems [9]. Examples of data-intensive systems are business information systems and database systems. LCM is based on a combination of dynamic logic and process algebra, and contains features to specify control structures as well as data structures. LCM comes with a method for conceptual modeling (MCM), which provides a set of heuristics to find models of external system behavior, and for validating the quality of these models [16][17]. Methodologically, MCM is closely related to the Jackson System Development (JSD), which has been applied to the specification of control-intensive systems [12]. Examples of control-intensive systems are industrial process control systems and robot control systems. This paper is intended to show that LCM/MCM is suitable for the specification of control-intensive systems as well. However, it will become clear from this paper that the utility of LCM/ MCM would be enhanced if automated support would be provided for reachability analysis of systems specified in LCM, and if LCM would be extended with constructs to deal with real time and with exceptions such as device failure. It will be argued in this paper that safety analysis is a special case of reachability analysis.

In section 19.2, MCM is explained by applying it to the development of a model for the production cell control system. Section 19.3 introduces LCM 3.0, again using the production cell control system as running example. Section 19.5 discusses the lessons learned from this application, and section 19.6 concludes the paper with a list of topics for further work.

19.2 Method for Conceptual Modeling (MCM)

MCM is a method to produce formal and informal conceptual models of observable system behavior. An important reason why MCM produces both formal and informal models is that MCM is designed with the aim of allowing both the possibility of formal *verification* of an implementation against a conceptual model, and of *validation* of the conceptual model against informal requirements that arise from discussions with the customer. The goal of verification is to show that a system specified in one way has the same behavior as a system specified in another way. The goal of validation is to show that a specification conforms to the intentions of the domain specialists. In other words, verification is concerned with the question whether we implement a system right, whereas validation is concerned with the question whether we model the right system.

Verification can proceed by formal proof or by testing; validation is essentially an informal affair, because the intentions of the domain specialists are themselves informal. In MCM in order to make formal verification possible, the conceptual model is specified in a formal language (LCM 3.0). In addition, in order to make validation possible, the model is specified informally, by means of diagrams and structured text. The informal presentation techniques have a precise correspondence with the formal specification.

Another design aim of MCM is to integrate useful elements of existing methods, ranging from Entity-Relationship modeling and Data Flow modeling to object-oriented modeling techniques. One reason for this is that there should be progress in system development methods. Inventing something new every five years does not constitute progress. Instead, we should take whatever is good from existing methods and try to improve on it. Another reason for this approach is that the acceptance of a method is likely to be increased when it uses techniques that are familiar to practitioners. This is incidentally an additional reason why formal and informal methods are combined in MCM: it enhances the usability of the formal method.

MCM models a system as consisting of a collection of communicating objects. Each object has a local state and a local behavior. Objects communicate with each other by means of synchronous communication events. Objects are subject to integrity constraints, which are static or dynamic constraints on allowable object states or behavior. Constraints may be local to one object or global to the entire system. In a data-intensive system, constraints represent business rules. In a control-intensive system, they can be used to express safety constraints.

When we model a data-intensive system in MCM, we first make a model of the *universe of discourse* (UoD) of the system, which is the part of the world about which the system registers data. For example, in a model of a library database system, we would model the UoD as a set of objects such as DOCUMENT, MEMBER, RESERVATION and LOAN instances. A borrow event in the UoD would be a synchronous communication event that in one atomic state transition deletes a RESERVATION object and creates a LOAN object. UoD objects would be represented in the system by records, that act as *surrogates* for the corresponding UoD objects. The state of the surrogates is updated whenever the corresponding UoD objects change state. For example, the borrow event in the UoD would be registered by a database transaction, that deletes a RESERVATION record and creates a LOAN record in one atomic state transition. Thus, each database system transaction corresponds to an event in the UoD, in which one or more UoD objects participate. Database systems are essentially registration systems.

The basic idea to transfer and extend this to a method for modeling control systems is to model the UoD as a set of communicating devices, each of which is represented in the system by a device surrogate. In real-time methods, such a surrogate is often called a *virtual device* [11]. Devices communicate by participating in a synchronous event. The control function of the system is provided by defining control objects in addition to the device objects. A control object does not correspond to a UoD object. It encapsulates device communications and enforces the required behavior of the devices. The problem of controlling devices in the UoD is thus reduced to the problem of defining the required synchronizations between the devices. This idea will become clear by the illustrations in the following paragraphs, taken from the production cell control system.

19.2.1 The class model

As explained above, one class is defined for each device in the UoD. It turned out to be convenient to define one class for each sensor device and for each actuator device. Examples of sensor object classes are TABLE_SWITCH, TABLE_POTMETER, and ARM1_POTMETER. Examples of actuator object classes are TABLE_H_MOTOR, TABLE_V_MOTOR and ARM1_MOTOR. Each object class has only one existing instance in the system.

In addition to the device objects, control objects are defined, that enforce the required behavior of the devices. Corresponding to the modular structure of the production cell system, the control system contains the following control objects: TABLE_CONTROL, ARM1_CONTROL, ARM2_CONTROL, ROBOT_CONTROL, PRESS_CONTROL and CRANE_CONTROL. Again, these are objects classes that each have exactly one existing instance. There are no control objects for the conveyor belts, because it is assumed that they move continuously. This means that it is also assumed that the blanks are spaced on the feed belt with sufficient distance so that the robot and rotary table have the time to return to the positions in which they are ready to receive the next blank from the feed belt.

The central component of models of data-intensive systems is a class diagram, which usually is some form of enhanced Entity-Relationship diagram. In the case of control-intensive systems, such a diagram is usually quite simple. Figure 1 shows a fragment of the class diagram representing the table switch and and the table control object classes.

A class is represented in Coad & Yourdon style [6] by a rectangle partitioned into three areas, listing, from top to bottom, the class name, names of attributes and predicates applicable to class instances, and names of events applicable to class instances. By convention, a predicate name starts with an upper-case letter and an attribute name consists only of lower-case letters. The TABLE_SWITCH object has three predicates, Exists, Table_lower_position and Table_upper_position, and two events, table_lower_position and table_upper_position. No object in the model has attributes.

Formally, the extension of a class is the set of all possible identifiers (oids) of the class instances. All predicates, attributes and events declared in the class are applied to oids. Each class has at least the Exists predicate, which is set to true when the object starts its existence and set to false when it ceases to exist. All in-

TABLE_SWITCH
Exits Table_lower_position Table_upper_position
Table_lower_position Table_upper_position

TABLE_CONTROL
Exits
start blank_drops_on_table table_stops_high table_stops_low table_stops_unload table_stops_load move_arm1_to_table remove_blrank_from_table

Figure 1   Class diagram of the table switch and the table control

tegrity constraints only concern existing class instances. The ability to create and delete objects is essential in data-intensive systems, but tends to be less important in control-intensive systems. In the production cell example, all classes have only one instance that eternally exists, and no object is ever created or deleted. One way to introduce objects that are dynamically created and deleted would be to model each incoming blank as an object. However, to express the parallelism between the different device and control objects, we must model these objects as separate processes anyway. Defining a BLANK object class then does not add any information, even in the case that different blanks would require a different treatment by the production cell. (Different treatments of different kinds of blanks could be specified by adding tests and choices to the specification of control object behavior.)

Returning to the class specifications in Figure 1, the Table_lower_position predicate in the TABLE_SWITCH class is needed to be able to specify a safety constraint.

MCM allows the specification of attributes of objects and of relationships between them. A relationship can itself have attributes and behavior. In addition, there is a special *is_a* relationship, that expresses that one class is a subclass of another one, and that defines inheritance of attributes, behavior and constraints from the superclass to the subclass. In our production cell model, there are no attributes and no relationships and there is no taxonomic structure — or more accurately, the attributes, relationships and taxonomic structure that the actual devices and control objects have in the real world, are not represented in the model. Attributes and relationships are typically needed in data-intensive systems to be able to store the necessary data to answer queries. In control-intensive systems, all data needed to be able to perform the control function is usually present in the state of the object life cycles.

The events in the TABLE_SWITCH class are events generated by this sensor. As explained below, these events are forced by the TABLE_CONTROL object to synchronize with other events in the UoD, such as table_stop_v. All events in the TABLE_CONTROL class are synchronization events between events in the UoD.

The life cycle model

The behavior of an object is called an *object life cycle* in MCM and we will follow this practice here. Figure 2 shows two equivalent representations the life cycle of the table control object. The start event is a synchronization event between control objects that is needed because of different speeds with which the parts of the system move. When a blank drops on the table, the control object tells the table motors to move upward and right, until the table reaches the top position and has the direction needed to be unloaded by arm1. It then tells arm1 to move to the table. When arm1 picks up the blank from the table, the table is moved downward and returned to its starting position.

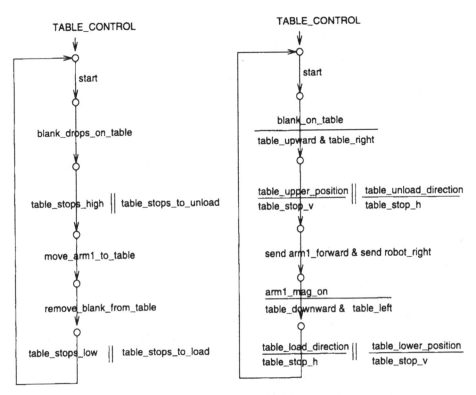

Figure 2 Life cycle of table control

The notation a ‖ b for two events a and b stands for the process ab + ba (a choice between the sequential processes ab and ba). More in general, we can label an arrow in a life cycle diagram with a multiset of processes. A transition along that arrow is then equivalent to a parallel execution of the processes in this multiset. These life cycle diagrams are also called *recursive process graphs* and are defined formally in [13].

On the right-hand side in figure 2, each event in the control life cycle has been replaced by the events that it synchronizes. The notation

blank_on_table
_____
table_upward & table_right

is a fancy way of writing blank_on_table & table_upward & table_right, used to express informally that blank_on_table triggers the other two events. This notation is similar to the well-known stimulus/response notation in Mealy machines [14]. The difference between stimulus and response is not represented in the formal specification; all that is represented is that certain events occur synchronously. Stimulus/response pairs have thus been modeled using Esterel's *synchrony hypothesis*, which says that the response to a stimulus occurs simultaneously with the stimulus [2][5].

The communication structure of the model can be shown by means of context diagrams and by a transaction decomposition table. For each control object, we can draw a *context diagram* such as the one shown in Figure 3.

Just like in data flow (DF) diagrams [22], the system of interest is drawn as a circle, and the systems with which it communicates are drawn as rectangles. These external objects are all devices (sensors or actuators). Note that the table control communicates with devices that are not part of the table, such as ARM1_MOTOR and ROBOT_MOTOR. This is to enforce synchronization between different parts of the production cell. Unlike DF diagram conventions, an arrow represents a synchronous communication rather than a flow of data, control, energy or material. No distinction is made between these different kinds of communications in the diagram. The direction of the arrow suggests initiative. A double-headed arrow represents a continuous communication. Continuous communications must be translated by an event recognizer into the relevant discrete events. The behavior of the event recognizer can easily be specified by a life cycle (e.g. as a polling algorithm).

A *transaction decomposition table* is a simple way to represent the decomposition of communications (called transactions) into their component events. Figure 4 shows a fragment of the transaction decomposition table of the table control.

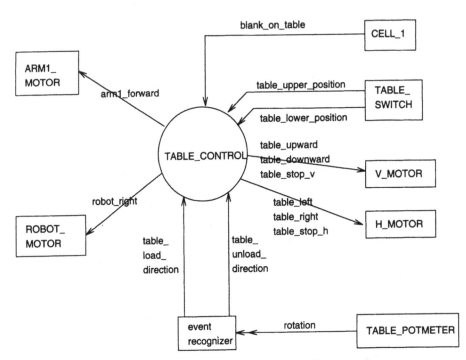

Figure 3   Context diagram of the table control

TABLE_CONTROL	start	blank_drops_ on_table	table_stops_ high	table_stops_ to_unload	move_arm1_ to_table	remove_blank_ from_table
CELL1		blank_on_table				
TABLE_SWITCH			table_upper_ position			
TABLE_V_MOTOR		table_upward	table_stop_v			table_downward
TABLE_H_MOTOR		table_right		table_stop_h		table_left
TABLE_POTMETER				table_unload_ direction		
ARM1_MOTOR					arm1_forward	
ARM1_MAG						arm1_mag_on
ROBOT_MOTOR					robot_right	

*Figure 4   Part of the transaction decomposition table containing trans-
actions of the table control object. The table control transaction omitted
are table_stops_low and table_stops_to_load*

The leftmost column contains object classes, and the top row shows all transactions of the control object. Each transaction is a communication event involving two or more events in the life of device objects. Each column thus shows in the context of which other events a local event is executing, and each row shows what the local events of the objects of a class are. One local event may be part of more than one communication event. The empty column below the *start* event is explained by the fact that the start event is a synchronization event between the different control objects and not between the devices of the system.

The control objects not only synchronize events in the life of sensors and actuators, they also synchronize events among themselves. in other words, global control is *distributed* over the control objects. For example, the move_arm1_to_table event in TABLE_CONTROL synchronizes the rotary table with ARM1 and ROBOT. Distributing these synchronization events over the different control objects makes the specification easier to understand but it makes it less reusable. As an alternative, one could define a PRODUCTION_CELL_CONTROL object, to which global control events are allocated. The lower-level control objects such as TABLE_CONTROL would thereby make less assumptions about the context in which they are employed, and would therefore become more reusable in other contexts.

19.3 Language for Conceptual Modeling (LCM 3.0)

19.3.1 Syntax and intuitive semantics

An LCM 3.0 specification consists of three components:

- A value type specification that defines all abstract data types needed for the model, such as natural numbers or rationals, in an order-sorted conditional equational specification. The intended semantics of the value type specifications is initial.

- A class specification that defines all object– and relationship classes in the model.

- A service specification that defines the system transactions. In a data-intensive system, these are registration events in which one or more system surrogates are updated because their corresponding UoD objects changed state.

In our model of the production cell control system, it turned out that the service specifications had to be enhanced to control object specifications, which have a life cycle that consists of system transactions. This does not involve a change in the underlying logic of the language, but it does require an extension of the syntactic sugar in which this logic is presented to the user.

For each object class to be defined in the class specification, the value type specifications must define an *identifier sort*, that provides all oids of the objects of that class. The identifier sort has the same name as the corresponding class. In the production cell example, the identifier sorts are the simplest sorts possible: they each contain only one constant as element. The identifiers of the table switch and table control objects are defined as shown in Figure 5. In data-intensive systems, there will in general by infinitely many identifiers of a class.

```
begin value type TABLE_SWITCH
    functions
        ts : TABLE_SWITCH;
end value type TABLE_SWITCH;

begin value type TABLE_CONTROL
    functions
        tc : TABLE_CONTROL;
end value type TABLE_CONTROL;
```

Figure 5   Specification of two identifier sorts

There are two kinds of class specifications in LCM 3.0, *object class* and *relationship class* specifications. For each object class, there is a corresponding identifier sort declaration of the same name. For each relationship class, the identifier sort is defined to be the cartesian product of the component identifier sorts. For example, if LOAN is a relationship class between DOCUMENT and MEMBER, then LOAN identifiers have the form $\langle d, m \rangle$, where d is a DOCUMENT identifier and m a MEMBER identifier. There are no relationship classes in the production cell example.

The *is_a* relationship between classes can be defined by defining a partial order on identifier sorts. For example, in a library database, we may want to declare the subclass relationship BOOK $\leq$ DOCUMENT. Semantically, this means that the class of book identifiers is a subset of the class of document identifiers. There is no significant taxonomic structure in the production cell (this is discussed in more detail in Section 19.5.)

The TABLE_SWITCH and TABLE_CONTROL classes are specified in Figures 6 and 7.

```
begin object class TABLE_SWITCH
    predicates
        Exists                              initially true;
        Table_upper_position;
        Table_lower_position               initially true;
    events
        table_lower_position;
        table_upper_position;
    life cycle
        TABLE_SWITCH = table_upper_position . table_lower_position . TABLE_SWITCH;
    axioms
        [table_upper_position(ts)] not Table_lower_position(ts);
        [table_lower_position(ts)] not Table_upper_position(ts);
end object class TABLE_SWITCH;
```

Figure 6 Specification of the TABLE_SWITCH object class

For each class, attributes, predicates and events are declared. The TABLE_SWITCH object contains two predicates, which are both initialized to true. In general, the Exists predicate will be set to true for an identifier by a creation event and set to false by a deletion event.

The production cell specification does not contain any attribute declarations. As an example of what an attribute declaration looks like, the following attribute of TABLE_CONTROL would count the number of blanks that has dropped on the table:

```
attributes
    nr_of_blanks : NATURAL           initially 0;
```

An attribute is a unary function on the identifier sort of the class. Only the co-domain of the function is shown in the specification. Thus, nr_of_blanks is a function TABLE_CONTROL → NATURAL.

The events applicable to the instances of the class are declared in the events section. Each event is a function with codomain EVENT and may have several argument sorts. The first argument sort is not shown in the declaration, because it is always the identifier sort of the class. Thus, table_upper_position is a function TABLE_SWITCH → EVENT and blank_drops_on_table is a function TABLE_CONTROL x CELL1 x TABLE_V_MOTOR x TABLE_H_MOTOR → EVENT.

All communication events are *transactions* of the control system with its environment, and they are declared as such in TABLE_CONTROL. Their decomposition into local events is defined in the transaction decomposition section of the class specification where the transactions are specified.

```
begin object class TABLE_CONTROL
    predicates
        Exists                      initially true;
    transactions
        start;
        blank_drops_on_table(CELL1, TABLE_V_MOTOR, TABLE_H_MOTOR);
        table_stops_high(TABLE_SWITCH, TABLE_V_MOTOR);
        table_stops_low(TABLE_SWITCH, TABLE_V_MOTOR);
        table_stops_to_load(TABLE_POTMETER, TABLE_H_MOTOR);
        table_stops_to_unload(TABLE_POTMETER, TABLE_H_MOTOR);
        move_arm1_to_table(ARM1_MOTOR, ROBOT_MOTOR);
        remove_blank_from_table(TABLE_H_MOTOR, TABLE_V_MOTOR);
    transaction decompositions
        blank_drops_on_table(c1, tv, th) = CELL1.blank_on_table(c1) &
                    TABLE_V_MOTOR.table_upward(tv) &
                    TABLE_H_MOTOR.table_right(th);
        table_stops_high(ts, tv) = TABLE_SWITCH.table_upper_position(ts) &
                    TABLE_V_MOTOR.table_stop_v(tv);
        table_stops_low(ts, tv) = TABLE_SWITCH.table_lower_position(ts) &
                    TABLE_V_MOTOR.table_stop_v(tv);
        table_stops_to_load(tp, th) = TABLE_POTMETER.table_load_direction(tp) &
                    TABLE_H_MOTOR.table_stop_h(th);
        table_stops_to_unload(tp, th) = TABLE_POTMETER.table_unload_direction(tp) &
                    TABLE_H_MOTOR.table_stop_h(th);
        move_arm1_to_table(a1m, rm) = ARM1_MOTOR.arm1_forward(a1m) &
                    ROBOT_MOTOR.robot_right(rm);
        remove_blank_from_table(th, tv) = TABLE_H_MOTOR.table_left(th) &
                    TABLE_V_MOTOR.table_downward(tv);
    life cycle
        TABLE_CONTROL = start .
                    blank_drops_on_table .
                    (table_stops_high II table_stops_to_unload) .
                    move_arm1_to_table .
                    remove_blank_from_table .
                    (table_stops_low II table_stops_to_load) .
                    TABLE_CONTROL;
end object class TABLE_CONTROL;
```

Figure 7   Specification of the TABLE_CONTROL object class

The life cycle of the class instances is defined in a recursive process specification in the style of ACP [1]. The class name is used as main variable of this specification. The & operator in the transaction decomposition specification is the communication operator from ACP. It is commutative and associative. If $t = e_1$ & e_2, then t is considered to be different from e_1 and e_2, and the effect of t is the joint effect of that of e_1 and e_2. The process executed by the specified system is the parallel composition of all object life cycles, in which all local events that are not transactions are encapsulated (renamed to deadlock). The only events that can occur, are transactions, whose effect is the same as the joint effect of the component events.

A class specification may contain axioms in order-sorted dynamic logic with equality, that constrain the values of attributes, define the effect of the events on the attributes, and define event preconditions. All axioms are universally quantified, but the quantifications are not shown in the example. Axioms must either be static integrity constraints, effect axioms, or precondition axioms.

A *static integrity constraint* is a formula without modal operators. It is an invariant of the state space of the system. An example of a static integrity constraint would be

nr_of_blanks(t) < 1000;

Safety constraints are all specified as static integrity constraints.

An *effect axiom* has the form $\phi \rightarrow [\alpha]\psi$, where ϕ and ψ are conjunctions of atoms of the form $a = c$ or literals, and α is an event. The meaning of $[\alpha]\psi$ is that after all possible executions of α, ψ is true.

A *precondition axiom* defines a necessary precondition of success for an event and has the form $\langle\alpha\rangle true \rightarrow \psi$. The meaning of $\langle\alpha\rangle\phi$ is that there is a possible execution of α that terminates and after which ϕ is true. The meaning of $\langle\alpha\rangle true$ is that there is a possible execution of α that terminates. The meaning of the precondition axiom $\langle\alpha\rangle true \rightarrow \psi$ is therefore that if we are in a state where there is a possible execution of α that terminates, then currently, ψ is true. For all non-creation events, there are implicit preconditions of the form

<table_lower_position(ts)>true → Exists(ts);

That is, only existing objects can execute (non-creation) events.

There are implicit *frame axioms* that define the non-effects of an event, i.e. they say what attributes and predicates do not change during the event.

19.3.2 Outline of declarative semantics

The declarative semantics of a class specification consists of three parts, an abstract data type, process algebra and a Kripke structure (Fig. 8).

Details on the declarative semantics of LCM specifications are given elsewhere [15][20]. Here, a brief outline is given of the basic ideas.

The value specification part of an LCM specification consists of a data type specification, which is interpreted in an *abstract data type A*, and a process type

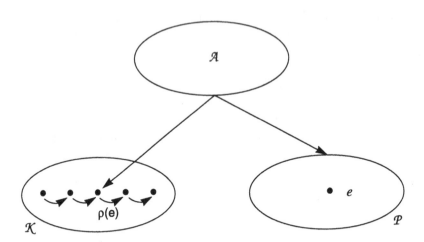

Figure 8   The structure of a model for LCM specifications

specification, explained below. As stated before, the intended semantics of the value type specification is the initial algebra semantics. This means that the abstract data type contains only data elements that can be named by closed terms in the specification, and that data elements are identified if and only if the closed terms denoting them can be proven equal in the value specification [7][10].

In addition to a value type specification, an LCM specification consists of class and transaction specifications. The class specification declares unary functions (called attributes) and unary predicates, both of which can be updated. The axioms of the specifications give static constraints on the attributes and predicates, and give effect and precondition axioms, that state how they are updated. These are interpreted in a *Kripke structure K* that consists of a set of possible worlds. All the possible worlds share the same domain, which is the abstract data type defined by the value specification. Different possible worlds may however assign a different interpretation to the attributes and predicates.

The class and transaction specifications in an LCM specification also declare atomic events, that are combined into object life cycles. Formally, we have sorts EVENT and PROCESS with EVENT ≤ PROCESS. The value specification defines process combinators as functions on PROCESS. For example, choice is a function declared with infix notation as

$$\_ + \_ \ : \ \text{PROCESS} \times \text{PROCESS} \rightarrow \text{PROCESS}.$$

The axioms for the process combinators are taken from ACP [1]. These declarations and axioms jointly form a process theory, which can be viewed as a specification of a process type. The process type specification is interpreted in the *process algebra P* of the model. The intended semantics is here the standard graph model of processes, slightly enhanced to include recursive process graphs [13]. The process algebra gives a meaning to processes independently of their effect on attributes and predicates. This means basically that *P* defines an equivalence relation on processes, that formalizes an observational equivalence notion. For example, the terms $e_1 + e_2$ and $e_2 + e_1$ are interpreted as the same process, because choice is commutative.

To define the effect of events and terminating processes on the attributes and predicates in the Kripke structure, a function ρ is defined that for each event *e* in *A*, defines an accessibility relation

$$\rho(e) \subseteq PW \times PW,$$

where *PW* is the set of possible worlds in *K*. ρ is extended by structural induction to terminating processes. There are many such functions ρ compatible with the axioms in the LCM specification. The intended semantics is that ρ assigns a minimal accessibility relation to *e*. In particular,

- ρ assumes that the conjunction of the completed preconditions of each event is necessary and sufficient for the event to lead to a next world. The completed precondition of an event *e* is the conjunction of all preconditions of *e* listed in the specification, all static integrity constraints (i.e. nonmodal axioms) and a nonmodal formula that guarantees the *e* leads to a next possible state.

- ρ(*e*) leads only to worlds that differ minimally from the current world.

There are still many different formalizations of this minimal change semantics, some of which are computationally tractable. This is subject of current study [18].

19.3.3 Axioms and inference rules

An axiom system that is complete for the loose semantics is given in [20]. A complete axiom system for the intended semantics given above is not yet known but in [19], we give a system that contains some of the needed frame axioms. To give an impression of the system, I list the modal logic axioms in Figure 9.

The R axiom says that if two actions are equal in the process algebra, then they have the same effect on the attributes and predicates of objects. It corresponds with the congruence condition on the function ρ in the semantics and it makes the axiomatization (without the frame axioms PosFr and NegFr) complete with respect to the loose semantics [20]. Axioms not listed above include first-order logic axioms,

(K) $[e] (\phi \rightarrow \psi) \rightarrow ([e] \phi \rightarrow [e]\psi)$

(R) $$\frac{\forall D :: e_1 = e_2}{\forall D :: [e_1]\phi \leftrightarrow [e_2] \phi}$$

(Barcan) $\forall x: s :: [e]\phi \rightarrow [e]\forall x:s :: \phi$ for $\notin Var(e)$

(PosFr) $\forall D :: P(t_1, ..., t_n) \rightarrow [e]P(t_1, ..., t_n)$
             where P is nonupdatable and t_i contains only nonupdatable function symbols ($i = 1, ... , n$).

(NegFr) $\forall D :: \neg P(t_1, ..., t_n) \rightarrow [e] \neg P(t_1, ..., t_n)$
             where P is nonupdatable and t_i contains only nonupdatable function symbols ($i = 1, ... , n$).

(N) $$\frac{\phi}{[e]\phi}$$

Figure 9   Modal logic axioms

axioms for substitution and congruence, equality axioms and the usual inference
rules for first-order logic.

19.4 Verification of Safety and Liveness

Safety constraints are easily formalized as static integrity constraints in LCM. An
example of a safety constraint is

```
    not Robot_zero_angle(rm)                    and
    not Arm1_zero_extension(a1m)                 and
    not Table_lower_position(ts)
->  not (Blank_on_table(c1) and Arm1_holds_blank(a1))
```

Here, rm identifies the robot motor, a1m the motor of arm1, ts the table switch,
c1 the photoelectric cell that signals whether a blank has arrived at the table, and
a1 the magnet of arm1. The constraint says that if arm1 comes close to the table,
then there must not be both a blank on the table and a blank in arm1. Specification
of this constraint requires the addition of a number of predicates to the specification, and a number of effect axioms that update these predicates at the appropriate
events. In order to prove that this constraint is respected during any possible behavior of the system, it was necessary to add some synchronization points. This is
the reason why the events start and move_arm1_to_table are present in the table
control life cycle. Since there is currently no automatic theorem-prover for LCM,

the constraints were proven manually. This could not be done as rigorous and detailed as would have been possible with the aid of a theorem-prover.

The idea of the proof of safety is very simple. The safety constraint is an axiom that should be true in all states of the model (i.e. in all possible worlds of the Kripke structure). The system is a parallel composition of a number of cyclic processes that in each of their transactions should respect the safety constraints. This is a classic integrity constraint verification problem known from databases: For each possible transaction, we should be able to prove that if the system currently satisfies the constraint, then it satisfies the constraint after executing the transaction. For any given constraint, only a few transactions are able to violate a constraint, viz. those transactions that update a predicate or attribute that occurs in the constraint. For example, only a few transactions update the safety constraint listed above, such as move_arm1_to_table. The proof that these transactions do not cause violations of the safety constraints is an elementary (but tedious) application of the axioms outlined above.

The specification makes a number of assumptions about the environment, that cannot currently be expressed in LCM. In general, it is assumed that the control system is ready to receive an event from its environment when it occurs. For example, it is assumed that the elevating rotary table is ready to receive the next blank when it arrives. This assumption cannot be expressed without constructs to specify and reason about real-time properties.

The safety constraints are currently formulated in an overly restrictive way, and as a consequence, the synchronization points introduced in the current specification cause an unnecessary reduction in concurrency between the movement of the parts of the production cell. This makes the system less efficient than it could be.

One of the results of this experiment in formal specification is that it became clear that the designer needs a tool to explore the design space effectively and efficiently. The tool should allow the designer to search for a specification that is optimal according to a set of criteria. One such criterion is that the parts of the production cell should not collide. Another is that the speed with which a blank is put through the system is as high as is possible, given the constraints on the system. Putting this differently, what we need is a tool that allows the designer to translate assumptions about the speed with which blanks arrive at the elevating rotary table and the speed with which the parts of the production cell moves into an optimal design of the control system. Safety analysis is only one part of the capability of such a tool. Real time analysis and, possibly, real space analysis (to reason about locations and speed) is another.

We are currently designing a tool for *reachability analysis*, that allows the designer to evaluate the reachability properties of the specified system [8]. In general, the system will be able to answer questions of the form "Starting from a state satisfying ϕ_1, is there a sequence of transactions to a state satisfying ϕ_2 such that the path satisfies constraints Φ?". If there are such paths, the system should exhibit one of these and be able to show why it is compatible with the constraints. If there is no such path, the system should be able to explain why. Answering these queries requires theorem-proving as well as planning capability. We believe that such a system would aid the designer in the exploration of the design space in search of a system design that is safe as well as efficient. If a path of transactions is found that leads from an initial state to some desired state, it would allow the designer to present a constructive proof that a liveness property is satisfied by the system.

In general, reachability queries are unsolvable, but since our specifications have a simple form, there is hope that we can solve some interesting subcases using theorem-proving techniques. The techniques for finding a reachability (dis)proof borrows ideas from plan generation and theorem-proving in AI. To make the specification more amenable to these techniques, it is first translated into situation calculus, where for each updatable predicate, the events that can lead to its update, and the necessary and sufficient preconditions for each event, are listed. This is used to reason from the desired final state ϕ_2 to conditions ψ on an initial state that is compatible with the given conditions ϕ_1 on the initial state. A tableaux technique is used to find a model of the formula $\psi \wedge \phi_1$. A path from this model to the final state is then generated by applying the events in forward direction. There is a prototype implementation of this procedure in Prolog, using techniques from the Satchmo theorem prover [4][8].

19.5 Discussion of the Specification

19.5.1 Extending MCM to control-intensive systems

The application of MCM to control-intensive systems required one change in the method. In data-intensive systems, all control is present in the UoD. Most data-intensive systems are *registration systems*: they merely register the events that occur in the UoD (and answer queries about the registered data). Consequently, the structure of a model of system behavior of data-intensive systems is very simple: We define a number of classes, one for each UoD object. Each UoD object may perform events, and some of them perform communications. These events and communications are registered by the system. The system model is therefore a copy of the UoD model, with the difference that the events and communications of system objects are initiated by the events and communications of the corresponding UoD objects.

Just as for registration systems, the transactions of the production cell control system correspond with transactions in the UoD. There are however two differences between control systems and registration systems:

- The initiative of the transaction may be with the control system as well as with objects in the UoD. Usually, a transaction is initiated by a device in the UoD. Because the control system enforces a synchronization with another device, the second device is then also forced to perform an event.

- The transactions of the control system are encapsulated in a control object, which has its own life cycle, by which it enforces meaningful behavior on the objects in the UoD.

No assumption is made about how many sensors or actuators participate in one transaction. In addition, we may specify different transactions to occur synchronously, so that several stimulus/response pairs may occur at the same time.

These extensions to MCM do not require a change in the logic of LCM. They merely reflect a different use of this logic. Note that control objects are very similar to the interactive function processes of JSD [12].

19.5.2 Extending LCM to control-intensive systems

LCM was found to be suitable for the specification of control-intensive systems. Some parts of the language were not used at all in this example. Thus, the specification uses mainly the constructs from ACP. Attributes, relationships between objects and taxonomic structures were not defined. In addition, object classification is not an issue in the production cell control system. However, the facilities of the language to specify these structures did not decrease the ease of use of the language for the specification of control structures.

There are some obvious extensions to the language, which would make it more useful for the specification and validation of such systems:

- *Real time* constructs that allow one to specify timing properties of events and states of the system. For example, in the spirit of time Petri nets [3], we could label each transition e with an interval $[t_1, t_2]$ that expresses that *e* cannot be performed before time t_1 has elapsed from the moment that *e* is enabled, and must be performed before time t_2 has elapsed. This allows one to specify the time that a system must wait in a state as well as a time-out before which a certain event must occur. It also allows us to derive, by means of a reachability analysis, the maximum time needed for the production cell before it is ready to receive the next blank from the feed belt.

- *Exceptions* such as device failures have not been expressed at all in the current model. This requires the specification of time-outs as well as, more generally, the occurrence of abnormal events and recovery from abnormal behavior.

As a first move towards the realization of these extensions to LCM, we have started a project that extends dynamic logic with real time and with deontic logic (the logic of actual and ideal behavior) [21].

19.5.3 Implementation

The system has not been implemented in executable code. In the past, students have manually translated LCM specifications of database system behavior into SQL database schemas embedded into C, and into a persistent version of C++. The goal of these projects was to find out whether these translations could be done at all, and what could be preserved of the structure of the LCM specification. Verification of the implementation has not been performed, but it is clear that the use of dynamic logic offers the possibility to verify whether transactions have been implemented correctly. Each transaction is a terminating program that should realize the effects as specified in the effect axioms of the LCM specification. This remains a topic for further research.

19.5.4 Specification and verification effort

The specification was found by first searching for an informal model, represented in a number of diagrams and accompanying (unstructured) notes and comments. The major tool to come to grips with the informal model turned out to be the event trace diagram (also called message sequence diagram in some methods), showing all local events in the objects of the system as well as the synchronizations between the processes. It took me three iterations to arrive at the current architecture of the control system (i.e. this is the fourth version of the system). Altogether, this took me about 12 hours, distributed over one week. By dry-running the system, I convinced myself that I had modeled a system that should be able to respect the safety constraints. To increase my confidence, I specified a number of safety constraints formally and proved them manually.

Having satisfied myself that I had found a stable model of a safe system, I wrote the formal specification, including the predicates that I discovered would be necessary for the formal proof of system safety. Excluding the crane specification, the result is about 8 pages and took about a day to write. Detailed proof of the safety constraints, down to the most detailed propositional logic manipulations, was not performed because this requires a theorem-prover.

Writing down the formal specification did not involve much creative work (other than inventing informative names for parts of the specification). What is sorely needed in this kind of clerical work is a workbench with intelligent text editing facilities for the specification as well as graph editing facilities for the diagrams, and the ability to cross-check the different parts of the formal and informal specification. Current research includes the incremental specification of such a workbench in LCM, and an implementation in C++.

19.6 Conclusions

With the minor extension of a construct to define control objects, LCM 3.0 is suitable for the specification of control-intensive systems. However, it does not contain any facilities for the specification of real-time properties that are usually important for control-intensive systems. Future work will therefore include extension of the language and its logic with real-time constructs. Current work includes the design and implementation of workbench for building integrated formal and informal specifications and of a tool for reachability analysis (without real time) of systems specified in LCM 3.0.

Acknowledgements

This paper benefited from comments made by Remco Feenstra and Claus Lewerentz on an earlier version.

References

[1]   J.C.M. Baeten and W.P. Weijland. *Process Algebra*. Cambridge Tracts in Theoretical Computer Science 18. Cambridge University Press, 1990.

[2]   G. Berry and G. Gonthier. The ESTEREL synchronous programming language: design, semantics, implementation. *Science of Computer Programming*, 19:87–152, 1992.

[3]   B. Berthomieu and M. Diaz. Modeling and verification of time dependent systems using time Petri nets. *IEEE Transactions on Software Engineering*, SE-17:259–273, March 1991.

[4]   F. Bry, H. Decker, and R. Manthey. A uniform approach to constraint satisfaction and constraint satisfiability in deductive databases. In *Proceedings of the International Conference on Extending Database Technology (EDBT)*, pages 488–505, Venice, 1988. Springer-Verlag.

[5]   R. Budde. ESTEREL Applied to the case study production cell. In *Case Study "Production Cell"*, FZI-Publication 1/94, pages 51–75. Forschungszentrum Informatik an der Universität Karlsruhe, 1994.

[6]   P. Coad and E. Yourdon. *Object-Oriented Analysis*. Yourdon Press/Prentice-Hall, 1990.

[7]   H. Ehrig and B. Mahr. *Fundamentals of Algebraic Specification 1. Equations and Initial Semantics*. Springer, 1985. EATCS Monographs on Theoretical Computer Science, Vol. 6.

[8]   R.B. Feenstra and R.J. Wieringa. Validating database constraints and updates using automated reasoning techniques. Submitted for publication.

[9]   R.B. Feenstra and R.J. Wieringa. LCM 3.0: a language for describing conceptual models. Technical Report IR-344, Faculty of Mathematics and Computer Science, Vrije Universiteit, Amsterdam, December 1993.

[10]   J.A. Goguen, J.W. Thatcher, and E.G. Wagner. An initial algebra approach to the specification, correctness, and implementation of abstract data types. In R.T. Yeh, editor, *Current Trends in Programming Methodology*, pages 80–149. Prentice-Hall, 1978. Volume IV: Data Structuring.

[11]   H. Gomaa. *Software Design Methods for Concurrent and Real-Time Systems*. Addison-Wesley, 1993.

[12]   M. Jackson. *System Development*. Prentice-Hall, 1983.

[13]   P.A. Spruit and R.J. Wieringa. Some finite-graph models for process algebra. In J.C.M. Baeten and J.F. Groote, editors, *2nd International Conference on Concurrency Theory (CONCUR'91)*, pages 495–509, 1991.

[14]   P.T. Ward and S.J. Mellor. *Structured Development for Real-Time Systems*. Prentice-Hall/Yourdon Press, 1985. Three volumes.

[15]   R.J. Wieringa.   A formalization of objects using equational dynamic logic.   In C. Delobel, M. Kifer, and Y. Masunaga, editors, *2nd International Conference on Deductive and Object-Oriented Databases (DOOD'91)*, pages 431–452. Springer, 1991. Lecture Notes in Computer Science 566.

[16]   R.J. Wieringa.   A method for building and evaluating formal specifications of object-oriented conceptual models of database systems (MCM).   Technical Report IR-340, Faculty of Mathematics and Computer Science, Vrije Universiteit, December 1993.

[17]   R.J. Wieringa and R.B. Feenstra.   The university library document circulation system specified in LCM 3.0. Technical Report IR-343, Faculty of Mathematics and Computer Science, Vrije Universiteit, Amsterdam, December 1993.

[18]   R.J. Wieringa and W. de Jonge. Object identifiers, keys, and surrogates. *Theoretical and Practical Aspects of Object Systems*, To be published.

[19]   R.J. Wieringa, W. de Jonge, and P.A. Spruit.   Roles and dynamic subclasses: a modal logic approach. In M. Tokoro and R. Pareschi, editors, *Object-Oriented Programming, 8th European Conference (ECOOP'94)*, pages 32–59. Springer, 1994. Lecture Notes in Computer Science 821. Extended version to be published in *Theory and Practice of Object Systems (TAPOS)*.

[20]   R.J. Wieringa and J.-J.Ch. Meyer.   Actors, actions, and initiative in normative system specification. *Annals of Mathematics and Artificial Intelligence*, 7:289–346, 1993.

[21]   R.J. Wieringa and J.-J.Ch. Meyer. DEIRDRE (deontic integrity rules, deadlines and real time in databases). Faculty of Mathematics and Computer Science, Vrije Universiteit., 1993.

[22]   E. Yourdon. *Modern Structured Analysis*. Prentice-Hall, 1989.

XX. Modula-3

Modelling and Implementation

Andreas Rüping, Emil Sekerinski

Forschungszentrum Informatik, Karlsruhe

Abstract

We present the modelling, implementation, and verification of a software system for the control of an industrial production cell. We use techniques of object-oriented and of parallel programming for both modelling and implementation. The implementation is done in Modula-3. We demonstrate the verification of safety requirements for the production cell.

We discuss how well Modula-3 is suited for developing the control software in this case study. In detail, we analyse the benefits of object-oriented and parallel constructs and how both can be integrated with each other.

20.1 Introduction

This report is part of a comparative study analysing different modelling, specification, verification, and implementation techniques when applied to a realistic example: an industrial production cell, consisting of six interacting devices (cf. [5]). Several approaches from the field of formal methods have been applied to this example, with the focus on discussing their practicability.

The different approaches include this one which deals with the modelling, implementation, and verification of a control software system for the production cell using Modula-3 (cf. [3][6]), a successor to Modula-2. Modula-3 adopts the techniques of structured and modular programming well known from Modula-2, and adds constructs for object-orientation and parallel programming.

Since the production cell is likely to be seen as several parallel objects, Modula-3 is a language predestined for its modelling and implementation. Furthermore, Modula-3 provides the basis for the verification of safety requirements placed on the production cell.

The aims of this case study are therefore the following:

- *Modelling*

  We set up a model of the production cell while keeping the object-oriented and parallel constructs of Modula-3 in mind. We analyse how well the object-oriented and parallel constructs can be integrated with each other.

- *Verification*

  On the basis of abstract program obtained in the modelling stage, we verify that the control software system fulfils certain safety requirements of the production cell.

- *Implementation*

  We present a Modula-3 implementation for controlling the production cell. We analyse in detail how well Modula-3 is suited for implementing a reactive system.

- *Evaluation*

  We evaluate the integration of techniques of object-orientation and parallel programming. We conclude with some remarks on the effort that was necessary for carrying out this case study.

Before going into details of the specification, we briefly introduce the main features of Modula-3 (cf. [3][6]):

- *Basic features*

  Basic features include declarations, statements, basic types, structured types, and procedures.

- *Modules*

  As indicated by its name, Modula-3 provides a modularisation concept. Above all, modules are compilation units. Also, modules serve for the structuring of large programs and thus contribute to the comprehensibility and the reusability of software.

- *Objects*

  Modula-3 supports object-oriented programming with a specific set of features that is integrated with the rest of the language. The basic object-oriented concepts include encapsulation, inheritance, and polymorphism.

- *Threads*

  Concurrent programming is also supported in Modula-3. A Modula-3 program may have multiple points of execution: multiple threads, each of which is an individual sequential program. The concurrent facilities of Modula-3 are lightweight, since all threads in a program live in the same address space. To guarantee mutual exclusion with entering critical sections, Modula-3 provides mechanisms for the synchronization of threads.

20.2 Modelling

For a detailed task description of the production cell we refer to [5]. The task description presented in [5] serves as a requirement specification.

The basic idea of our approach is to model each major device of the production cell as an individual object. Hence we have six major objects: the feed belt, the elevating rotary table, the robot, the press, the deposit belt, and the travelling crane. All devices have to fulfil their tasks autonomously, and they communicate with each other to ensure the correct processing. Therefore, the control of each device is organized in parallel threads that communicate via signals.

Next we discuss in more detail the static and dynamic aspects of all objects.

20.2.1 Object states

We set up a state-based model of the production cell. At every point in time, each object is in a certain state. The state space of an object is determined by the Cartesian product of its attributes. The current state of an object is represented by the current values of these attributes. The state space for each object is finite.

For instance, the press can be in a position for either loading, unloading, or pressing. This is reflected by one attribute which can take three different values. Hence there are three different states of the press.

For more complex objects, however, we only consider essential combinations of attribute values. For the robot for instance, we identify states such as loading, unloading, or waiting, which all cover several attribute values.

The states of different objects are independent of each other, thus allowing for an arbitrary combination of states for several objects. Given n devices, the objects of the entire production cell span an n-dimensional state space. The state of the whole production cell is represented by a vector of this state space.

To ensure the safety requirements for the production cell, we have to restrict the state space of the entire system, in order to rule out unsafe combinations of the states of several objects. An example of such a safety requirement will be presented later on in this case study.

20.2.2 Signals and synchronization

The objects representing the six major devices run through their states in a cyclic manner. For instance, the press cyclically loads, presses, and unloads. Since several devices have to cooperate, synchronization is necessary between them. For instance, before the press can load a metal plate, the robot must be in a state allowing for unloading. Thus, both devices are required to be in appropriate and corresponding states if metal plates shall be handed over from one to the other.

For each pair of neighboured devices there is such a corresponding state in either device. Until both devices have reached this particular state, neither device can continue and change into the next state.

Synchronization requires that two objects communicate via signals. Objects can send a signal and continue their thread. They can also wait for a signal and not continue until the signal is received. We use CSP-like communication constructs (cf. [4]), with *!!* representing the sending of a signal and *??* representing the waiting for it.

In Figure 1 we exemplarily present the possible states of the robot and the press and their communication. Each state is represented by a round cornered box, giving the name of the state and saying which signals are sent as soon as a state is entered. The state transitions are annotated with the signals for which an object has to wait before it can enter the next state.

Figure 1 shows that for the neighboured devices press and robot there are two pairs of corresponding states in which metal plates are handed over: Load (press) and Unload (robot) and, respectively, Unload (press) and Load2 (robot).

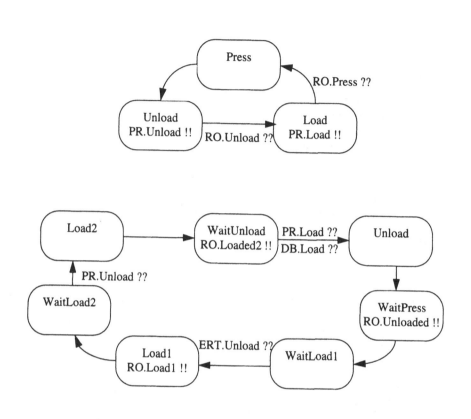

Figure 1   States and synchronization of the press and the robot

The press only communicates with the robot since it sends signals only to the robot and also receives signals only from the robot. However, the robot also has to communicate with the elevating rotary table and the deposit belt since it also exchanges metal plates with these devices.

In some cases, objects have to reach specific states in a particular order. As Figure 1 shows, special waiting states are introduced to ensure this order. For instance, the robot may rotate into the position for loading (WaitLoad2), but then has to wait for the press to be ready before it can extend its arm and switch its magnet on (Load2).

We obtain the exemplary specification of the press and the robot presented in Figure 2 which demonstrates their synchronization. A preliminary specification of the entire production cell is presented in [2].

```
PROCEDURE ControlPress () =
BEGIN
    < initialize >
    WHILE TRUE
    DO PR. Unload !!;
        < provide metal plate to the robot >;
        RO.Loaded2 ??;
        < position for loading >;
        PR.Load !!;
        < receive metal plate from the robot >;
        RO.Unloaded ??;
        < press metal plate; position for unloading >
    END
END


PROCEDURE ControlRobot () =
BEGIN
    < initialize >
    WHILE TRUE
    DO  PR.Unload ??;
        < extend arm 2 into the press >;
        < switch on magnet2 >;
        < retract arm2 >;
        RO.Loaded2 !!;
        < rotate for unloading arm1 >;
        PR.Load ??;
        < extend arm 1 into the press >;
        < switch off magnet1 >;
        < retract arm1 >;
        < unload arm2 at the deposit belt >;
        RO.Unloaded !!;
        < load metal plate on arm1 from the elevating rotary table >;
        < rotate for loading arm2 >
    END
END
```

Figure 2 Synchronization of the robot and the press

20.2.3 Sensors

One problem, however, remains. Some of the actions included in the above specification require sensors to work correctly. For instance, the robot must read a sensor in order to bring its one arm next to the press. Typically, after an actuator such as the press has been activated, its control software has to wait for a sensor to return a particular result before de-activating the actuator.

To avoid active waiting which would result in the purely sequential controlling of the entire production cell, we introduce a particular sensor control, dealing with all sensor values at every point in time. A special process cyclically reads all sensors and, if a sensor returns a relevant value, sends a signal to the device that needs this sensor value. We therefore introduce a special sort of signals thus allowing for a working implementation of sensor depending tasks.

20.2.4 Controlling the production cell

Controlling the entire production cell consists of controlling all single devices plus controlling the sensors. Inside the single threads all processes are executed sequentially and, as we have seen, in a cyclic manner. The threads of all single devices are executed in parallel.

20.3 Verification

It is an important issue to prove that the model obtained in the previous section fulfils certain safety requirements. Safety requirements can be expressed as invariants that have to hold throughout the execution of the control system. We are able to verify that the abstract program controlling the production cell (cf. Figure 2) preserves such invariants and thus guarantees the safety requirements.

As an example of such verification, we prove that the robot's arms cannot be damaged by the press.

20.3.1 Safety requirements as invariants

As mentioned before, safety requirements are fulfilled by appropriately restricting the state space of each object. This is done by introducing invariants on the object's states. Since safety requirements often describe how several devices may or may not behave, invariants typically have to cover the states of several objects.

For instance, for the press and the robot we obtain the following states, describing the position of the press, and the robot's position, arms, and magnets:

- Press: PR.position ∈ { press, load, unload }

- Robot: RO.position ∈ { arm1_to_elevating_rotary_table, arm1_to_press, arm2_to_deposit_belt, arm2_to_press },

    RO.arm1 ∈ { in, out }, RO.arm2 ∈ { in, out },

    RO.magnet1 ∈ { on, off }, RO.magnet2 ∈ { on, off }

The following invariant expresses the safety requirement which says that the robot's arms cannot be damaged by the press:

(INV-SR)   ((RO.position = arm1_to_press ∧ RO.arm1 = out)   ⇒
            (PR.position = unload ∨ PR.position = load))   ∧
            ((RO.position = arm2_to_press ∧ RO.arm2 = out)   ⇒
            (PR.position = unload))

This invariant is required to hold at all *observed* points in time during the parallel execution of all processes. This means that we may consider several actions together to be atomic, and do not check the invariant in between them.

We also assign an invariant to each signal used as a means of synchronization. Such a signal invariant describes a condition which has to hold if a process sends a signal, and which another process can assume to hold when receiving this signal. For the signals used for the synchronization of the robot and the press (cf. Figure 2), we obtain the following signal invariants:

(INV-PR-Load)   PR.position = load

(INV-PR-Unload)   PR.Position = unload

(INV-RO-Loaded2)   RO.arm2 = in ∧ RO.magnet2 = on

(INV-RO-Unloaded)   RO.arm1 = in ∧ RO.magnet1 = off ∧
                       RO.arm2 = in ∧ RO.magnet2 = off

In Figure 3 we present the procedures controlling the press and robot. These procedures are assigned to a module that controls all single devices and their synchronization. The abstract programs in Figure 3 also include conditions (in curly parentheses) which hold at certain times. Each condition is composed of the safety requirement's invariant and certain signal invariants, with the safety requirement's invariant always holding and the signal invariants holding at appropriate times.

```
PROCEDURE ControlPress () =
BEGIN
    PR.position := unload;

    WHILE TRUE

    DO {INV-SR ∧ INV-PR-Unload} (PR-C₀)
        PR. Unload !!;
        RO.Loaded2 ??;
        {INV-SR ∧ INV-RO-Loaded2} (PR-C₁)
        PR.position := load;
        {INV-SR ∧ INV-PR-Load} (PR-C₂)
        PR.Load !!;
        RO.Unloaded ??;
        {INV-SR ∧ INV-RO-Unloaded} (PR-C₃)
        PR.position := press; PR.position := unload
    END
END

PROCEDURE ControlRobot () =
BEGIN
    RO.position := arm2_to_press; RO.arm1 := in; RO.arm2 := in;
    RO.magnet1 := on; RO.magnet2 := off

    WHILE TRUE

    DO {INV-SR} (RO-C₀)
        PR.Unload ??;
        {INV-SR ∧ INV-PR-Unload} (RO-C₁)
        RO.arm2 := out; RO.magnet2 := on; RO.arm2 := in;
        {INV-SR ∧ INV-RO-Loaded2} (RO-C₂)
        RO.Loaded2 !!;
        RO.position := arm1_to_press;
        {INV-SR} (RO-C₃)
        PR.Load ??;
        {INV-SR ∧ INV-PR-Load} (RO-C₄)
        RO.arm1 := out; RO.magnet1 := off; RO.arm := in;
        < unload arm2 at the deposit belt >;
        {INV-SR ∧ INV-RO-Unloaded} (RO-C₅)
        RO.Unloaded !!;
        < load metal plate on arm1 from the elevating rotary table >;
        RO.position := arm2_to_press; RO.arm1 := in; RO.arm2 := in
    END
END
```

Figure 3   Conditions for controlling the robot and the press

20.3.2 Proving invariants

Using the weakest precondition calculus (cf. [1][7]), we can now formally prove
that the conditions presented in Figure 3 indeed hold. The proof obligations
include the following:

- after the initialization the initial condition C_0 hold:

  wp (init, C_0) = true

- given the fact that the condition at a certain point inside the control loop
  holds, then the condition at the (cyclically) next point inside the loop also
  holds:

  $C_i \Rightarrow$ wp (m, C_{i+1})

Such a proof has to be performed for each process. As an example, we demonstrate that the initialization of the press establishes the initial condition PR-C_0 (cf.
Figure 3). We apply the following transformation rule for weakest precondition
expressions:

  wp (x := e, Pred) = Pred [x := e]

This rule allows for replacing a variable *x* by an expression *e* in the predicate
Pred. The proof looks as following:

  wp (position := unload, INV-SR $\wedge$ INV-PR-Unload) =

  wp (position := unload,
        ((RO.position = arm1_to_press $\wedge$ RO.arm1 = out) $\Rightarrow$
        (PR.position = unload $\vee$ PR.position = load)) $\wedge$
        ((RO.position = arm2_to_press $\wedge$ RO.arm2 = out) $\Rightarrow$
        (PR.position = unload)) $\wedge$
        PR.position = unload) =

  ((RO.position = arm1_to_press $\wedge$ RO.arm1 = out) $\Rightarrow$
  (unload = unload $\vee$ unload = load)) $\wedge$
  ((RO.position = arm2_to_press $\wedge$ RO.arm2 = out) $\Rightarrow$
  (unload = unload)) $\wedge$
  unload = unload) =

  true   ❐

The remaining proofs (showing that PR-$C_i \Rightarrow$ wp (m, PRC$_{i+1}$) and considering
the robot) can be performed in a similar way. Thus, we have verified that the
invariant representing the above safety requirement holds whenever the state of the
production cell is observed. Further safety requirements (cf. [5]) can be verified
according to the same principle.

20.4 Implementation

We only outline the principles of the implementation in Modula-3 since the implementation details are beyond the scope of this paper. For details we refer to [2] where a preliminary version of the implementation is presented.

20.4.1 Objects and classes

We introduce one object for each of the six devices. Thus we obtain the classes *FeedBelt*, *ElevatingRotaryTable*, *Robot*, *Press*, *DepositBelt*, and *TravellingCrane*. Since we need only one object of each of these classes, they are only instantiated once. The classes supply procedures for controlling the device and attributes describing the subobjects of which a device consists.

For instance, the class *Robot* has procedures to rotate clockwise or counterclockwise and to extend or to retract one of its arms. Attributes of *Robot* include its motor and the sensors that are used to measure how far each arm is extended.

However, not all of these features are public. Applying the principle of information hiding, only procedures representing the major tasks of an object are made public. For instance, the robot provides the task of picking up a metal plate from the press. This task is implemented by calling private procedures as well as sending and receiving signals.

Attributes are public only if this is necessary for the state change of an object to be invoked from outside. For instance, the control class has to have write access to the position of the press.

The attributes of the six major objects all belong to one of the following two categories: they are either actuators or sensors. The motor, for instance, is an example of an actuator. We therefore introduce two abstract classes *Actuator* and *Sensor* from which the various actuators and sensors inherit.

There is also an inheritance hierarchy for the classes representing the six major devices. An abstract class *Device* is introduced serving as a common superclass for all of them. There is also a class *Belt* which is a superclass of both *FeedBelt* and *DepositBelt*. The inheritance hierarchy is presented in Figure 4.

Finally, we introduce a class *Controller* for controlling the whole production cell by invoking the public methods of the single devices. Within the control system, there is only one instance of this class, a controller object that incorporates the complete control over the production cell.

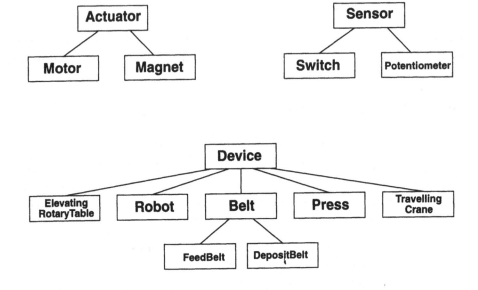

Figure 4   Inheritance hierarchy

20.4.2 Threads

We apply three mechanisms provided by the thread concept of Modula-3: thread creation, mutual exclusion, and waiting for events.

Each device has its own thread which, after being created, controls this device in an infinite loop. Mutual exclusion is used when accessing the actuators. If an actuator is accessed by one thread, it cannot also be accessed by another one. Hence, mutual exclusion is used to implement the synchronization of the entire system. Finally, threads may have to wait for events, in our case for certain sensor values to be received via signals, to synchronize themselves.

20.4.3 Signals

Signals are implemented in Modula-3 as so-called conditions. A thread can be blocked until a certain condition is fulfilled and is then unblocked. This works well for all sensors since signals are never received before the thread is blocked.

However, a problem occurs with signals used for the synchronization of objects. In some cases, the condition to continue a thread is fulfilled before the thread is actually blocked. To deal with this problem, we use a mechanism to put received signals into a queue. For the details we refer to [2].

20.5 Evaluation and Conclusions

The case study of applying Modula-3 to the industrial production cell has two major characteristics, both stemming from the language features of Modula-3:

- object-orientation,

- concurrency.

Object-oriented features were successfully ₁applied in the case study. The features used include encapsulation, abstraction, subclassing. Since the production cell consists of several devices, it was natural to model and implement these devices as single objects. In detail, object-orientation turned out to be beneficial with respect to the following aspects:

+ The encapsulation inherent of object-orientation allowed us to hide details of objects that are irrelevant to other objects. For instance, the robot provides as public methods only its major tasks such as picking up a metal plate from the press, rather than switching its motor on or off, waiting for certain sensors, and the like.

+ Object-orientation improved the overall structure of both the model and the implementation. We have set up a class hierarchy with subclasses being specializations compatible with their superclass. Applying the principle of conceptual abstraction, properties and behaviour shared by several classes are introduced in a common superclasses. For instance, features typical of all sensors are extracted to the class *Sensor*. Features are generally introduced at exactly one place in the system, thus the structure of the entire control system became systematic and well comprehensible.

+ Finally, inheriting features from superclasses helped with the reuse of design. Introducing concepts as high as possible in the class hierarchy reduced the effort for the development of the whole system and thus made the development more efficient.

Concurrency is generally useful for this case study for the very obvious reason that the single devices of the production cell are indeed able to work in parallel. As to the threads of Modula-3, we identified the following advantages and disadvantages:

+ The Modula-3 threads allowed us to implement a parallel control software system with little overhead. The threads turned out to be easily applicable in this case study.

- In our example, the different threads do not access the common address space. However, the separation of more complicated threads is likely to be difficult if they have to communicate to a large extent via the common address space.

- Finally, the communication mechanisms provided are not very powerful. First, it is impossible to send messages along with signals. Second, there is no queuing mechanism provided for signals. Such a mechanism is needed whenever a signal may have already been sent out before the receiving thread begins to wait for it.

The integration of object-orientation and parallel programming worked well. Each device is implemented as an autonomous object with its own thread. Furthermore, objects are provided with certain properties with respect to parallel programming, such as mutual exclusion, by inheriting from a superclass which supplies these properties. We conclude that object-oriented and parallel mechanisms were integrated with little implementational overhead.

Being a safety-critical system, the industrial production cell is a good example of the verification of safety requirements. On the basis of an appropriate calculus, we verified that the program controlling the production cell indeed guarantees the safety requirements. We applied the following techniques which turned out to work well:

+ Safety requirements can well be expressed as system invariants. Applying the weakest precondition calculus, these invariants were verified with comparatively little effort.

+ The parallel processes of the control system are synchronized via signals. We incorporated parallel constructs into the verification by assigning an invariant to each communication signal. Thus, safety requirements were successfully proven which were placed on several processes working in parallel.

We conclude with some remarks on the effort that was placed on the case study presented. The preliminary version of the implementation (cf. [2]) includes 1.400 lines of code, organized in 20 classes and 21 modules. The development of the whole case study took 80 hours, in detail:

- 15 hours on specification and design,
- 60 hours on the preliminary implementation,
- 5 hours on the exemplary verification of safety requirements.

Acknowledgements

Thanks are due to Matthias Felger for the modelling and implementation of a preliminary version of the control software system, and to Thomas Lindner and Walter Zimmer for the discussion of this paper.

References

[1]   E. W. Dijkstra. *A Discipline of Programming*. Prentice Hall, 1976.

[2]   M. Felger. *Spezifikation und Implementierung einer Fertigungszelle mit Modula-3*. Studienarbeit, Forschungszentrum Informatik, Karlsruhe (German language), 1994.

[3]   S. P. Harbison. *Modula-3*. Prentice Hall, 1992.

[4]   C. A. R. Hoare. *Communicating Sequential Processes*. Prentice Hall, 1985.

[5]   T. Lindner. *Task Description*. Technical Report, Forschungszentrum Informatik, Karlsruhe, 1993.

[6]   G. Nelson (Ed.). *Systems Programming with Modula-3*. Prentice Hall, 1991.

[7]   A. U. Shankar. *An Introduction to Assertional Reasoning for Concurrent Systems*, in ACM Computing Surveys, Vol. 25, No. 3. ACM Press, September 1993.

XXI. TROLL *light*

Specification with a Language for the Conceptual Modelling of Information Systems

Rudolf Herzig, Nikolaos Vlachantonis

Technische Universität Braunschweig

Abstract

TROLL *light* is a language for conceptual modeling of information systems. It is designed to describe the Universe of Discourse (UoD) as a system of concurrently existing and interacting objects. TROLL *light* objects have observable properties modeled by attributes, and the behavior of objects is described by events. Possible object observerations may be restricted by constraints, whereas event occurrences may be restricted to specified life cycles. TROLL *light* objects are organized in an object hierarchy established by subobject relationships. Communication among objects is supported by event calling.

TROLL *light* was employed for the specification of a production cell. In this paper we show some characteristic points of that specification and conclude with a discussion of lessons learned from this work.

21.1 Goals and Concepts of TROLL *light*

TROLL *light* is a language for the conceptual modeling of information systems. The development of information systems can be roughly split up into two major phases. The aim of the *requirements engineering* phase is to obtain a conceptual model of a certain Universe of Discourse (UoD). On the basis of the conceptual model and by consideration of further non-functional constraints, a working system is constructed in the *design engineering* phase [1]. A conceptual model is an abstract mathematical structure which is a model of a theory. The theory, given by a set of sentences in a formal language, is usually called a *schema* [2].

A natural approach to information system modeling consists in viewing an information system as a community of concurrently existing and interacting objects [3]. As a language for the specification of such an object community we propose TROLL *light* [4]. The origins of TROLL *light* date back to the paper of A. and C. Sernadas and H.-D. Ehrich [5] in which principal ideas for object-oriented specification of information systems were formulated. This was the starting point for language proposals like OBLOG [6] and TROLL [7].

In particular TROLL is a voluminous language which offers a large variety of modeling concepts to satisfy the users' needs for expressing facts and laws of the UoD. In contrast, the aim of the development of TROLL *light* was to obtain a handy language with a modest number of basic constructs. One could say that TROLL *light* is grounded on a single specification concept called *object descriptions* (templates), having the following general structure.

```
TEMPLATE name
    DATA TYPES         import of data types
    TEMPLATES          import of other object descriptions
    SUBOBJECTS         declaration of local subobjects
    ATTRIBUTES         declaration of attributes
    EVENTS             declaration of event generators
    CONSTRAINTS        static integrity conditions (invariants)
    DERIVATION         rules for derived attributes
    VALUATION          effects of events on attributes
    INTERACTION        synchronization of events in different objects
    BEHAVIOR           restriction of possible event sequences
END TEMPLATE
```

A template describes static and dynamic properties, i.e. valid *states* and valid *state evolutions*, of a prototypical object. In an actual object community this prototypical object may be instantiated by any number of actual objects. In order to provide objects with identities, every object description is associated with a corresponding object identifier sort. There is the following convention: object descriptions always start with an upper case letter while the corresponding object identifier sorts start with a lower case letter.

The objects described by TROLL *light* are objects with visible attribute states. Items with data abstraction should be specified outside TROLL *light* by means of abstract data types. The specifications of standard and user-defined data types can be referred in the DATA TYPES clause of a template. An object description may also depend on other object descriptions which can be made known in the TEMPLATES clause of a template.

The declaration of local subobjects, attributes, and event generators makes up the interface of an object to other objects.

- By subobject relationships objects are arranged into an object hierarchy. Thereby the basic structure of an object community is fixed.

- Attributes build the observable state information of objects. By means of predefined sort constructors like SET and TUPLE arbitrary complex, data-valued, and object-valued attributes can be specified.

- Events are abstractions of state-modifying operations on objects.

Subobject, attribute, and event symbols define a signature over which axioms concerning static and dynamic properties can be specified. Static properties include:

- The possible states of an object can be restricted to admissible states by specifying static integrity conditions.

- Attributes are divided into explicit and derived attributes. While the values of explicit attributes are determined by event occurrences, the values of derived attributes are computed from other stored or derived information.

For stating static integrity constraints and derivation rules, TROLL *light* includes a SQL-like query calculus. This calculus is also used in the formulation of dynamic properties which include:

- The effect of event occurrences on attributes is specified by valuation rules of the kind $\{\phi\}[e]\, a = \tau$. This rule says that whenever an event corresponding to the event generator e occurs and the precondition ϕ is fulfilled, then the attribute a is set to the value of the term τ evaluated in the previous state.

- Events in different objects can be synchronized by interaction rules of the kind $\{\phi\}\, o_1.e_1 \gg o_2.e_2$. This rule says that whenever an event corresponding to the event generator e_1 occurs in object o_1 and the precondition ϕ is fulfilled, then the event e_2 must occur simultaneously in object o_2.

- Possible event sequences can be restricted to admissible ones by giving simple *process descriptions*. These process descriptions include events as elementary processes, and sequence -> and choice I as process building operators. Only right recursive processes are allowed so that process descriptions can be translated into an event-driven sequential machine. The occurrence of events can be restricted by preconditions which may refer to the overall state of an object.

A detailed informal description of the language concepts can be found in [4]. Semantic issues are addressed in [8]. The role of TROLL *light* in a development environment for information systems is discussed in [9].

Although a production cell is not a typical information system, it can be viewed as a community of concurrently existing and interacting objects, too. Hence in [10] we applied TROLL *light* to describe possible states and state evolutions of such a system. The following section gives a brief overview about how this was done.

21.2 The Specification

The general proceeding to describe an object community with TROLL *light* is as follows:

1. Identify objects and place them into an object hierarchy.

2. Describe attributes and events as state-modifying operations as well as the effects of events on attributes.

3. Restrict possible attribute states by constraints and possible event sequences by behavior patterns.

4. Synchronize events in different objects by interaction rules.

Steps 1 and 4 effect the global structure of an object community while steps 2 and 3 refer to local object properties.

Step 1: Both the objects for which object descriptions had to be written and the hierarchical organization of these objects were clearly prescribed by the task description provided by the FZI. We give a detail from the object description for the topmost element of the object hierarchy.

```
TEMPLATE ProductionCell
      ...
   TEMPLATES
       FeedBelt, ElevRotaryTable, Robot, Press,
       DepositBelt, PhotoelectricCell, TravellingCrane;
   SUBOBJECTS
       FeedBelt1            :    feedBelt;
       ElevRotaryTable1     :    elevRotaryTable;
       Robot1               :    robot;
       Press1               :    press;
       DepositBelt1         :    depositBelt;
```

DepositSensor1	:	photoelectricCell;
TravellingCrane1	:	travellingCrane;

...

END TEMPLATE;

The template reflects the fact that a production cell is composed of a feed belt, an elevating rotary table, a robot and other parts. Since all these components are objects with different properties, separate object descriptions had to be given for most of them. Note that FeedBelt denotes an object description while feedBelt is the name of the corresponding object identifier sort. In order to distinguish between subobjects and their object descriptions, a number was added in the construction of subobject names.

Step 2: As an example for the specification of attributes, events and effects of events on attributes we cite the object description of a feed belt.

TEMPLATE FeedBelt
   DATA TYPES
      Nat;
   ATTRIBUTES
      Blank#:nat;
   EVENTS
      BIRTH install; start; stop;
      getBlank; dropBlank;
      DEATH destroy;
   VALUATION

[install]	Blank#=0;
[getBlank]	Blank#=Blank#+1;
[dropBlank]	Blank#=Blank#-1;

   ...

END TEMPLATE;

For the specification of the production cell most attributes only served to report the status of some component.

Step 3: Attribute states can be restricted by constraints. Here we assume that a feed belt may not transport more than nine blanks at a time.

TEMPLATE FeedBelt

   ...

   CONSTRAINTS
      Blank#<10;

   ...

END TEMPLATE;

TROLL *light* specifications give no hint how to proceed when certain conditions are not fulfilled. It is only specified that when there are nine blanks on a belt then the event getBlank will lead to an inconsistent state.

The possible event sequences can be restricted by process descriptions.

TEMPLATE FeedBelt
   ...
  BEHAVIOR
    PROCESS FeedBelt =
      (install -> StoppedBelt);
    PROCESS StoppedBelt =
      (start -> RunningBelt);
    PROCESS RunningBelt =
      (getBlank -> RunningBelt |
      {Blank#>0} dropBlank -> RunningBelt |
      stop -> StoppedBelt);
END TEMPLATE;

The specification reflects the fact that a feed belt may only get or drop blanks when it is running. The fact that a belt may only drop blanks when there is at least one blank on the belt is described by an event precondition.

Step 4: Events in different objects are synchronized by interaction rules. The following interaction rules describe the interaction between the most important system components.

TEMPLATE ProductionCell
   ...
  INTERACTION
    ...

feed	>> FeedBelt1.getBlank;
FeedBelt1.dropBlank	>> ElevRotaryTable1.getBlank;
Robot1.graspBlankfromTable	>> ElevRotaryTable1.dropBlank;
Robot1.dropBlankforPress	>> Press1.getBlank;
Robot1.graspBlankfromPress	>> Press1.dropBlank;
Robot1.dropBlankforBelt	>> DepositBelt1.getBlank;
DepositSensor1.dropBeam	>> DepositBelt1.stop,
	TravellingCrane1.getOrder;
DepositSensor1.detectBeam	>> DepositBelt1.start;
TravellingCrane1.openGripper	>> FeedBelt1.getBlank;

    ...
END TEMPLATE;

The complete specification contains about 10 pages and can be found in [10].

21.3 Evaluation

A template describes the admissible states and state evolutions of an object or an object community like the production cell. Hence safety constraints on attributes can be easily described as invariants and safety conditions on event occurrences by event preconditions. In this way a template restricts object behavior but it normally does not stipulate a particular object behavior.

Object descriptions abstract from the initiative of events, i.e., it cannot be distinguished whether a given event occurrence is caused by an object itself or by the environment. This distinguishes TROLL *light* from object-oriented (concurrent) programming which assumes an implicit processor concept, and shows that TROLL *light* is tailored for the specification of information systems as typical interactive systems. So it is not desired to derive a self-running system from TROLL *light* object descriptions. In contrast to that, object descriptions are used to check for a given evolution whether it matches the specified behavior patterns.

State evolutions can be simulated by using a validation tool for TROLL *light* specifications called *animator* which allows to query object community states and to perform state transitions by interactively triggering event occurrences. Animating object descriptions may help to assure informal correctness. Issues of formal correctness are addressed by a verification calculus with which propositions about object descriptions may be formally proved. The animator and a verification tool are important components of the TROLL *light* development environment we are working on [9].

A production cell is not a typical information system. For instance, the specification of observable attributes was of minor importance for the description with TROLL *light*. Nevertheless, some aspects of the production cell could be described in a straightforward way:

- The information about what templates were to be written and how objects had to be organized in an object hierarchy could be directly taken from the task description (or the real model of the production cell) provided by the FZI.

- The specification of events and interaction rules could be directly taken from the task description as well.

- For the specification of allowed life cycles the simple process descriptions built up on sequence and choice satisfied the needs in many cases.

However in some cases an additional process building operator like *parallel composition* (cf. [11]) would have been helpful.

- Safety constraints could be specified by constraints and event preconditions.

It is important to note that besides abstracting from the initiative of an event, there is no notion of time in TROLL *light*. So real-time conditions cannot be expressed in a direct way. For a related object-oriented specification language with real-time aspects see ALBERT [12].

References

[1]   P. Loucopoulos and R. Zicari, editors. *Conceptual Modeling, Databases, and CASE: An Integrated View of Information Systems Development*. John Wiley & Sons, Inc., 1992.

[2]   R.J. Wieringa. Three Roles of Conceptual Models in Information System Design and Use. In E. Falkenberg and P. Lindgreen, editors, *Information System Concepts: An In-Depth Analysis, Proc. IFIP WG8.1 Working Conference, Namur (Belgium)*, pages 31-51. North-Holland, Amsterdam, 1989.

[3]   H.-D. Ehrich, G. Denker and A. Sernadas. Constructing Systems as Object Communities. In M.-C. Gaudel and J.-P. Jouannaud, editors, *Proc. TAPSOFT'93: Theory and Practice of Software Development*, pages 453-467. Springer LNCS 668, 1993.

[4]   S. Conrad, M. Gogolla and R. Herzig, TROLL light: A Core Language for Specifying Objects. Informatik-Bericht 92-02, Technische Universität Braunschweig, 1992.

[5]   A. Sernadas, C. Sernadas, and H.-D. Ehrich. Object-Oriented Specification of Databases: An Algebraic Approach. In P.M. Stocker and W. Kent. editors, *Proc. 13th Int. Conf. on Very Large Data Bases VLDB*, pages 107-116. Morgan-Kaufmann, 1987.

[6]   J.-F. Costa, A. Sernadas and C. Sernadas. OBL-89 Users Manual (Version 2.3). Internal report, INESC, Lisbon, 1989.

[7]   R. Jungclaus, G. Saake, T. Hartmann and C. Sernadas. Object-Oriented Specification of Information Systems: The TROLL Language. Informatik-Bericht 91-04, Technische Universität Braunschweig, 1991.

[8]   M. Gogolla, S. Conrad and R. Herzig, Sketching Concepts and Computational Model of TROLL light. In A. Miola, editor, *Proc. 3rd Int. Conf. Design and Implementation of Symbolic Computation Systems DISCO*, pages 17-32. Springer LNCS 722, 1993.

[9]   N. Vlachantonis, R. Herzig, M. Gogolla, G. Denker, S. Conrad, and H.-D. Ehrich. Towards Reliable Information Systems: The KORSO Approach. In C. Rolland, F. Bodart and C. Cauvet, editors, *Advanced Information Systems Engineering Proc. 5th CAiSE'93, Paris*, pages 463-482. Springer LNCS 685, 1993.

[10]   R. Herzig and N. Vlachantonis. Spezifikation einer Fertigungszelle in TROLL light. Interner Bericht, Technische Universität Braunschweig, 1992.

[11]   R.J. Wieringa, *Equational Specification of Dynamic Objects*. In R.A. Meersman, W. Kent, and S. Khosla, editors, *Object-Oriented Databases: Analysis, Design & Construction (DS-4), Proc. IFIP WG2.6 Working Conference, Windermere (UK) 1990*, pages 415-438. North Holland, Amsterdam, 1991.

[12]   E. Dubois, P. Du Bois, and M. Petit. O-O Requirements Analysis: an Agent Perspective In O.M. Nierstrasz, editor, *ECOOP'93 — Object-Oriented Programming*, pages 458-481. Springer LNCS 707, 1993.

A. Simulation

Implementation and Usage of a Simulation with Graphical Visualization of the Production Cell

Artur Brauer, Thomas Lindner

Forschungszentrum Informatik, Karlsruhe

Abstract

In this chapter, we describe the implementation and operation of a production cell simulation with graphical visualization.

The simulation imitates the behaviour of the production cell and can be controlled by a simple ASCII protocol. The status of the production cell is computed and visualized in realtime, sensor values are returned accordingly.

Furthermore, we describe how to get and how to install the simulation. The protocol for reading sensor values or error codes, and for giving commands is documented.

A.1 Problem

Our motivation is to provide all contributors to the case study with an enviroment by which they can enhance their first understanding of the problem by watching a demo controller. Additionally, we find it indispensable to be able to validate constructed software by connecting it with the simulation. Without this, the notion of a *specification error* would make no sense at all, as there is nothing to test specification or implementation against it.

A central requirement is to make controlling the simulation by a program as simple as possible. This is solved by allowing a control program to read sensor values from stdin and write actuator commands to stdout. Therefore, we are independent from the choice of the control software's implementation language.

A.2   Simulation

We developed a simulation of the production cell [2], which imitates the important behaviour of the real production cell. The simulation runs a graphical window which displays the substantial elements of the production cell. In order to achieve good performance, all the devices are drawn simplified (see Figure 1).

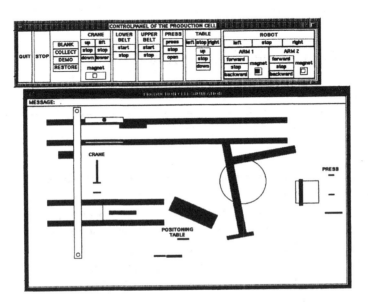

Figure 1   Screen dump of the visualization (together with a control panel)

The simulation is controlled by receiving sensor information from it and transmitting commands to it (the commands are described in subsection A.5). Both can be done using UNIX-pipes: any control software can just read the sensor information from stdin and write commands to stdout. The simulation takes care about the movement of the devices and blanks, detects collisions, and reports them. The simulation displays a top view of the production cell. For the elevating rotary table, the press, and the crane information about the third dimension is necessary. For that reason we have introduced scales, which represent the height of the corresponding devices. The scales contains height marks which represent the height of the neighbour devices. For a simple identification, the marks use the same colour as the devices they stand for.

The blanks processed by the production cell are initially stored in the left bottom corner of the window. To put a blank onto belt 1, one can use the corresponding command. During testing it might be useful to put a blank directly on a particular device. This can be done by selecting the blank pressing the left mouse button and dragging it over one of the following devices: both belts, elevating rotary table, or press.

The simulation is distributed together with a control panel. It can be used to become familiar with the simulation by manual control. It also provides a *demo* button which causes the simulation to perform a demonstration handling only one blank.

The simulation can be used in two modes. In the *synchronous* mode the control program has to provide the clock for the simulation. After a command has been given, a *react* command must be sent each time you want the devices to move one more step. In this way the simulation can be forced to run synchronous to your controller. In the (default) *asynchronous* mode the simulation will use its own clock, updating the simulation is done automatically. If your control program is quick enough, this is the most comfortable choice. If you think that your control program may be too slow and may occasionally miss the right time to swich off devices, we recommend you to use the synchronous mode.

Tcl/Tk

The simulation was implemented using the Tool Command Language (Tcl) and the widget Toolkit (Tk), both developed by J. Ousterhout at the University of California at Berkeley [1]. Tcl is a C-like, string-based interpretative language. It was designed for tool integration, therefore the interpreter is extendable. One of the most useful extensions is Tk, where a powerful graphical widget set was made available for X11. Together, Tcl/Tk is a very promising approach for both tool integration and the construction of graphical user interfaces.

Tcl/Tk is free software and can be obtained via ftp from sprite.berkeley.edu.

A.3   Installing the Simulation

We presuppose a running Tk/Tcl installation on your system to start the simulation. The simulation is available as the compressed tarfile visualization.tar.Z on the ftp server ftp.fzi.de [141.21.4.3] in the directory pub/korso/fzelle/simulation. To use the simulation you first have to uncompress and then to un-tar the file into a directory of your choice. The README file contains a list of files which should be now in your simulation directory. Change the file startsimu by entering the correct path of your wish interpreter (this is the above-mentioned Tcl/Tk-interpreter) in the first line and set the variable wishpath in the third line also to the path for wish. Now, you have a running version of the simulation.

A.4   Starting the Simulation

For starting the simulation the executable file startsimu is provided. It takes several parameters and starts the simulation coupled together with your control program or with the (default) control panel. If you call startsimu without any parameters, it will come up together with the manual control panel, both will use colour display mode and english language. The simulation will use the asynchronous mode. Below, the parameters are listed which can be used to configure the simulation.

-**con** <controllername>

> The simulation will be started together with the specified controller; in case you want to give some parameters to the controller you have to quote them together with controllername, like e.g.

> > startsimu -con "mycontroller -option1 -option2".

-**snc**   The simulation will use the synchronous mode; the controller has to give a react command each time it wants the simulation to do one more cycle.

-**bw**   The simulation will use monochrome display.

-**grm**   The simulation will use german language.

-**eng**   The simulation will use english language (the default).

A.5   Definition of the Protocol

To couple control program and simulation UNIX-pipes are used. This allows the program for controlling the simulation just by writing ASCII commands to stdout and reading status information of the simulation from stdin.

The simulation provides two kinds of commands. First, there are commands for switching devices on and off. Secondly, there are some commands to get information from the simulation, which then can then be read from stdin. It is of great importance that after both kinds of commands the control program flushes the pipe, since otherwise the commands will be buffered and will have no immediate effect. This can be done e.g. in C with the fflush command.

A.5.1   System commands

These commands provide general facilities to manipulate the simulation which do not belong to certain devices.

- **system_quit**   quits the simulation
- **system_stop**   stops any movement in the simulation
- **blank_add**   puts a blank from the stock on the feed belt
- **blanks_collect**   puts all the blanks dropped irregularly back to the stock
- **system_demo**   starts a demonstration of the simulation
- **system_restore**   restores the simulation
- **react**   this command is only available in the synchronous mode; it is then used to force the simulation to do one update cycle; before this command is transmitted the simulation will not perform a change

A.5.2   Feed belt commands

These commands allow to toggle the feed belt.

- **belt1_start**   starts the feed belt to move right
- **belt1_stop**   stops the feed belt

A.5.3   Deposit belt commands

These commands allow to toggle the deposit belt.

- **belt2_start**   starts the deposit belt to move left
- **belt2_stop**   stops the deposit belt

A.5.4   Elevating rotary table commands

These commands can be used to rotate and move up and down the elevating rotary table.

- **table_left**   starts rotation to the left of the elevating rotary table
- **table_stop_h**   stops rotation of the elevating rotary table
- **table_right**   starts rotation to the right of the elevating rotary table
- **table_upward**   starts upward movement of the elevating rotary table

- **table_stop_v**   stops vertical movement of the elevating rotary table
- **table_downward**   starts downward movement of the elevating rotary table

A.5.5   Robot commands

These commands allow for rotating the robot and move the arms forward and backward.

- **arm1_forward**   starts forward movement of arm 1
- **arm1_stop**   stops movement of arm 1
- **arm1_backward**   starts backward movement of arm 1
- **arm1_mag_on**   activates magnet of arm 1
- **arm1_mag_off**   deactivates magnet of arm 1
- **arm2_forward**   starts forward movement of arm 2
- **arm2_stop**   stops movement of arm 2
- **arm2_backward**   starts backward movement of arm 2
- **arm2_mag_on**   activates magnet of arm 2
- **arm2_mag_off**   deactivates magnet of arm 2
- **robot_left**   starts rotation of the robot to the left
- **robot_stop**   stops rotation of the robot
- **robot_right**   starts rotation of the robot to the right

A.5.6   Crane commands

This commands control the movement of the crane.

- **crane_to_belt2**   starts the crane moving towards the deposit belt
- **crane_stop_h**   stops movement of the crane
- **crane_to_belt1**   starts the crane moving towards the feed belt
- **crane_lift**   starts the crane's magnet moving upward
- **crane_stop_v**   stops movement of the crane's magnet
- **crane_lower**   starts the crane's magnet moving downward
- **crane_mag_on**   activates the crane's magnet
- **crane_mag_off**   deactivates the crane's magnet

A.5.7   Press commands

This commands are used to control the press.

- **press_upward**   starts press moving upward
- **press_stop**   stops movement of the press
- **press_downward**   starts press moving downward

A.5.8   Commands to get status information from the simulation

The simulation provides two commands which can be used to obtain information from the production cell. Both cause the simulation to produce output which can be read by the control program from stdin. The get_status command provides the control program with information on device sensors. The get_passings command was introduced to achieve some synchronization possibilities. It should not be necessary when the synchronous mode is used.

- **get_status**

  writes the status information of the simulation to the standard output

  This command causes the simulation to write all the status information to stdout as a 15-element vector. Each element of the vector is separated by a newline. The status vector contains all the information that the real production cell can provide and additionally a list of errors that occurred since the last status report. The format of the status vector is described in Table 1.

- **get_passings**

  writes the number of passings through the main loop to stdout

  This command causes the simulation to write the number of passings through the main loop to stdout. The number of passings is computed modulo 10000. This can be used to synchronize the control program with the simulation, since the performance of the simulation depends much on the host it is running on.

Index	Corresponding Device	Short Description	Range/ Meaning
1	press	Is the press in bottom position?	1=y/0=n
2		Is the press in middle position?	1=y/0=n
3		Is the press in top position?	1=y/0=n
4	robot	Extension of arm 1	$0..1^a$
5		Extension of arm 2	$0..1^a$
6		Angle of rotation of the robot	$-100..70^b$
7	elevating rotary table	Is the elevating rotary table in bottom position?	1=y/0=n
8		Is the levating rotary table in top position?	1=y/0=n
9		Angle of rotation of the table	$-5..90^b$
10	crane	Is the crane above the deposit belt?	1=y/0=n
11		Is the crane above the feed belt?	1=y/0=n
12		Height of the crane's magnet	$0..1^c$
13	feed belt	Is a blank inside the photoelectric barrier?	1=y/0=n
14	deposit belt	Is a blank inside the photoelectric barrier?	1=y/0=n
15	general	A list of errors, that occurred since the last status report (enclosed in curly braces; elements separated by space.)	see Table 2

Table 1   The status vector

a. Value 1 means that the arm is fully extended.
b. Value 0 means that the device is in the initial position given by the simulation.
c. Value 0 means that the magnet is in top position.

Number	Meaning
0	No error occurred; if this is true, no other elements are in the error list.
1	A blank dropped irregularly during the passage from the feed belt to the elevating rotary table.
2	Collision between elevating rotary table and the feed belt.

Table 2   Error codes

Number	Meaning
3	Elevating rotary table reached the right stop.
4	Robot reached the left stop.
5	Robot reached the right stop.
6	Arm 1 dropped a blank invalidly.
7	Collision between arm 1 and press.
8	Arm 2 dropped a blank invalidly.
9	Collision between arm 2 and press.
10	A blank dropped from the end of the deposit belt.
11	Collision between crane and the deposit belt.
12	Collision between crane and the feed belt.
13	Crane dropped a blank irregularly.
14	Crane reached the stop over the feed belt.
15	Crane reached the stop over the deposit belt.

Table 2   Error codes

A.5.9   Guard command

This command has been introduced to allow for controlling the simulation on a higher level of abstraction. Using the guard command would completely change the nature of the problem, therefore, we do not recommend this. The command is intended to be used by control programs which performance is so bad that they are not able to react quick enough and do not use the synchronous mode.

- **new_guard sensor_number operator destination_value command**

  a new guard is created.

  The simulation allows to create guards, which assures that a particular action is performed at the right time. A guard tests a condition until it becomes true and than call a simulation command and removes himself. The condition to be tested consists of a sensor, an operator and a destination value. The **sensor_number** describes which sensor should be compared with the **operator** to the **destination_value**. When this expression is true, **command** will be executed.

A.6 Operating the Cell

To transport a blank through the production cycle of the simulation one has to know some more detailed information about it. There are various kinds of collisions that can occur during operating the cell and which of course should be avoided.

On the other hand one has to know constants describing the appropriate angles and extentions for loading and unloading various devices. Both kinds of informations are provided below.

A.6.1 Collisions

There are three kinds of irregular behaviour detected by the simulation. First, two devices can clash against each other. Secondly, a device should be stopped before it reaches the borders of it's movability, if the controller misses to do so this is treated as a failure. And thirdly, a blank that is not transported correctly from one device to another one, is dropped an causes an error report, too.

The following list enumerates the collisions between devices and gives conditions which assures that the production cell works safely. Note that the conditions below are not the only possible which assure collision avoidance. There may be weaker conditions, but if the ones below are obeyed, no collisions will occur.

1. The rotary elevating table can clash against the feed belt. This can be avoided, if the angle of the table is kept greater or equal than 0.

2. The rotary elevating table and the robot can collide, if both arm 1 and the table hold a blank and the table is in top position.[1] Then the two blanks can touch each other, what is reported as a collision. To prevent this, one of the following conditions must be satisfied:

   — the angle of the robot is less or equal to 0,

   — the extension of the robot is 0,

   — the table is in bottom position.

3. The robot can reach into the press with both arms and cause a collision, if the press's height is in the range of the corresponding arm. Collisions between arm 1 and the press may be avoided by ensuring one of the following conditions:

---

1. More precisely: the two blanks will collide, not the robot and the table.

— the angle of the robot is greater or equal to -70,

— the extension of arm1 is less or equal to 0.3708,

— the press is in bottom or middle position.

To prevent arm 2 from clashing with the press one the following conditions must be satisfied:

— the robot's angle is greater or equal to 55,

— the robot's angle is less or equal to 15,

— the extension of arm 2 is 0,

— the press is in top or bottom position.

4. Collision can occur between the crane and the belts. The following condition is sufficient to prevent failures:

— during the crane's movement its height should be less or equal to 0.6593.

Besides collisions between devices a controller can cause conflicts by moving some devices out of their moving range. So one has to regard some constraints when moving the devices.

1. The robot's angle should be kept between 65 and -95.

2. The elevating rotary table must not rotate over 85 degrees.

3. The crane should not move outside the region which is limited by the two switches.

The third kind of possible conflict occurs if a device transfers a blank irregularly to the next device. That is e.g. when arm 1 of the robot releases a blank while not beeing above the press or the table. There are various situations where a blank May not be released. Instead of describing all possible wrong situations, we rather give a guide how to transfer blanks by giving the right constants in the following section.

A.6.2 Constants

During operating the cell blanks move from one device to the next. It is obvious that for such a transition the concerned devices should be in the right position, characterized by the constants given below.

1. When a blank changes from the feed belt to elevating rotary table, the table has to be in bottom position and it's rotation angle must o be 0.

2. When arm 1 wants to pick up a blank from the elevating rotary table, the following conditions should be satisfied:

   — robot's angle: 50

   — arm 1 extension: 0.5208

   — table in top position

   — table's rotation: 50.

3. To put the blank with arm 1 into the press, one has to ensure the following conditions:

   — robot's angle: -90

   — arm 1 extension: 0.6458

   — press in middle position.

4. To get a blank from the press with arm 2 one has to make sure that:

   — robot's angle: 35

   — arm 2 extension: 0.7971

   — press in bottom position.

5. Arm 2 can release the blank above the deposit belt if:

   — robot's angle between -90 and -45

   — arm 2 extension: 0.5707

6. The crane can pick up the blank from the deposit belt if:

   — the crane's horizontal position is adequate (cf. sensor 10)

   — crane's height: 0.9450.

7. The crane can deposit the blank on the feed belt if:

   — the crane's horizontal position is adequate (cf. sensor 11)

   — crane's height: 0.6593.

References

[1]   J. K. Ousterhout. An X11 Toolkit Based on the Tcl Language. *Proceedings of the 1991 Winter USENIX Conference*, 1991.

[2]   A. Brauer, C. Lewerentz, T. Lindner. Implementing a Visualization of an Industrial Production Cell Using Tcl/Tk. *Proceedings of the First Tcl/Tk Workshop*, Berkeley, June 1993.

Lecture Notes in Computer Science

For information about Vols. 1–814
please contact your bookseller or Springer-Verlag